Selenium in Medicine and Biology

Selenium in Medicine and Biology

Proceedings of the Second International Congress on Trace Elements in Medicine and Biology
March 1988 · Avoriaz, France

Editors
Jean Nève · Alain Favier

Walter de Gruyter · Berlin · New York 1989

Editors

Jean Nève, Assistant Professor, Dr. Pharm. Sc.
Université Libre de Bruxelles
Unité de Toxicologie et de Chimie Bioanalytique
Campus Plaine 205-1
Boulevard du Triomphe
B-1050 Bruxelles
Belgium

Alain Favier, Professor, Dr. Pharm. Sc.
Centre Hospitalier Universitaire de Grenoble
Hôpital A. Michalon
Laboratoire de Biochimie C
BP 217X
F-38043 Grenoble cedex
France

Library of Congress Cataloging in Publication Data

International Congress on Trace Elements in Medicine and Biology
(2nd : 1988 : Avoriaz, France)
Selenium in medicine and biology.

Includes indexes.
1. Selenium--Physiological effect--Congresses.
2. Selenium in human nutrition--Congresses. 3. Selenium--Health aspects--Congresses. I. Nève, Jean, 1951– . II. Favier, Alain, 1945– . III. Title.
[DNLM: 1. Selenium--adverse effects--congresses. 2. Selenium--analysis--congresses. 3. Selenium--physiology--congresses.
QU 130 I612s 1988]
QP535.S5I45 1988 612'.01524 88-30976
ISBN 0-89925-503-5 (U.S.)

Deutsche Bibliothek Cataloging-in-Publication Data

Selenium in medicine and biology : proceedings of the 2. Internat. Congress on Trace Elements in Medicine and Biology, March 1988, Avoriaz, France / ed. Jean Nève ; Alain Favier.
[Organized by: Soc. Française d'Etude et de Recherche sur les Elements Trace Essentiels (SFERETE) ... Organizing committtee: A. Favier ...]. – Berlin ; New York : de Gruyter, 1988
ISBN 3-11-011770-3
NE: Nève, Jean [Hrsg.]; International Congress on Trace Elements in Medicine and Biology <02, 1988, Avoriaz>; Société Française d'Etude et de Recherche sur les Eléments Trace Essentiels

 Printing: Gerike GmbH, Berlin. – Binding: Dieter Mikolai, Berlin. – Printed in Germany.

Preface

For four days during March 1988, 200 scientists with various qualifications from 20 countries met in Avoriaz, the splendid site in the French Alps already famous for its fantastic film festival. They were participating in the Second International Congress on Trace Elements in Medicine and Biology devoted to selenium. This event was the follow-up to a first successful symposium on zinc which was held during May 1986, in Villard de Lans, in the French massif du Vercors. Numerous winter sport possibilities together with weather varying from beautiful to abominable harmoniously tempered the intense scientific work of participants. Twenty exhibitors and several institutions and firms whose support is acknowledged elsewhere also contributed to the success of the event. The present book, the rapid publication of which is due to Walter de Gruyter, is a collection of plenary lectures and other contributions that were delivered on that occasion.

Having to take into consideration the quasi simultaneous occurrence of other international congresses devoted to the selected trace element, the organisers decided to contact the largest possible panel of specialists known not only for their competence but also for their different specializations and opinions. Young scientists who have been publishing in the field for less than 10 - 15 years were also selected on the basis of their original work and use of modern investigation techniques. The confrontation of these experts with other investigators who contributed with oral or poster presentations gave rise to a very exciting event that was appreciated by the participants for the high quality of contributions and discussion in a congenial and relaxing atmosphere.

From a scientific point of view, it was astonishing even for those who have closely followed the selenium saga, how much the knowledge on selenium has progressed these last few years. Several important gaps were filled in strengthening the favourable opinion on the beneficial properties for health of this essential trace element.

In the field of biochemistry, many contributions were devoted to glutathione peroxidase, this powerful protective enzyme whose dysfunction is associated with many symptoms of selenium deficiency. However, other biological aspects of the element were also carefully considered such as its interaction with xenobiotics or heavy metals as well as its anticarcinogenic and immunomodulatory properties. The precise role of other selenium containing biological molecules, particularly the numerous selenoproteins, is now under investigation. The metabolism of selenium as well as its availability from the environment were other thrilling subjects for discussion. Some mechanisms for selenium homeostasis have been elucidated, but much effort remains to be made concerning the fate or organic forms. Assessment of selenium status including its determination in biological materials has been refined by the identification of sensitive and specific functional indices. Satisfaction of selenium requirement for some functions can now be quantified: as a consequence, marginally low selenium status appeared more widespread than previously thought. Under these conditions, selenium supplementation, if not undertaken on a large scale in some countries, is increasingly becoming considered in several population groups as a protective factor against chronic diseases. The possible toxicological consequences of selenium exposure were also largely discussed, but it was admitted that few problems actually occur. Epidemiology has made considerable progress, however there is still a lack of data particularly for developing countries. Large international studies were therefore encouraged.

The implication of selenium in human pathology was extensively considered during the congress, with a special focus on those susceptible to selenium deficiency such as children, pregnant women, the elderly, subjects with restricted regimens, and inhabitants of low selenium areas... Diseases such as cystic fibrosis, cirrhosis, inflammatory diseases, cancer, cardiovascular diseases, renal insufficiency or AIDS... were more particularly discussed showing the growing interest of clinicians in the pathological consequences of selenium deficiency. New and refined laboratory methods were described in order to demonstrate the effects of selenium deficiency and supplementation more precisely. Intervention studies and the possible beneficial effects for health of selenium supplementation were commented upon on different occasions.

We sincerely hope that this book in which varied and independent opinions are collected will constitute an agreeable and exhaustive way of keeping the scientific community aware of the latest data on the role of selenium in biology and medicine.

August, 1988

A. Favier, L. Molle, J. Nève

Acknowledgements

Abbott Laboratories
Laboratoires Aguettant
Becton Dickinson
Biomerieux
Biotrol
Boehringer Mannheim
Cahiers de Nutrition et de Diététique
Cell Life Int Inc
Crédit Lyonnais
Dietaroma
Société des eaux d'Evian
Finnigan Matt GmbH
Gif Diagnostics
GN Pharm GmbH
Laboratoires des Granions
Laboratoires Herbaxt
Laboratoires Holistica
Société La Roche Posay
Huhtmaki Oy Novamed
Jobin Yvon Instruments
Johnson Matthey SA
Laboratoires Labcatal
Laboratoire Lafon
Laboratoires Millot Solac, Sanofi
Merck SA
Nestlé SA
Nycomed AS
Perkin Elmer Instruments
Philips Instruments
Laboratoires Richelet
Selenium Tellurium Development Association
Société Sodipro
Technicon Instruments
Université de Strasbourg Louis Pasteur
Université de Grenoble Joseph Fourier
Varian Instruments
VG Instruments

We greatly appreciate the support given by these institutions and firms.

Selenium in Medicine and Biology

Second International Congress on Trace Elements in Medicine and Biology
Avoriaz, France, March 1988

Organized by:

Société Française d'Etude et de Recherche sur les Eléments Trace Essentiels (SFERETE), Société Française de Biologie Clinique (SFBC), and Association des Anciens Internes du CHU de Grenoble.

Organizing Committee: A. Favier, President, A. Alcaraz, J. Arnaud, C. Boujet, B. Dardelet, V. Ducros, H. Faure, J.L. Lafond, J.C. Potie, M.J. Richard, J. Rocipon, M. Rual (Grenoble).

Scientific Committee: J. Nève, President (Brussels), J. Arnaud (Grenoble), F. Baruthio (Nancy), P. Chappuis (Paris), A. Favier (Grenoble), O. Guillard (Poitiers), M. Lamand (Clermont-Ferrand), F. Pierre (Nancy), A. Pineau (Nantes), R. Zawislak (Strasbourg).

Under the Auspices of:

Université J. Fourier, Grenoble
Conseils Régionaux Rhone - Alpes et Haute Savoie
Société Belge de Chimie Clinique

CONTENTS

SECTION I : SELENIUM INTAKE, METABOLISM AND HOMEOSTASIS

Plenary lectures given by : G. Gissel - Nielsen
A. Favier
D. Behne

Chairmen : L. Molle
J.M.C. Gutteridge
J. Clausen
R. Bourdon

*

*   *

*

* *

SECTION III : ASSESSMENT OF SELENIUM STATUS

Plenary lectures given by : J. Nève
Y. Thomassen
P. Chappuis

Chairmen : M. Lamand
D. Behne
Y. Thomassen
R. Bourdon

*

* *

SECTION IV : SELENIUM IN HUMAN DISEASES

Plenary lectures given by : I. Lombeck
J.M.C. Gutteridge
A. Peretz
G.N. Schrauzer

Chairmen : J.K. Huttunen
J.F. Narbonne
J. Nève
M. Tolonen

*

* *

SECTION V : SELENIUM SUPPLEMENTATION AND TOXICITY

Plenary lectures given by : J.K. Huttunen
A. Pineau
F. Baruthio

Chairmen : A. Favier
M. Lamand
D. Behne

*

* *

SECTION VI : SELENIUM IN ANIMALS

Plenary lecture given by : M. Lamand
Chairman : J. Fontaine

Section I

Selenium intake, metabolism and homeostasis

SELENIUM INTAKE BY PLANTS, ANIMALS, AND HUMANS.

Gunnar Gissel-Nielsen

Agricultural Research Department, Risø National Laboratory
DK-4000 Roskilde, Denmark.

Introduction

Selenium has been recognized as an essential trace element for animals and humans for 25-30 years, and in many countries lack of Se in food is a common example on mineral imbalance due to intensive plant production. Several attempts have been made to demonstrate essentiallity for plants, too, by depleting the growth medium for Se, but so far without success. Furthermore, of all the Se-containing biologically active components found in animals none are found in plants. The interest for Se in plants is consequently related to the quality of the plants as animal fodder and as human food.

Geographical distribution of Se

The minimum Se-concentration of the total fodder that will meet animal requirements is considered to lie in the range of 0.05-0.1 ppm, and toxic effects can be expected at chronic intakes of fodder that exceed about 1 ppm Se. On the basis of these limits, different areas of the world are characterized as Se-deficient, Se-adequate, and Se-toxic areas. Outside Europe crops containing toxic Se-concentrations are found in the midwestern regions of the US and Canada, in Venezuela, India, and China. Se-deficiency areas are more widespread. It is reported from both the western and the eastern coastal areas of North America, from Venezuela, Australia, New Zea-

Selenium in Medicine and Biology

land, Japan, and China (1). From most countries of the world no information is available about the Se-status.

The situation in Europe is illustrated in figure 1. There is a geographical pattern in this map showing Scandinavia as a natural low-Se area, while central Europe is balancing between deficiency and sufficiency. From southern Europe information is infrequent. A few samples from Italy point to a level ranging from somewhat deficient to sufficient (2). Se-toxicity is only observed spot-wise from Wales and Eire.

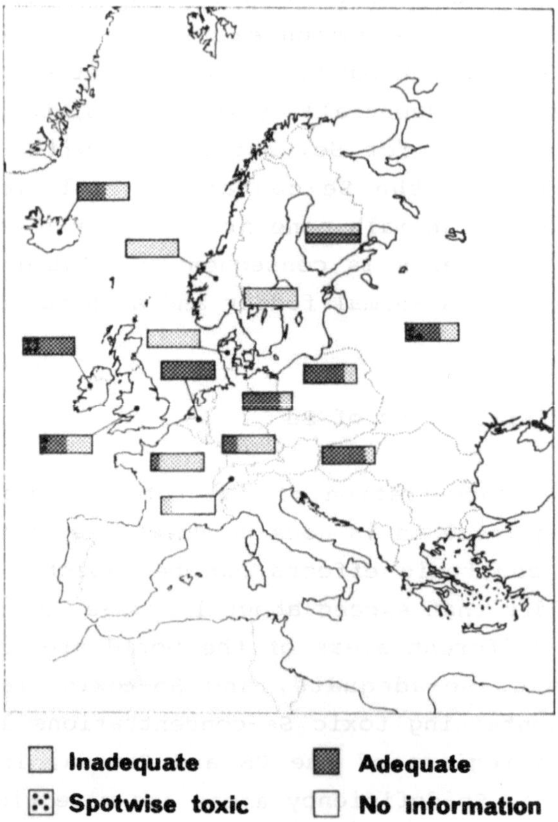

Figure 1. Distribution of Se in fodder crops in Europe. The two symbols for Finland refer to before and after 1985 (3,4).

How bad the Se situation is in the Scandinavian countries can be seen in Table 1. Finland (3) was most seriously affected with only about one-tenth of the minimum required Se in barley before 1985, the year when the Finns started adding Se to all compound fertilizers. Then follows Norway (5), Sweden (6), and most fortunate is Denmark (7). The Se-content of the grass is twice as high as that of the barley grain, but even the grass from Denmark has only 0.040 ppm Se on average.

Table 1. Se Concentrations in Scandinavian Crops. (ppm Se)

		Grass	Cereals
Finland	1968-69	0.014	0.007
Sweden	1968	0.027	0.011
Denmark	1972-73	0.040	0.018
Norway	1975-78	0.025	0.009

Se-intake by humans and animals

These low Se-concentrations in plants are reflected in the daily Se-intake by humans and in their blood Se-content. Table 2 gives some data on this from a few selected countries (1).

Table 2. Human Dietary Se-intakes and Blood Se-levels.

	Se-intake µg/day	Blood Se µg/l
Belgium	55	123
Canada	98-224	182
China	11-4990	8-3180
Denmark	40	86
Finland	30	56-87
New Zealand	28-56	59-83
USA	62-216	157-265

These values for Se indicate a relationship between the Se content of food produced in different countries and the blood Se-content of the inhabitants. A great number of publications

indicate a similar correlation between the Se content of animal food and animal blood Se.
The scources behind the intake of Se are many, and so are the publications giving the results of surveys on Se-content of foodstuffs. One of the first comprehensive surveys was that of Koivostoinen (8), and it gives the results from the low-Se area of Finland, while, e.g., that of Robberecht et al. (9) from Belgium illustrate the situation in more Se-sufficient countries. The differences in Se-content of the same foodstuff from different surveys are obvious when comparing the reported results, but arranged in order of decreasing Se-content the similarities are obvious, too. Concequently, I shall not give exact values for the Se-content of the different foodstuffs here, but list the foodstuff groups in a relative order (Table 3).

Table 3. Groups of Foodstuff in Order of Relative Se-content.

Relatively high	Seafood
	Meat
	High-sulphur vegetables
	Dairy products
	Cereals
	Other vegetables
Relatively low	Fruits

Because of the correlation between the Se-intake and blood-Se as shown Table 2, the Se-intake of humans and animals living in a certain area can be estimated to a great extent by evaluating the Se-uptake by the fodder crops grown in the particular area.

Se-uptake by plants

A straight correlation between soil Se and plant content excists only when comparing the situation for areas with extreme differences in Se-status. Within low to moderate Se-areas no general correlation is present. This is due to the large number of factors influencing the availability of soil Se to plants (1).

In soils with high pH, inorganic Se will be present mainly as selenate, which is hardly fixed at all in the soil. If the precipitation and the leaching is low, most of the Se will be available for the plants, and the crops will be Se-rich. This is known in some places in India. Contrary to this, low pH favours the selenite form, which is fixed strongly in the soil. Even with the same total Se content of the soil as in the above-mentioned example, the Se concentration of the crop might be ten times lower.

This situation is refered to in figure 2, where a fixation of selenite by clay minerals and organic matter is indicated, along with a very strong fixation by iron hydroxides. Through microbial activity volatile Se is lost to the atmosphere. However, Se also returns to the soil from the atmosphere. All this leaves only a minor part of the Se in the cycle to pass through the plant-animal system.

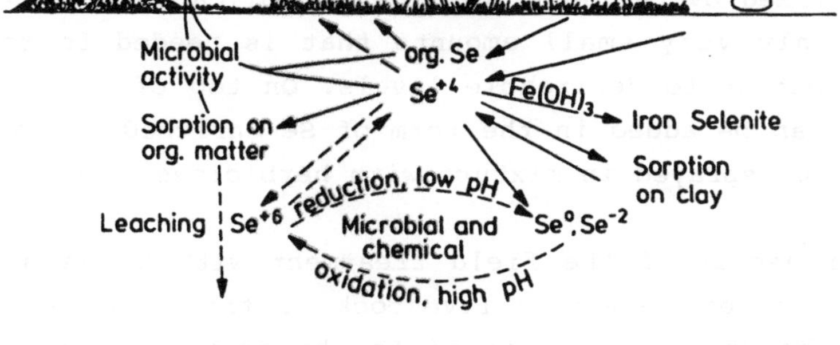

Figure 2. Possible Cycles of Se under Field Conditions.

Futhermore, the Se-concentration of plant samples depends on the time of sampling (10). The release of Se from clay and from organic matter is a very slow process. The Se released during the winter is available for the crops in the spring when the yield of, e.g., grass is low, giving a relative high Se concentration. However, when the growth of the grass increases during the summer, less Se will be taken up and distributed in a much bigger yield, and the concentration drops as a result of dilution.

Field treatment with Se

To remedy the low natural crop Se-concentration, some ways of raising it have been developed during the years, and those that are most possible have been summarized in a field experiment from which the main results are given in Table 4 (11).

Table 4. Se Treatment of Barley in Field Experiments.

Control	0.013	ppm
10 g Se/ha as selenite in PK-fertilizer	0.069	-
20 - " -	0.137	-
120 g Se/ha as selenite in PK-fertilizzer	0.088	-
5 - " - sprayed on plants	0.069	-

From these results it is seen that when using selenate which is not fixed by the soil, or foliar application to the plants it is only very small amounts that is needed to raise the plant content to desireable levels. On top of this, the selenate can be added in the form of Se-enriched compound fertilizer or sprayed in mixture with herbicides or other pesticides.

Another aspect of the field treatment with Se as a mean to prevent Se-deficiency in livestock is the chemical form in which the Se is administered to the animals. Whether Se is supplemented through pre-mixed fodder, concentrate, mineral supplement, water, lickstone, or ruminant pellet the Se is offered as inorganic selenite, while Se in crops predominantly is present as free seleno-amino acids or in proteins. The impact of this is still subject to many discussions.

Bioavailability of Se

However, there is no doubt that the bio-availability of different Se-compounds varies strongly. This has been shown

Table 5. Relative biological values of Se

Se scource	% survival
chick diet	34
+ dried fish	38
+ raw bread	53
+ raw beef	69
+ baked bread	72
+ selenite	81

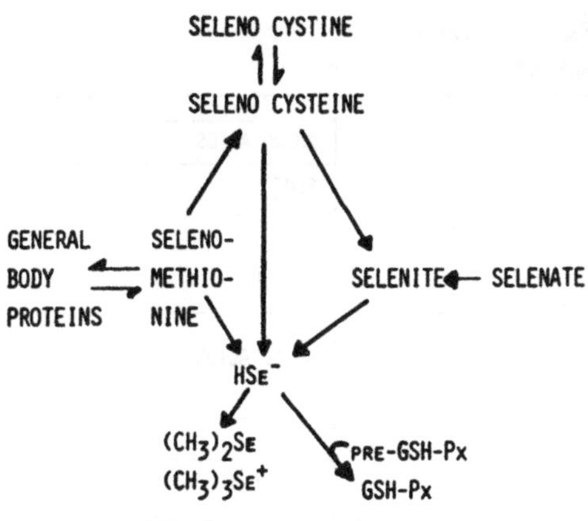

Figure 3. Possible Pathways of Se-compounds in the Formation of GSH-Px.

in several experiments, and Table 5 gives the results from just one of them. Laws et al. (12) fed Se-deficient chickens with a number of Se-sources, and Table 5 shows the percentage of surviving birds fed the same amount of Se.
Table 5 indicates a marked difference in the bio-availability even between organic scources of Se. This question was discussed by R. A. Sunde at the 12th Linderström-Lang symposium in 1982. He suggested the pathways for different Se-compounds in the production of Glutathione-peroxidase (GSH-Px) shown in figure 3.
Figure 3 indicate that surplus of Se-methionine can be stored in general body protein, while surplus of Se-cystine, Se-cysteine, and inorganic Se will be excreted with the urine.
Such differences stress the importance of the form of Se in animal fodder and human feed, and some work has been done to evaluate the possibilities of influencing the chemical form of the Se in plants. Earlier experiments (13) have shown that Se is translocated in the xylem-sap of maize as selenate when added to the nutrient solution as selenate, while selenite is metabolized immediately to Se-amino acids and translocated as such (figure 4).

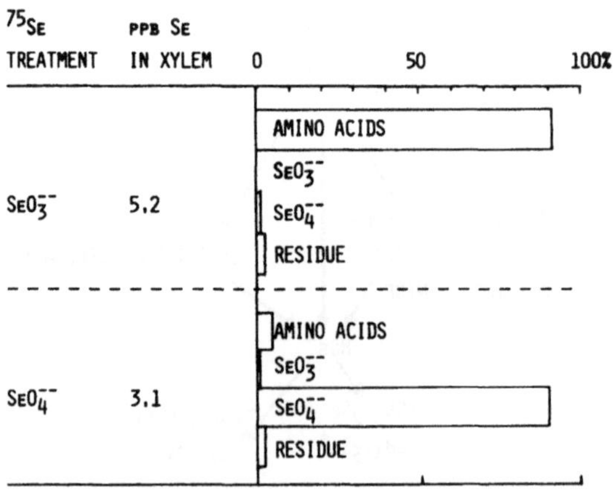

Figure 4. The Distribution of Se into Fractions of Xylem Sap from Maise when Added as Selenate or Selenite.

These results indicate a possible way of affecting the metabolic pathway of Se in plants, so we carried out some experiments in which rye-grass and barley are given Se as selenite or selenate through the roots or by foliar application. By a series of fractionations of the Se-compounds of the grass and the barley grain the compounds were separated into total and water-soluble Se-proteins, Se-amino acids, selenite and selenate. Neither the oxidation state of the added Se nor the way of application had any significant effect on the distribution of the Se between the four possible compounds. In all four cases only about 10% of the Se was present in an inorganic form (13). This implies that any means of supplementation of crops with inorganic Se can be evaluated on the basis of the uptake in per cent of the total added.

Conclusion

The human and animal intake of Se from natural sources varies over the world from a few cases of toxicity to rather widespread incidences of moderate to strong deficiency. Only China has reported severe cases of deficiency leading to the death of many humans, but the moderate deficiency as seen in many countries might be more dangerous to human health than often considered. In Finland they have taken the consequences of high rates of cardiovascular diseases and certain forms of cancer and by law introduced Se in all compound fertilizers in amounts that equal 10 g Se/ha as selenate. In New Zealand selenized compound fertilizers are available. In many other countries Se-enriched pre-mixed fodder for the livestock and Se-pellets for human supplementation are available, but most often the Se is present in an inorganic form. This means that in many cases the Se nutrition of animals is taken much better care of than that of people, even if people in general eat a much more mixed variety of food including more of the relatively Se-rich foodstuff as fish and some vegetables. In the future increased attention to the importance of a sufficient Se-content in human food - as is already done for livestock - can be expected to be paid. Hopefully, it will be accompanied by the realization that a reasonable uptake of Se by our crops is a cheap, sure, and easy way of ensuring a desirable human and animal intake of Se.

References

1. Gissel-Nielsen, G., U.C.Gupta, M.Lamand, and T. Westermarck. 1984. Advances in Agronomy, 37, 397-460.
2. Bordini, A., M. Di Vincenzo, C. Sinisalchi, and A. Valeriani. 1985. Riv. Soc. Ital. Sci. Aliment, 14, 357-360.
3. Oksanen H.E. and M. Sandholm. 1970. J. Sci. Agric. Soc. Finl., 42, 250-253.
4. Ylåranta, T. (Personal communication.)
5. Frøslie, A., J.T. Karlsen, and J. Rygge. 1984. Acta Agric. Scand. 30, 17-25.
6. Lindberg, P. and S. Bingefors. 1970. Acta Agric. Scand. 20, 133-136.
7. Gissel-Nielsen, G. 1975. Acta Agric. Scand., 25, 216-220.
8. Koivistoinen, P. 1980. Acta Agric. Scand., suppl. 22.
9. Robberecht, H., H. Deelstra, D. vanden Berghe, and R. van Grieken. 1982. Revue des Fermentations et des Industries Alimentaires, 37, 188-201.
10. Gissel-Nielsen, G. 1975. Acta Agric. Scand., 25, 216-220.
11. Gissel-Nielsen, G. 1986. Biol. Trace Element Research, 10, 209-213.
12. Laws, J.E., J.D. Latshaw, and M. Biggert. 1986. Nutrition Reports Intern., 33, 13-24.
13. Gissel-Nielsen, G. 1987. Journal of Plant Nutrition, 10, 2147-2152.

SELENIUM INTAKE OF THE SWISS POPULATION - PRELIMINARY RESULTS

B. Zimmerli
Federal Office of Public Health, Division of Food Control, CH-3000 Bern 14

A. Wyttenbach
Paul-Scherrer-Institute (PSI), CH-5303 Würenlingen

Introduction

Selenium (Se) is known to be an essential element for birds, mammals and humans but not for plants (1,2). Mainly by extrapolations from the Se requirement of mammalian species it has been suggested that a daily intake in a range of 50-200 µg is adequate and safe for adults (3). In recent years a relative Se-deficiency of humans has been suggested to be associated with an increased risk of developing cancer and cardiovascular diseases (1,4).

The Se intake of a population depends on the Se concentration of the soil and its bioavailability to plants. According to the geological history of Switzerland it can be assumed that the concentration of Se in Swiss soil is rather low, especially south of the Alps. This assumption has recently been supported by the results of some bed-rock analysis (5). The results of forage and cattle blood analyses as well as the sporadic occurrence of "white muscle disease" (north of the Alps), suggest that the Swiss roughage feed does not always contain enough Se to cover the requirements of cattle (6).

However, data concerning the Se content of Swiss food are very scarce and no intake study was available hitherto. On the other hand there exists a growing demand for registration of Se-enriched health food. The purpose of this study, which is a part of a general survey of the Swiss diet (7), was to estimate and assess the dietary Se intake of the population. The results obtained for nutrients, Pb, Cd, Hg, Ni and pesticides from this study are discussed in (8,9,10), those for Se and other 18 trace elements are summarized in (11).

Selenium in Medicine and Biology

Experimental

Ten complete daily rations, including breakfast, lunch and dinner, were sampled at 4 different food service kitchens (Jan./Feb. 1983) on 10 consecutive days. The specific locations were: a personnel restaurant (A), a hospital (B), a lacto-ovo-vegetarian restaurant (C) and a military canteen (D). Additional beverages not served with the meals and snacks were not included. The mean mass and energy of the daily rations amounts to 2.02 kg resp. 9.54 MJ (2260 kcal.). Although some deviations from the per capita consumption figures are evident (e.g. less meat and fruits), the mean composition can be taken as typical for the Swiss population (7).

All Se determinations have been performed by purely instrumental neutron activation analyses [Se-75, irradiation: 24 h with thermal neutrons ($1.2 \cdot 10^{13} n/cm^2/s$), decay time 60 d] on 100 mg aliquots of the homogenized and freeze-dried samples (10). Accuracy was tested by analyzing aliquots of NBS SRM 1571 (orchard leaves).

Results and Discussion

Not unexpectedly the mean Se intake at the lacto-ovo-vegetarian restaurant (C) is the lowest (table 1). Nevertheless the reason might not be the absence of meat, seafood and offals, but rather the very low incidence of pastas (1 from 10 rations). Although about 1/3 (13 from 40 rations) of the daily rations are below 50 µg/d (6 from C, 3 from A, 2 from B and D), the overall mean daily intake is about 70 µg/d. Taking further into account that snacks were not included and that meat amounted to only about half of the estimated per capita consumption figure (7), the results suggest that the mean Se intake of the general population should be well within the proposed adequate and safe range.

Recent intake figures from other countries revealed by chemical analyses of 24 h duplicate portions of individuals, indicate that the estimated mean nutrition density of Se seems to be comparable to the one of FRG, NL, S (12-14), whereas those of the U.S.A. and Canada are higher (15-17). Some recent results (18) of Se determination in blood serum of 10 healthy Swiss women ($\bar{x}$ = 93 ng/ml, 80-108 ng/ml) are also comparable to the mean found in other European countries (e.g. S, NL, B, UK). For the U.S.A. and Canada somewhat higher figures have been reported (19).

Table 1:. Daily Se Intake and Nutrition Density at Different Locations

Measures		Location (n = 10 each) A	B	C	D	All
$\bar{x} \pm s$	(µg/d)	82 ± 31	68 ± 20	53 ± 29	79 ± 33	71 ± 30
Range	(µg/d)	37 -127	39 - 96	32 -125	30 -135	30 -135
Se/dry matter	(µg/g)	0.18	0.16	0.13	0.16	0.15
Se/energy	(µg/MJ)	8.58	7.62	6.17	7.35	7.51
Se/protein	(µg/g)	0.98	0.83	0.88	0.84	0.88

Taking into account that Switzerland tends to be a region poor in Se, and that only about 65 % (in 1986/87 about 80 %) of the cereals for bread-making are grown in the country (20), and further that Se is not routinely added to feed or fertilizers (21), our results can be explained by the share of imported food and feed. It has been estimated that the net self-suffiency in food is 50-60 % on an energy basis and 60-70 % on a protein basis (22).

By comparing the composition of the different daily rations, a preliminary rough estimate of the mean Se content of food can be made (fresh weight): "meat" 0.06 µg/g, egg 0.2 µg/g, "liver and fish" 0.3 µg/g and dry pastas (egg-noodles) 0.5 µg/g. Pastas are exclusively manufactured from hard wheat (durum) imported from the U.S.A. and Canada mainly. Therefore the estimated Se concentration in pastas is in a fairly good agreement with the reported values in those countries, whereas the concentrations in meat seem to be much lower (23). The following tentative conclusions can be drawn: 1. There is no evidence that the dietary Se intake of the Swiss population in general is outside the suggested adequate and safe range. 2. Imported food and feed seem to be a substantial direct and indirect source of dietary Se. 3. Since Switzerland seems to be a region poor in Se, it cannot be excluded that certain population groups have a rather low Se intake. This question as well as the Se concentration in different home-grown foods warrants further studies.

References

1. Gissel-Nielsen, G., U.C. Gupta, M. Lamand, T. Westermarck, 1984. Adv. Agronomy 37, 397-460.
2. Brown, T.A., A. Shrift. 1982. Biol. Rev. 57, 59-84.
3. National Academy of Science (NAS). 1980. Recommended Dietary Allowences. 9th ed.. National Academy of Science, Washington D.C..
4. Levander, O.A.. 1986. In: Trace Elements in Human and Animal Nutrition. Vol. 2, 5th ed. (W. Mertz, ed.). Academic Press, New York, pp. 209-279.
5. Jäger, E., D. Kurz, U. Krähenbühl (University of Berne): to be published.
6. Mathis, A., H. Horber, H. Jucker. 1983. Schweiz. Arch. Tierheilk. 125, 317-328.
7. Zimmerli, B., R. Knutti. 1985. Mitt. Gebiete Lebensm. Hyg. 76, 168-196.
8. Stransky, M., P. Scheffeldt, A. Blumenthal. Mitt. Gebiete Lebensm. Hyg. 76, 197-205.
9. Knutti, R., B. Zimmerli. 1985. Mitt. Gebiete Lebensm. Hyg. 76, 206-232.
10. Corvi, Cl., C. Majeux, J.Vogel. 1986. Trav. chim. aliment. hyg. 77, 583-597.
11. Wyttenbach, A., S. Bajo, L. Tobler, B. Zimmerli. 1987. In: Trace Element - Analytical Chemistry in Medicine and Biology (P. Brätter, P. Schramel, eds.). W. de Gruyter, Berlin, New York, pp. 169-178.
12. Schelenz, R. 1983. In: Essentielle und toxische Inhaltsstoffe in der Gesamtnahrung. BFE-R-83-02 (R. Schelenz, Hrsg.). Bundesforschungsanstalt für Ernährung, Karlsruhe, pp. 192-200.
13. Ellen, G.. 1987. Dietary Studies in the Netherlands: Duplicate Portion Approach. Paper Presented at the EC Workshop on Total Diet Studies, Zeist, November 9-11.
14. Kumpulainen, J., M. Mutanen, M. Paaki, J. Lehto. 1987. Var Föda, Suppl. 1/87, 75-82.
15. Welsh, S.O., J.M. Holden, W.R. Wolf, O.A. Levander. 1981. J. Am. Diet Assoc. 79, 277-285.
16. Levander, O.A., V.C. Morris. 1984. Am. J. Clin. Nutr. 39, 805-815.
17. Gibson, R.S., C.A. Scythese. 1984. Biol. Trace Elem. Res. 6, 105-116.
18. Bajo, S., A. Wyttenbach, 1986. EIR, Würenlingen, Report, TM-44-86-13.
19. Thorling, E.B., U. Overhard, J. Geboers. 1986. Ann. Clin. Res. 18, 3-7.
20. Brand, F., Eidg. Getreideverwaltung, Berne. 1988. Personal communication.
21. Morel, J., Eidg. Forschungsanstalt für viehwirtschaftliche Produktion, Grangeneuve, Posieux. 1988. Personal Communication.
22. Grütter, R., Schweiz. Bauernsekretariat, Brugg. 1988. Personal Communication.
23. Schubert, A., J.M. Holden, R.W. Wayne. 1987. J. Am. Diet. Assoc. 87, 285-299.

SELENIUM IN FOOD AND NUTRITION IN FRANCE

M. Simonoff, P. Moretto, G. Simonoff

C.E.N.B.G.-U.A. 451 du C.N.R.S.-Chimie Nucléaire, Le Haut Vigneau,
33170 Bordeaux-Gradignan, France

C. Conri, B. Cornaille, B. Fleury, Y. Ducloux,
Service de Medecine Interne, Hôpital de Haut Lévèque,
33605 Bordeaux-Pessac, France.

Introduction

A nutritional deficiency in selenium may be a risk factor for cardiovascular diseases (1). The anticarcinogenic properties of selenium, in addition, have been demonstrated in numerous animal experiments, recently reviewed by Shamberger (2), and in several epidemiological studies (3,4,5). A study involving 750 hospital patients in Bordeaux (6,7) showed that the serum selenium level was significantly lower in most pathological states (cancer, diabetes, cardiovascular diseases, alcoholism ...), compared to healthy controls. Whether these decreased levels were cause or effect has not been established.

The most obvious medical use of selenium at the present time is to prevent the development of a deficit during total parenteral nutrition (8,9). It however seems that supplementation should be more frequent to correct nutritional deficits in man, particularly those arising in breast fed infants because of inadequate maternal intake of selenium: indeed, the level of selenium in human milk is directly related to maternal intake (10). A somewhat more speculative use would be as a prophylactic anticarcinogenic agent (11).

To the best of our knowledge, there are no published measurements of the mean daily selenium intake in France, or of the selenium content of those food which are generally consumed. In the present study, we have measured the selenium levels in various samples of meat, fish, vegetables, fruits, and other foodstuffs by a proton induced X-ray emission (PIXE) mtehod.

Using the INSEE (Institut National de la Statistique et des Etudes Economiques) data, we calculated the average daily intake according to the different French regions and the social and professional classes. Moreover, the selenium content of several nutritional regimens was also determined.

Selenium in Medicine and Biology

Materials and Methods

Food samples

The choice of samples was based on their popularity according to information received from various government agencies and from the directors of supermarkets.

The samples were cooked or left raw, according to French custom, and placed in a 100 mL Pyrex flask, then weighed directly (wet weight). The sample weights were generally in the range 1-2 g, except for vegetables where ~ 5g samples were taken because of the low selenium levels encountered.

Chemical pre-concentration of selenium

The method is described in detail elsewhere (12). Selenium is determined in the samples by PIXE after wet ashing and coprecipitation on tellurium. The metallic residue is dissolved in nitric acid. A known volume of a 25 mg/L yttrium solution is added as an internal standard. Recovery for this preconcentration step is measured with the use of carrier-free ^{75}Se.

PIXE irradiation

A beam of 3.2 MeV protons from the 4MV Van de Graaff accelerator at the "Centre d'Etudes Nucléaires" in Bordeaux is used to generate X rays on the subsequent targets. Photon spectroscopy was carried out by means of a 100 mm^2 intrinsic germanium X ray detector.

The limit of detection for selenium under the present experimental conditions is of the order of a nanogram per target.

INSEE data

These data (13) give the dietary intake per year for each product in France (regions, social and professional classes).

Diets

The mean daily intake of selenium was calculated for different diets :

- Normal civilian and military hospitals, school cafetarias.
- Therapeutic for specific disorders.

The selenium intake was calculated as the product of the measured levels in the samples and the amounts of the foodstuffs for each of the different diets.

Results and Discussion

Mean results

The results of the measurements in various foodstuffs have been presented elsewere (14) and can be summarized as follows (Table I).

Product	Mean selenium content ng/g
Meat	80
Fish	200
Cereals	80
Milk	⩽13

Table I

Data per regions and professions

Using the INSEE tables (13) giving the dietary intake for each product in France along with the region and the social and professional class of the people we have calculated the mean dietary intake. It ranges from 40 to 60 µg/day. The details are given in tables II and III.

Selenium (μg/day) for different professions	
Farmers	59.5
Agricultural workers	52.2
Liberal professions	43.9
High grade employees	39.8
Employees	42
Working class	45

Table II

French Regions	Se μg/day
South West	52
Paris	40
North	48
East	49
Mediterranean	48
National Mean	45.9

Table III

Our values lie in the lower range of data from european countries except for some parts of Finland. No major differences are noted according to the professional classes. The lowest contribution of selenium in the dietary intake corresponds to the region of Paris, resulting mostly from a lower consumption of bread, eggs, cereals and fats.

Regimens

Selenium contents of several nutritional regimens were also determined (15) in normal and therapeutic diets.

Normal diets

* Diets from the Haut-Leveque Hospital (Bordeaux)

Mean 57 μg/day
Range (49.5-65)μg/day

* Diets from the Military Hospital

Mean 72.5μg/day
Range (63.8-81.1)μg/day

* Open-air schools

Mean 43 μg/day
Range (34-57)μg/day

* Vegetarian diets 42 μg/day

Therapeutic diets

* Calcium calculus 24 μg/day (Restricted : milk, cheese, tripe, shellfish, fish, dried fruits)
* Oxalate calculus 31 μg/day (Dairy products, meats, eggs, fish restricted but not excluded)
* Urate calculus 48 μg/day (excludes ; delicatessen, game, meat, mushrooms, offal, but the satisfactory Se intake is obtained from dairy products, meats, lean fish and cereals)
* Diabetic 30 μg/day (with 120 g of carbohydrates)
* Hypocaloric 44 μg/day
* Hypercholesterolemia 66 μg/day

Selenium intake in these different regimens lies within the national mean range except for recommended oxalic and calcic lithiasis diets.

Acknowlegments
The authors thank Mrs Regnier, Mrs Barrot and Pr. Cuisinier-Raynal for their contribution with the regimens.

References

1.Salonen, J.T., Alfthan, G., Huttunen, J.K., Pikkarainen, J., Puska. P. 1982. Lancet. 2, 175-179.

2-Shamberger, R.J. 1986. Trace Elements in Medicine. 3, n°3, 105-111.

3-Schrauzer, G.N., White, D.A., Schneider, C.J. 1977. Bioinorg. Chem. 7, 23-24.

4-Schrauzer, G.N., White, D.A., Schneider, C.J. 1977. Bioinorg. Chem. 7, 35-56.

5-Salonen, J.T., Alfthan, G., Huttunen, J.K., Puska, P. 1984. Am. J. Epidemiol. 3, 342-347.

-Salonen, J.T., Salonen, R., Lappete Lainen, Maenpaa, P.H., Alfthan, G., Puska, P. 1985. Br. Med. J. 290, 417-420.

6-Simonoff, M., Conri, C., Fleury, B., Ducloux, G., Berdeu, B., Llabador, Y.,Moretto, P. First Meeting of the International Society for Trace Element Research in Humans, Palm Springs, California, 1986.

7-Conri, C., Simonoff, M., Fleury, B., Ducloux, G. First Meeting of the International Society for Trace Element Research in Humans,Palm Springs, California, 1986.

8-Nève, J., Molle, L., Timmermans, J. 1985. Pharmakon. 62, 95-105.

9-Kein, C.L., Ganther, H.E. 1983. Am. J. Clin. Nutr. 37,319-328

10-Kumpulainen,J.,Salmenpera, L., Slimes, M.A. Koivistoinen, P. and Perkeentupa, J. 1985 Am. J. Clin. Nut. 42 , 829

11-Combs, G.F., Jr and Clark, L.C. 1985. Nutr. Rev. 43, 325

12-Simonoff, M., Hamon, C., Llabador, Y.,Moretto, P., G. Simonoff. 1988. Nucl. Inst. Meth. (in press)

13-Beyer, M.N., Mercier M.A. Collection de l'INSEE - Ménages M117/1986

14-Simonoff, M., Hamon, C., Moretto, P., Llabador, Y. 1988. Journal of Food Science (in Press)

15-Cornaille, B. 1988. Thèse médecine Bordeaux

SELENIUM IN WINE

H. Eschnauer

Gelnhäuser Str. 15, D-6463 Freigericht 1, FRG

F. Alt, J. Messerschmidt, G.Tölg

Institut für Spektrochemie und angewandte Spektroskopie, Bunsen-Kirchhoff-Str. 11, D-4600 Dortmund 1, FGR

Introduction

Little is known about selenium content in wine. The reason for this is, that up to the present time, no suitable determination method was available permitting the quantitative analytical detection of selenium in pg- and lower ng-regions. An appropriate procedure was therefore worked out which facilitated the reliable determination of the selenium at this concentration level.

Experimental

Selenium is a regularly and naturally occurring component contained in wine at the trace element level; the determination of selenium was required in widely varying samples such as in wine, in the skin of the grapes, in kernels, in the fruit pulp, in leaves, in stems, in yeasts, in soils, in taste-improving agents of wine and in fungicides. The concentration level investigated extended in this case over the pg/g- to the µg/g-region. Accordingly, the individual steps of the entire procedure had to be adapted to the appropriate operating conditions prevailing in each instance.

Selenium in Medicine and Biology

Open wet digestion of the samples

For all of the above kinds of samples, with the exception of the fungicides, an open wet chemical digestion method with nitric acid and perchloric acid was carried out. In this digestion method, it is ascertained that by maintaining a controlled time/temperature programme, no loss of selenium by volatilization occurs [1,2].

Soil samples and taste-improving agents of wine are not completely digested under these operating conditions. Determinations can only be made of those parts which are digestable by $HNO_3/HClO_4$. For the reduction of Se(VI) to Se(IV), the digestion residue is treated with 1 mL 5 M hydrochloric acid and heated to 95^0C in a boiling water bath for a period of 15 min.

Burning of the sample material in an oxygen flux (Trace-O-Mat$^{(R)}$)

200 mg of the fungicides were burnt in an oxygen flux. The volatile products of the burning process (including Se), frozen out in the cooling system of the apparatus, were collected in the stock vessel by refluxing with 5 M sulphuric acid and then transferred into a 10 mL volumetric flask by water [3]. By this technique, it is possible to separate selenium from matrix components (e.g. copper), which would cause interferences in the determination with the hydride technique of AAS.

Determination with the hydride-AAS technique, resp. with hydride-generation condensation-AAS (HGC-AAS)

The common hydride-AAS was employed as the only determination step in the ng-region and the hydride generation-condensation technique in the lower ng- and in the pg-regions (1,4). By the HGC-AAS the formed SeH_2 is first collected in an adsorption zone cooled with liquid nitrogen. By subsequently heating up of this zone, the SeH_2 vapourizes abruptly and by He carrier gas the SeH_2 is transported into the heated quartz cuvette of the spectrometer (for a more detailed description, see [4]).

Results and Discussion

With the described combined procedure (digestion and HGC-AAS), reliable determinations of selenium in the lower ng- and in the pg-region can be made. According to results attained, selenium is therefore a regularly and naturally occurring trace element contained in wine.

This was confirmed by the results of investigations of 103 various blends of wines of 1959 to 1984 vintages, i.e. 20red and white Ingelheim wines from the district of Rheinhessen, 42 Late Burgundy wines from Baden, 23 red and white wines from the district of Franken and 18 white wines from the Nahe district. The mean valuesof the above investigations lie between 0.29 and 0.79 µg/L; it thus appears, that the results attained can be correlated to the location of the respective vineyards and thus with the respective geological soil formation.

The red wines cultivated in comparable vineyard locations such as locations in Ingelheim (Rheinhessen) and locations in Franken contain more selenium than the white wines cultivated in the same locations. Perhaps, the cause for this lies in the higher Se concentrations in the skins of the red grape wines. The differences between the lower mean selenium contents in white wines from Franken (0.29 µg/L) and the higher contents in wines from the Nahe district might be explained by the different selenium contents in vineyard soils.

The relatively low selenium contents of the Late Burgundy wines from Baden might also be due to the lower selenium content in the vineyard soils cultivated in that area.

The total selenium content in commonly commercially available wines can thus be assumed to lie between 0.1 and 1.0 µg/L. And in this instance, we are speaking of the primary, naturally occurring content.

Further investigations involved the question, whether there would occur secondary contents of selenium in wines caused by external influences such as dust particle emissions from the air or by oenological contaminations. It was found that 13 different sulphur-containing fungicides used in vineyard cultivation contain amounts of selenium between 20 and 1260 µg/kg (mean value 144 µg/kg). But no secondary selenium contents in wines could be caused by such fungicides, even if used frequently during dry years and shortly before grape harvests.

German bentonites have unrisky selenium contents (2.7; 4.5 and 7.5 µg/kg). White wine bottles may be discoloured by SeO_2 and they may subsequently contain about 4 mg/kg Se. After storage for a period of 5, respectively 25 years of an Ober-Ingelheim red wine, comparable normal Se contents of 0.54 µg/L wine, respectively 0.58 µg/L wine were found. A selenium migration from the glass of the white bottle into the wine may thus be excluded. The Se-coloured ruby glass may contain up to 4000 mg/kg Se and during a period of storage of one year, 1 to 2 µg/L Se may migrate from it into the wine.

Phosphate-containing fertilizers can contain selenium and this may explain certain selenium contents in soils of long-term cultivation, including vineyard soils.

No selenium contaminations to grapes by brown coal power station could be established. However, dust particles emitted by a pigment factory were the cause that grapes showed selenium contents which were higher by several orders of magnitude.

The total content of selenium in 20 different vineyard soils in Ingelheim (Rheinhessen district) has been found to average 0.24 mg/kg Se, about 10% of which is water-soluble and is in that form available for plant consumption. It is likely that selenium reaches the grapes via vine root transport and not via air pollution on the grape surface. As regards soils, their total Se contents are generally stated to be 0.2 mg/kg; for limestone soils, it is stated to be only 0.08 mg/kg. This could be an explanation why selenium contents of wines from the district of Franken are so low.

Another studie involved the mobility and behaviour of selenium during the vegetation phase of the vines, subsequently labelled 'Red Portuguese' and 'White Silvaner' from Ober-Ingelheim. The water-soluble selenium available for plant consumption gains access to the leaves and the grapes and their respective parts via the roots and woods of the vine plant. There are no deposits of selenium on the surfaces of leaves and grape berries, because almost similar results in respect of Se contents have been obtained from washed and unwashed leave and berry samples. It can thus be concluded that Se transport moves via the root along plant body and enters into the fruit pulp and the kernels of the grapes. The skin of the grape berry of the 'Portuguese' grape shows the highest Se concentration, while in the 'Silvaner' grapes, the kernels have the highest Se content. The relatively high Se concentrations are leached out during mash-wash procedures; this is why high Se concentrations are found in new red wine while the must of the red and the white grapes are still containing similar concentrations of selenium.

After tapping of the wine, there are again almost similar selenium contents in the ready red and white wines. The reason for this is a substantial derichment of Se into the fermentation residue with derichment rates of between 30 to 60%, sometimes even of more than 95%.

Considering the Se distribution in the grape berries, it is the kernels which show high Se concentrations, as was observed in respect to other elements, too. The skins of the berries of the 'Red Portuguese' grape are considerably more Se-enriched than those of the berries of white grapes. As mentioned earlier, the cause for this is that red wines have higher Se contents, if they are mash-fermentated.

Acknowledgement

This research was suported by the 'Bundesministerium für Forschung und Technologie' and by the 'Ministerium für Wissenschaft und Forschung des Landes Nordrhein-Westfalen'.

References

[1] Piwonka, J., G. Kaiser, G. Tölg. 1985. Fresenius Z Anal Chem 321, 225

[2] Alt, F., J. Meserschmidt, K.H. Schaller. 1988. In: Analysis of Hazardous Substances in Biological Materials, Volume 2 (J. Angerer and K.H. Schaller, eds.). VCH Verlagesgesellschaft mbH., D 6940 Weinheim, p. 231

[3] Knapp, G., S.E. Raptis, G. Kaiser, G. Tölg, P. Schramel, B. Schreiber. 1981. Fresenius Z Anal Chem 308, 97

[4] Alt F., J. Messerchmidt, G. Tölg. 1987. Fresenius Z Anal Chem 327, 233

SELENIUM METABOLISM

A. Favier
Biochimie C, C.H.R.U. Grenoble, B.P 217 X, 38043 Grenoble Cédex, France

Among all trace elements, selenium possesses the more sophisticated metabolism. As a result many points remain unclear, even through a great many papers, reviews or books have been published on this subject (56, 9).

This fact is mainly due to electronic structure of selenium which gives to this metal chemical and physical properties very close to sulfur. Sulfur and selenium are elements of group VIA. They have very close electro-configuration, covalent radius, ionic radius, ionization potentials and electro-negativity.

This similarity explains that selenium compounds can be metabolized by the same enzymes as those devoted to their sulfur analogs, and that traces of selenium can be incorporated instead of sulfur in quite a lot of biological structures. So, when identifying selenium in a biological compound, it will be difficult to know if it has a specific role to play in this living cell, or if it reflects only sulfur labelling.

Another consequence of this similarity is that selenium exists in the body in different chemical forms : minerals like selenite or selenate or selenide, or organic like selenocysteine, selenomethionine. In the later form, selenium is covalently bound to carbon or sulfur, so that each one of the above-mentioned organic compounds has a specific metabolism in the body. But there exists also interconnective pathway between the different forms. Some of the main chemical forms are presented in table I. But sulfur and selenium are not entirely similar. Large differences exist in the acidic strenght of their hydride and in their oxydoreduction properties.

Selenium in Medicine and Biology

Table I : CHEMICAL FORMS OF SELENIUM

	OXYD.	FORMULA
Selenate	6+	SeO_4^{2-}
Selenite	4+	$Se O_3^{2-}$
Selenotrisulfide	0	R-S-Se-S-R
Selenocysteine	2-	$HSe\text{-}CH(NH_2)\text{-}COOH$
Se Methylselenocysteine	2-	$CH_3\text{-}Se\text{-}CH\text{-}NH_2\text{-}COOH$
Selenomethionine	2-	$CH_3\text{-}Se\text{-}CH_2\text{-}CH_2\text{-}CH\text{-}(NH_2)\text{-}COOH$
Selenocysthationine	2-	$COOH\text{-}CH(NH_2)\text{-}CH_2\text{-}Se\text{-}CH_2\text{-}CH_2\text{-}CH(NH_2)\text{-}COOH$
Trimethylselenonium	2-	$(CH_3)_3Se^+$
Dimethylselenide	2-	$CH_3\text{-}Se\text{-}CH_3$
Dimethyldiselenide	2-	$CH_3\text{-}Se\text{-}Se\text{-}CH_3$
Factor 3	?	

I - Biochemical reactions of selenium

These reactions are very important for the transformation of a dietary form of selenium into another one, as well as the incorporation of selenium in proteins (therefore for the synthesis of selenoenzymes), and for the synthesis of the chemical forms of excreted selenium.

Synthesis of selenoaminoacids in mammals : This is an important point of selenium metabolism.

Mc Connell in 1957 (35) had found selenomethionine and selenocysteine in animal proteins, but this may have been an artifact, because he used an hydrolysis step with hydrochloric acid that is now supposed to destroy selenoaminoacids. Morever, Schwarz in 1964 (52) found that, in vitro, selenite can bind to thioaminoacids and that complex formed presents the same retention time as selenoaminoacids. This reaction makes the presence of selenoaminoacid in an hydrolysate very difficult to prove. To remove this artifact Cummins (10) dialysed rabbit proteins before the hydrolysis step and found no selenoaminoacids. Fuss (16) had found some traces of selenomethionine and selenocysteine in the enzymatic hydrolysate of sheep proteins. Radioactive selenite^{75} had been previously injected (in a ruminant, selenoaminoacids may be synthetised by rumen flora). But Olson in 1976 (43), injected selenite^{75} in rats and found no selenocysteine nor selenomethionine in liver of kidney protein hydrolysate, but 2-7 diamino 4 thio 5 seleno octanoic acid.

Actually this subject remains controversial, but many authors argue that selenocysteine synthesis is possible (as discuss further) but not selenomethionine synthesis.

Figure 1

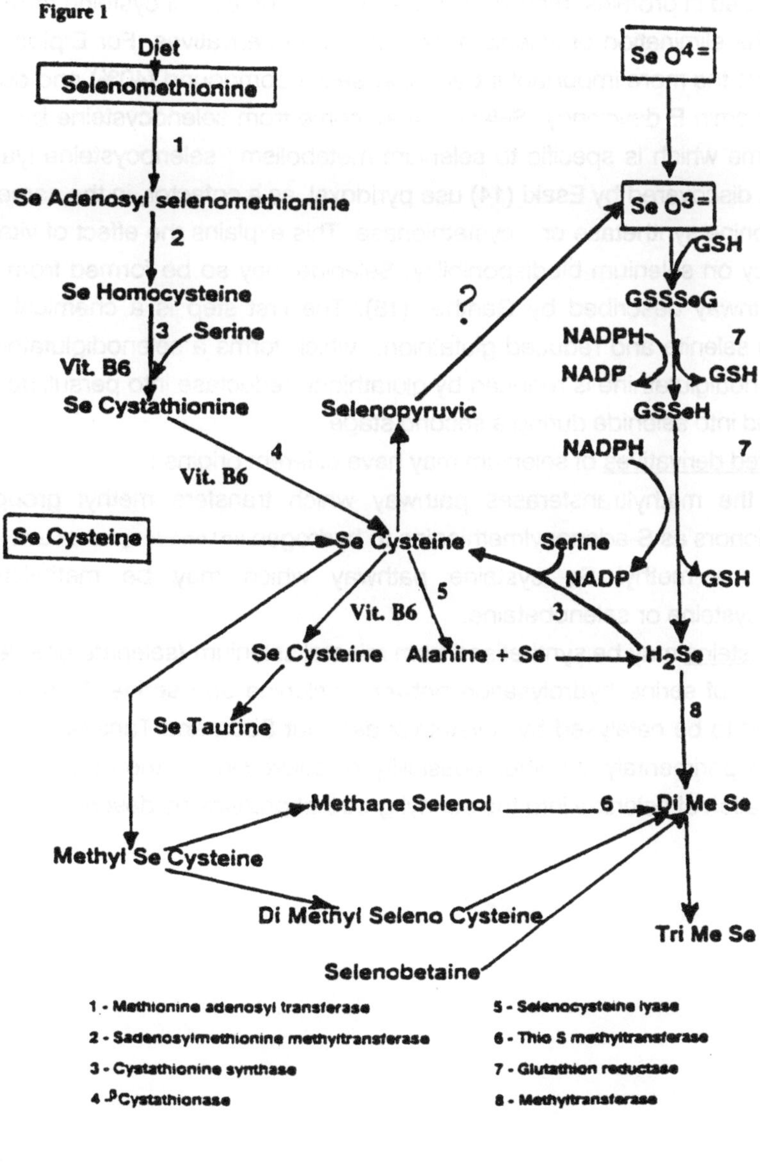

Reactions of transelenuration : As presented in figure 1, the first steps in selenomethionine metabolism are common with sulfur. Soda (58) demonstrated that enzymes of transulfuration, cystathionine β synthetase, β cystathionase are able to metabolize selenomethionine into selenocysteine. In human, inborn errors of metabolism of sulfur amino acids metabolism, disturb selenium metabolism, as showed by Belstein and Gahl (3) that found a decrease in the plasma selenium level and erythrocyte GPX, in patients with β cystathionase deficiency.

A very important metabolite is selenide (H_2Se). This compound may be incorporated in proteins, may lead as well to the synthesis of cysteine form selenite and to the elimination of selenium as methylated derivatives. For Diplock (12,13) selenide is the more important intracellular selenocompound (40%) and decreases during vitamin E deficiency. Selenide may come from selenocysteine by action of an enzyme which is specific to selenium metabolism : selenocysteine lyase. This enzyme, discovered by Esaki (14) use pyridoxal as a cofactor, in the same way as cystathionine synthetase or cystathionase. This explains the effect of vitamin B 6 deficiency on selenium biodisponibility. Selenide may so be formed from selenite by a pathway described by Ganther (18). The first step is a chemical reaction between selenite and reduced glutathion, which forms a selenodiglutathion (44). The selenodiglutathine is reduced by glutathione reductase into persulfide in a first stage and into selenide during a second stage.

Methylated derivatives of selenium may have different origins :

- the methyltransferases pathway which transfers methyl groups from methyl donors as S-adenosylmethionine to hydrogen selenide previously formed.
- the methyl Se cysteine pathway which may be methylated into dimethylcysteine or selenobetaine.

Selenocysteine may be synthetised from mineral selenium (selenate or selenite) by a reaction of serine hydrolysation between selenide and serine. This reaction is supposed to be catalysed by cystathionase, but Soda and Tanaka (58) failed to verify it experimentaly. Another possibility of conversion is the peritranscriptional incorporation of selenide into the serinyl group of proteins as described in a further chapter.

Analogies and differences between sulfur and selenium metabolism

Even if sulfur and selenium atoms are very close , they present some differences (33). The analogy between these two metals explains that some enzymes or transport proteins make no difference between their sulfur substrate and their selenoanalogs. This permits the absorption and metabolisation of selenomethionine but also explains antagonistic effects between sulfur and selenocompounds.

Antagonism between sulfate and selenate has been demonstrated by White (69, 70) and Pope (48) in sheep. Increasing the sulfate level of diet decrease urinary selenium and decrease selenium in blood and wool. Schubert (54) found in the same way that treatment of fields with gypsum generates White Muscle Disease in lamb. The same antagonism was found in rats by Ganther (20) and Bonhorst (4) : sulfate decreases the toxicity of selenate but not of selenite. This antagonism takes place at the level of urinary elimination not at intestinal absorption.

Selenate decreases the incorporation of sulfate into the cartilage mucopolysaccharides (23), but this is not a direct competitive mechanism because Campo (7) did not find radioactive selenium in the mucopolysaccharides of rats.

In vitro L-methionine competes with L-selenomethionine for absorption but paradoxically, a decrease in methionine level in diet, decreases SeGPX in rat (61). In the same way supplementation with methionine increases selenium level in blood and tissues (22). Some metabolic steps of selenium may depend on methionine (like the methylation of selenide that depends on adenosyl methionine). Inversely Burk (5) found an impairment of the conversion of methionine to cysteine in the selenium deficient chicken, and that addition of cysteine to selenium deficient diet protects against growth retardation. So there may be a selenium-dependant step for sulfur amino acid metabolism.

Main differences between sulfur and selenium related to oxydoreductive and acidic properties. Selenocompounds are reduced by their sulfur analogs in vitro as well as in vivo. This is clear in these two reactions :

In vitro $$\underset{(4+)}{H_2SeO_3} + \underset{(4+)}{2H_2SO_3} \longrightarrow \underset{(0)}{Se} + \underset{(6+)}{2H_2SO_4} + H_2O$$

In vivo $$\underset{(2-)}{4\,GSH} + \underset{(4+)}{H_2SeO_3} \longrightarrow \underset{(-1)(0)(-1)}{G\text{-}S\text{-}Se\text{-}S\text{-}G} + \underset{(-1)}{G\text{-}S\text{-}S\text{-}G} + 3H_2O$$

In the body sulfur tends globaly to be oxydised and selenium to be reduced as is reflected by the degree of oxydation of urinary excreted metabolites : sulfate (6+) as compared with selenonium (-2).

Selenides are more acidic than sulfides : pKa of selenol group of selenocysteine is 5.24, pKa of thiol group of cysteine is 8.25. So at physiological pH the selenol group is negatively charged (Se^-) and the thiol group positively charged (SH^{2+}). This explains the difference in properties of some enzymes when selenocysteine is incorporated at the place of cysteine in their structure.

Incorporation of selenium into proteins

Figure 2 : POSSIBLE MECHANISMS FOR INCORPORATION OF SELENIUM INTO PROTEINS

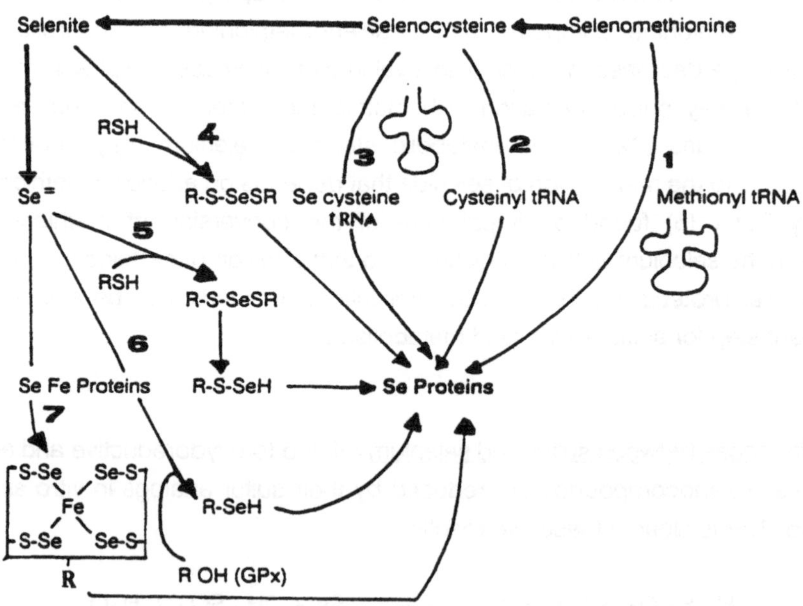

There exist many metabolic processes that may permit incorporation of selenium into proteins (Fig. 2) depending on the nature of the selenocompound. The study of these mechanisms is complicated by the possible interconversion between chemical forms and by the existence of two sorts of processes : a non specific remplacement of sulfur aminoacids by selenoaminoacids in general proteins and a specific incorporation of selenocysteine in glutathion peroxydase.

Selenomethionine may be randomly incorporated in place of methionine, by replacing methionine in methionyl t-RNA. Through this process incorporation of selenium is proportional to the selenomethionine level and is more important in methionine-rich proteins like thiolase, β galactosidase, and glutamine synthetase. A small rate of substitution has no effect on enzymatic activity, but high levels of substitution (at toxic intakes) may explain some of the toxic effects.

There exist other ways of selenomethionine incorporation, as proved by Millar and Shepphard (39), who obtained a different pattern of renal and hepatic proteins when injecting rats with radiolabelled methionine or selenomethionine. Selenomethionine may be partly catabolised and its metabolites incorporated into proteins.

Selenocysteine may be non specifically incorporated in place of cysteine by replacing cysteine on its t-RNA, but also by a specific mechanism using a specific t-RNA.

The non specific incorporation of selenocysteine will give very different properties to proteins due to the difference in pKa of the two aminoacids (24). This may be one of the mechanisms of selenium toxicity. In fact, it had been proved by Stadtman that the incorporation of selenocysteine is blocked by antibiotics inhibiting protein synthesis, but she failed to recover radioactivity when she used selenocysteine labelled on the carbon skeleton. More recently Sunde and Everson (63) have proved that the carbon skeleton of the selenocysteinyl residue on glutathion peroxydase, comes from serine. Selenide may be exchanged with the hydroxyl group of a proteic serine. This post traductional interconversion of serine into selenocysteine had been the first explanation of the experiment of Sunde and Everson (62). In fact, as described later, selenium may neither be incorporated before nor after traduction but simultaneously with traduction. Actually the most convincing hypothesis for the incorporation of selenium in the active site of glutathion peroxydase, is that Stadtman's (59). As described in figure 3 there exists a specific selenocysteine t-RNA formed from a serine t-RNA. This serine t-RNA is phosphorylated by a specific serinyl t-RNA kinase, then the phosphate group is exchanged spontaneously or enzymatically with selenide, leading to formation of a selenocysteine t-RNA. So selenium from selenite (or selenocysteine, or selenomethionine) is not incorporated after synthesis of a pre-glutathion peroxydase on the serine group, but one the serine t-RNA. Surprisingly the codon of this special selenocysteinyl t-RNA had been identified as UGA for the very close selenoenzyme formate deshydrogenase of Methanobacterium formicicum (57). This UGA codon is usually a stop codon in the genetic code, and is very close to, but different from the UGU codon of cysteine.

Selenite and selenide may be incorporated on the thiol group of proteins by the reaction described by Ganther (17) forming selenotrisulfide bridge, or selenopersulfides. Jenkins and Hiridoglou (27) had found a relation between the level of selenium fixation and the cystine content of proteins. Selenide may also replace sulfide in clusters or iron sulfur proteins. Mukai (42) prepared a selenoferrodoxin with the same activity as the sulfur enzyme. This non specific exchange can explain isolation of a selenoprotein on the respiratory chain.

Fig. 3 : INCORPORATION OF Se CYSTEINE IN GPX FROM HYPOTHESIS OF STADTMAN
(UGA Stop - UGU Cysteine - (UGA) Se Cysteine)

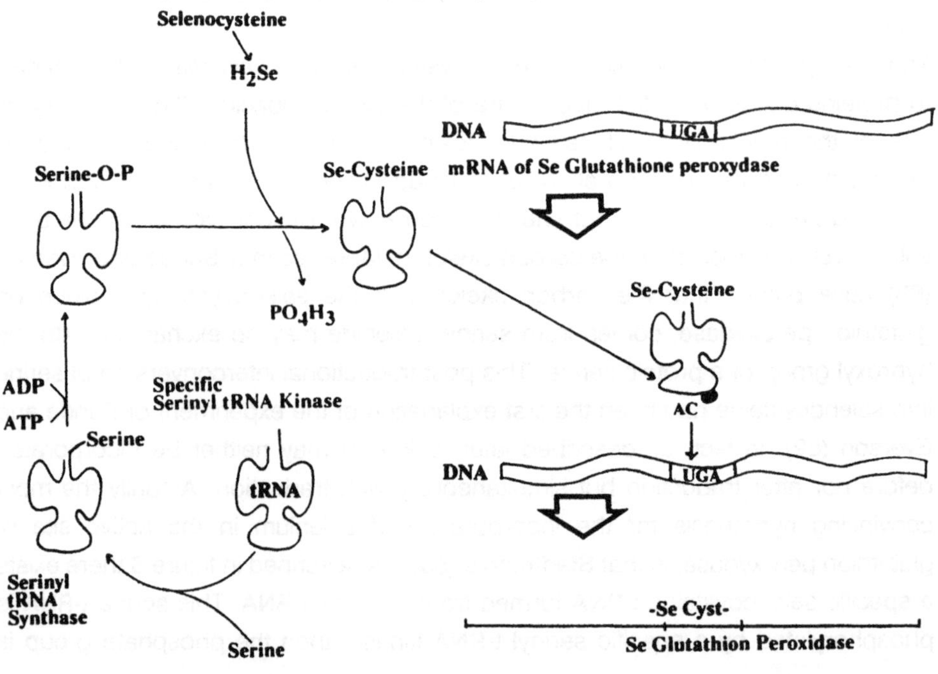

II - General metabolism

Intestinal absorption

Forms of selenium in diet : A great number of chemical forms of selenium have been identified in human diets. These forms are variable in species of plants or animals and depend on selenium level.

Selenocysteine and selenocystine have been identified in corn, selenomethionine in bacteria, grass or animal proteins. Selenocystathionine, seleno homocystine and methyl selenocysteine are abundant in seleno accumulator plants growing in selenium rich soil. Methyl derivatives are found in animals intoxicated with selenium. In animals some selenium exists in an unidentified form named factor 3 (56). A great number of others forms exist (selenotaurine, selenotrisulfides,...), but methods of identification suffer from artefacts. So quite a lot of work will be necessary to possess quantitative data on the nature of selenium in each component of human diet.

Mechanisms: Biochemical mechanisms of absorption are not entirely elucidated and depend on the chemical form of selenium in diet. Selenoaminoacids are liberated from animal or vegetal proteins by actions of digestive enzymes. L-selenomethionine is then transported across intestinal membrane by the same active mechanism as L-methionine. L-selenomethionine transport is half that of DL-selenomethionine, and is inhibited by methionine (36). Radioactive studies found no difference in cellular localisation between selenomethionine and methionine after absorption. Its absorption takes place along small intestine but more in duodenum. Absorption of selenite alone takes place in the small intestine and by a non active transfer. But physiologically selenite reacts first with cysteine or reduced glutathione, in the intestinal lumen, to form selenotrisulfides selenodiglutathione that is transported across the cellular membrane by gamma glutamyl transferase (2).

Selenate is absorbed in ileum by a binding process different from, and non inhibited by sulfate. Selenate for Thomson (66) is better absorbed than selenite, but had greater urinary elimination, and therefore a worse biodisponibility.

Factors influencing absorption or biodisponibility : They were reviewed by Combs (9).

Selenium status seems to have no effect on selenomethionine or selenite absorption, and so absorption is not the stage where selenium homeostasis is regulated. This makes selenium metabolism different from iron-zinc metabolism.

Heavy metals (As, Cd, Hg) decrease selenium absorption but they also modify selenium retention when injected. But reversibly Se decreases toxicity of heavy metals (As, Hg, Pb, Ag, Cd).

Vitamins level in diet modulate selenium biodisponibility : Vitamins B6, B2, E, A and C are more efficient. Vitamins A, C and E increase Se glutathion peroxydase (SeGPX) level but also intestinal absorption of Se. Pyridoxal and riboflavin seems without effect on intestinal absorption but their deficiency decreases Se GPX by decreasing the incorporation of Se into proteins.

Transfer by blood

The binding of selenium to plasma proteins is a difficult problem to understand due to fast redistributing mechanisms. In vitro selenite is not incorporated in plasma protein, but we have no proof that selenite is released, after absorption, by intestinal cells in the same chemical form. There are many proofs that selenium can be on a selenotrisulfide form (figure 4).

Figure 4 : GENERAL VIEW ON SELENIUM METABOLISM

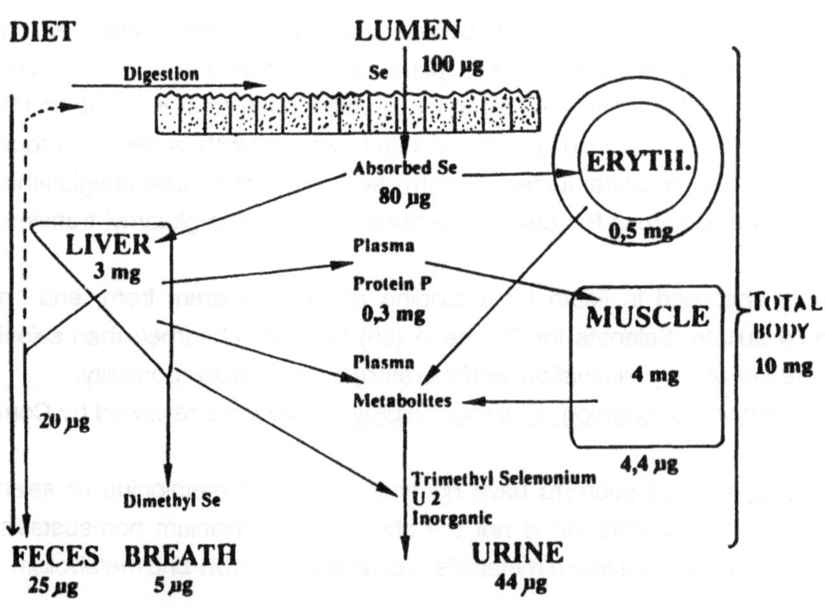

If some selenite exists in plasma it is very rapidly taken up by erythrocytes, and released in another form that Lee (31) identified as selenodiglutathione and Gasiewicz (21) as selenide fixed on plasma proteins.

In vivo experiments are conflicting. Many authors inject animals with selenite^{75} and measure radioactivity on plasma protein fractions. Schwarz (53) in dogs identified selenium in α_2 and β_1 globulins and albumin. Jenkins (28) found in chicken a different pattern after 2 hours (α_2 and α_3 globulins) than at 24 hours (α_2 and globulins), Millar (40) identified selenium in rat on α_2 globulins excreted by liver. Dickson (11) measured selenium in humans in α_2 and β globulins ; Burk (6) observed a net redistribution with time after injection of selenite^{75}. In the first hour selenium may be carried by VLDL, but at 6 H on LDL and α_2 globulin. Sandholm (51) found radioactivity on lipoproteins and on a peak between α_1 and α_2 globulins.

So many authors observe a variation of distribution with time, and a majority observe selenium on α_2 globulin. This is consistent with the results of Motzenbocker (41) who identify in rat a selenoprotein P excreted by liver a maximum 3 hours after injection of selenium^{75}. This protein contains selenium as selenocysteine, has a molecular weight of 80000, and migrates in the α_2 zone at electrophoresis. It is certainly the form of redistribution from liver to tissues as demonstrated by Motzenbocker by injection of this protein. This had been found also by Mc Murray (38) in sheep. He identified selenium in numerous proteins in the first hour but on a reduced form, labile after treatment with mercaptoethanol. In a second period he identified selenium on a stable form in only one protein with a molecular weight of 56750.

The redistribution is clearly visible on total radioactivity of plasma, which presents a decrease during first hours, then an increase with a maximum between 5 to 10 hours due to hepatic secretion of selenoprotein. We will see later that the same phenomen is found again during human studies with stable isotope.

Glutathion peroxydase don't play a role in transfer of selenium and represent only a small part of the plasma level.

<u>Transfer across membranes and uptake by tissue</u>

Selenium in different forms passes through placenta membrane, but some authors found that selenite when injected as ^{75}selenium has a lower transfer than selenomethionine or selenocysteine (67). The same difference seems to exist for the transfer from blood to milk through the mammary epithelial cells. Allen (1) found a greater entry of selenodiglutathion released by erythrocytes and fast incorporation of this form into milk proteins mainly casein that represente 70% of milk selenium. At the two levels, studies had been done by injecting radioselenite in blood of animals, but we have not any proof that this form still exists in blood after beeing absorbed. The mechanisms by which selenium is uptaken by cells remains to be elucidated.

Uptake by tissue depends on the mode of administration, on the selenium status, on the age and species of animals (8), on the form of selenium (i.e selenomethionine has a better uptake by pancreas and selenite by liver). Organs like the liver and pancreas attract selenium very fast but have a first rate of elimination, the brain lost selenium slowly, tests have a stedy uptake of selenium and a low elimination rate. Richer organs are testis, kidney, pineal gland, liver and pancreas, but the highest quantity of selenium is to be found in muscles with a content of 40% of the total.

Excretion of selenium

There exist many elimination routes but feces and urines are the main way at physiological intake, and breath the main way at toxic levels. Roughly we can estimate that urine excrete 60%, feces 35%, breath 1%, saliva 1%, and sweat 1% of normal intake.

Foster (15) in rat breath found less than 1% of a dose of 0.06 mg/Kg and 5% of a dose of 2 mg/kg, and Mc Connel (37) found 35% of a dose of 5.6 mg/Kg (as selenomethionine). In breath, selenium is present as dimethylselenium or dimethyldiselenium. So, methylation throught urine and breath is an important pathway of detoxification.

Urinary excretion depends on the form and intake of selenium. In rat Foster found that 30% of selenomethionine are excreted in urine as compared with 95% of trimethylselenonium, and Palmer (45) that 25% of selenite is excreted this way. Thomson (66) found excretion of selenate (58%) faster and more efficient than selenite (17%).

In man, depending of the dose, urinary excretion represents 21 to 25% of intake (50% for normal intakes) as can be seen in table II. The nature of the form of urinary selenium depends on the amount and form of intake. Trimethylselenonium ion is a specific form of urine but is little at normal intakes (4% in rat for Foster (15), 10% in man for Ganther (19)). The rest of urinary selenium remains to be identified. Urinary excretion seems to be one of the most important mechanisms of selenium homeostasis.

Table II : PERCENTAGE OF EXCRETION BY URINE OR FECES DEPENDING OF THE DOSE OR FORM OF SELENIUM INGESTED

INTAKE		URINE /ug	URINE % Dosis	FECES /ug	FECES % Dosis
STEWART 1978	24 /ug	13	62	11	45
LEVANDER 1981	24 /ug	29	129	17	70
	224 /ug Seleniferous wheat	112	50	86	38
SWANSON 1983	150 /ug	111	74	28	18
JANGHORBANI 1984	11 /ug 197 /ug	27 44	245 41		
LEVANDER 1985	90 /ug 74 /ug	48 39	53 52	34 23	37 31
PATTERSON 1987	Radiotracer Se^{75} 19 d. after 90 d. after		16 35		19 25
THOMSON 1986	1 mg selenite 1 mg selenate	210 680	21 68	380 60	38 6

Robinson (50) found a renal clearance at 0.2 to 0.3 ml/mn for American citizens (80 /ug intake), but at 0.1 to 0.2 ml/mn for New Zealand (30 /ug intake) ; in the later case by increasing intake to 150 /ug, clearance increases up to 0.4 to 0.7 ml/mn. Janghorbani (26) by decreasing intakes observed a net decrease in Se^{74}, injected seven days before.

Fecal excretion is around 20 to 30 /ug, but increases only with larger intakes. It represents non absorbed selenium but also excreted selenium from digestive secretions.

Saliva selenium is low (2.8 /ug/l) but may vary with intakes (34). Selenium in sweat is just as low (1.2 /ug/l) but does not depend on intakes.

Stable isotopes and quantitative study of selenium metabolism

Many isotopes can be used as a tracer to study selenium metabolism (Table III). Some are radioactives like Se^{75} but some stable isotopes like Se^{76} or Se^{74} have a small natural abundance and so are usable. Stable isotopes present important advantages, they are safe and so can be used for study on children or pregnant women.

Table III : SELENIUM ISOTOPES

STABLE		RADIOACTIVE	
	Nat.Ab %		Half-life
Se^{74}	0.81	Se^{75}	120 d.
Se^{76}	8.66	Se^{77m}	17.6 s.
Se^{77}	7.31	Se^{79}	3.89 mn
Se^{78}	23.21	Se^{81}	57 mn
Se^{80}	50.65	Se^{83}	70 s
Se^{82}	8.35		

Stable isotopes of selenium may be determined in biological materials by different methods :

- neutron activation as was done by Janghorbani (a time consuming and middle-range precision technique)
- thermoionisation mass spectrometry which presents a very high precision,
- gaz chromatography mass spectrometry of the piazselenol derivatives which is an easy to run but less precise technique (described by Reamed (49), Veillon and Levander (68)), that we have developped in our laboratory.

Metabolic studies can be done after ingestion or injection of labelled selenium by measuring disappearance rate of the isotope. By monitoring plasma enrichment it is possible to calculate turnover and half life of differents compartments. In figure 5 we present such a study for a normal subject that orally had absorbed selenite Se^{76}, which in plasma has a half life of 79 hours. By monitoring urine enrichment and plotting the log of enrichment versus time we obtain a succession of straight lines whose slopes are equal to the ratio intake on pool size.

Figure 5 : THE CHANGES IN PLASMA ^{74}Se WITH TIME AFTER INGESTION OF $^{74}SeO_3^{-}$ 100 µg (T 1/2 elimination = 79 h)

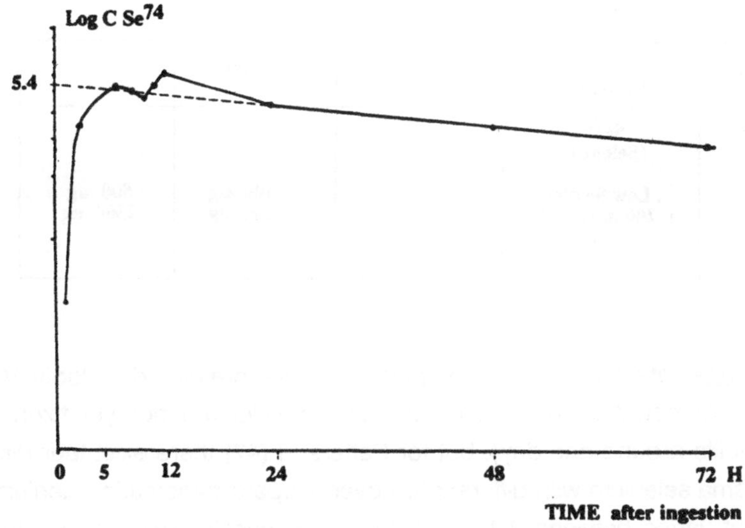

Stable isotopes can be used to compare biodisponibility of two different forms of selenium, each one labelled with a different isotope. Patterson (46) studied the incorporation into maternal milk of ^{74}Se selenomethionine and ^{76}Se selenite. In the same way we compared absorption of two drugs red ^{74}Se (Granions Laboratory) and ^{76}Se selenite.

<u>Size of selenium pools</u> measured by different authors are presented in table IV. Total body selenium is smaller in countries with normal intakes like the U.S.A. where total body pool size measured by labelling is close to those determinated by Schroeder (55). Stewart (60) found different results when the isotopic selenium is introduced as selenite or selenomethionine.

There exists a fast exchange pool (plasma and liver selenium) and a long time exchange pool (muscle selenium), but values for this pool differ with authors.

Table IV : SELENIUM POOL SIZE

	METHOD	TOTAL POOL Size	FRACTIONNAL POOL First Exchange	FRACTIONNAL POOL Long time
SCHROEDER	determination	14.6 mg		
STEWART (New Zealand)	Se^{75} (Se Methionine)	5.2 mg		
very low status 1978	Se^{75} (Selenite)	3.5 mg		
JANGHORBANI 1984	Se^{74} (Selenite)	10.3 mg	1 mg	
VEILLON 1985	Se^{76} (Selenite)			
	. Low Se diet . + 200 µg Se-Met. (21 days)		100 µg 200 µg	500 µg 2500 µg

Exchange rate for different compartments are presented in table V. The more easily eliminated is the fraction of plasma selenium not yet fixed upon proteins (half life around one day), but for Patterson (47) there exist four different forms of plasma selenium with different turnover. Hepato pancreatic selenium had a half life that varied between 2.4 to 46 days and muscle selenium around 150 days.

Table V : SELENIUM TURN-OVER IN PLASMA

		PLASMA	HEPATO PANCREATIC	TISSUES
RADIO TRACER Se^{75}				
THOMSON 1974	Selenite per os	1 d	8.3 d	115 d
JEREB 1975	Selenite IV		20 d	115 d
LATHROP 1982	Se Met IV	0.55	46 d	220 d
STABLE ISOTOPES Se^{74}				
JANGHORBANI 1984	Selenite per os		2.4 d	162 d low intakes
PATTERSON	Selenite per os	0.4 - 1.6 h 3 - 8 h 9 - 30 h 140 - 220 h	5 - 41 d	70 to 360 d

Another useful method to quantify selenium metabolism is balance method (Table VI), that may be combined with a tracer study. Using this method it had been showed by Stewart (60) that New Zealanders are in equilibrium with intakes for an intake that is four times lower as compared with that of Americans. This may be due to an adaptative mechanism to low intakes. Levander (34) had done numerous balance studies with different intakes, and calculated that zero balance is obtained with intakes between 60 to 80 μg.

Table VI : SELENIUM BALANCE (μg)

	INTAKES	FECAL Se	URINARY Se	APPARENT RETENTION
LEVANDER 1981	20 μg 200 μg (Seleniferous Wheat)	17 86	29 112	-26 +2
SWANSON 1983				
NW	150	28	111	+11
PW (14s)	154	33	100	+21
PW (34 s)	158	28	96	+34
LEVANDER 1981				
US Men	90	34	48	+8
US Women	74	23	39	+12
STEWART 1978 New Zealand Women	24	11	13	0

In concluding selenium is ambivalent, in part this metal has its own specific metabolism with enzymes as selenocysteine lyase, but in another part it may be a parasite of sulfur metabolism. This latter aspect is mainly greater at toxic level.
On the other side, at physiological level, selenium possesses a very precise and specific mechanism of incorporation into glutathion peroxydase. The selenium able to be incorporated in the process is selenide and may come from selenite as well as selenocysteine or selenomethionine. Different forms of selenium seem to have close the same biodisponibility on this point of view.

References

1. Allen J.C., W.J. Miller. 1981. Feedstuffs. 5, 22.

2. Anundi I., J. Hogberg, A. Sthahl. 1984. Acta Pharmacol. Toxicol. 54, 273.

3. Beilstein M.A., W.A. Gahl, P.D. Whanger.: TEMA 6 (in press)

4. Bonhorst C.W., I.S. Palmer 1957. J. Agric. Food Chem., 5, 931.

5. Bunk M.J., G.F Combs. 1981. Proc. Soc. for Exp. Biol. and Med., 167, 87.

6. Burk R.F. 1974. Biochim. Biophys. Acta. 372, 255.

7. Campo R.D., J.P.A. Wengert, C.D. Tourtelotte, M.A. Kirsh. 1966. Biochim. Biophys. Acta. 124, 101.

8. Chow C.K., J. Jeng. 1981. In : Selenium in Medicine and Biology (M.D Spallholz, H.E. Ganther and J.L. Martin, eds.). AVI Wesport, p. 477.

9. Combs G.F., S.B. Combs. 1986. The Role of Selenium in Nutrition. Academic Press, Orlando.

10. Cummins L.M., J.L. Martin. 1967. Biochemistry. 6, 3162.

11. Dickson R.C., R.H Tomlison. 1967. Clin. Chim. Acta. 16, 311.

12. Diplock A.T., C.P.J. Caygill, E.H. Jeffrey, C. Thoma. 1973. Biochem. J. 134, 283.

13. Diplock A.T., J.A. Lucy. 1973. FEBS Letter 29, 205.

14. Esaki N., T. Nakamura, H. Tanaka, K. Soda. 1982. J. Biol. Chem. 257, 4386.

15. Foster S.J., R. Kraus, H.E. Ganther. 1986. Arch. of Biochem. Biophys. 251, 77.

16. Fuss C.N., K.O. Godwin. 1975. Austr. J. Biol. Sci. 28, 239.

17. Ganther H.E. 1968. Biochemistry. 7, 2898.

18. Ganther H.E. 1971. Biochemistry. 10, 4089.

19. Ganther H.E. 1986. In : Proceeding of the Third International Symposium on Selenium in Biology and Medicine. (G.F Combs and Coll. eds.). AVI Wesport.

20. Ganther H.E., C.A. Bauman. 1962. J. Nutr. 77, 408.

21. Gasiewicz T.A., J.C. Smith. 1976. Biochim. Biophys. Acta. 428, 113.

22. Gu L., R. Zhou, S. Yin, Q. Yang. 1986. In : Proceeding of the Third International Symposium on Selenium in Biology and Medicine. (G.F. Combs and Coll. eds.). AVI Wesport.

23. Hilz H., F. Lipman. 1955. Proc. Natl. Acad. Sci. 41, 880.

24. Huber R.E., I.H. Segel, R.S Griddle. 1967. Biochim. Biophys. Acta. 141, 573.

25. Janghorbani M., M.J. Christensen, A. Nahapetian, V.R. Young. 1982. Am. J. Clin. Nutr. 35, 647.

26. Janghorbani M., L.J. Kasper, V. Young. 1984. Am. J. Clin. Nutr. 40, 208.

27. Jenkins K.J., M. Hiridoglou. 1971. Can. J. Biochem. 46, 468.

28. Jenkins K.J., M. Hiridoglou, J.F. Ryan. 1969. Can. J. Physiol. Pharmacol. 47, 459.

29. Jereb M., R. Falk, B. Jereb, C. Lindhe. 1975. J. Nucl. Med. 16, 846.

30. Lathrop K.A., R.E Johnston, M. Blau, E.O. Rothschild. J. Nucl. Med. 1982. 13, 7.

31. Lee M., A. Dong, J. Yano. 1969. Can. J. Biochem. 47, 791.

32. Levander O.A., B. Sutherland, V.C. Morris, J.C. King. 1981. In : Selenium in Biology and Medicine (M.M. Spallholz, J.L. Martin and H.E. Ganther eds.). AVI Wesport.

33. Levander O.A
Trace Elements in Human Health and Disease, (A.S. Prasad, eds). Académic Press Inc., p.135.

34. Levander O.A., V.C. Morris. 1985. In : Trace Elements in Man and Animals. TEMA 5, (C.F Mills, I. Bremner and J.K. Chester eds.). C.A.B., U. K., 498.

35. Mc Connell K.P., C.H. Wabnitz. 1957. J. Biol. Chem. 226, 765.

36. Mc Connell K.P., G.J. Cho. 1965. Am. J. Physiol. 208, 1191.

37. Mc Connell K.P., D.M. Roth. 1966. Proc. Soc. Exp. Biol. Med. 123, 919.

38. Mc Murray C.H., W.B. Davidson. 1985. In : Trace Elements in Man and Animals. TEMA 5, (C.F. Mills, I. Bremner and J.K. Chester eds.). C.A.B., U. K., 474.

39. Millar K.R., A.D. Shepphard. 1973. N-Z. J. Agr. Res. 16, 293.

40. Millar K.R. 1972, N-Z J. Agr. Res. 15, 547.

41. Motzenbocker M.A., A.L. Tappel. 1982. Biochim. Biophys. Acta. 719, 147.

42. Mukai K., J.J. Huang, T. Kimura. 1974. Biochim. Biophys. Acta. 336, 427.

43. Olson O.E., I.S. Palmer. 1976. Metabolism. 25, 299.

44. Painter E.P. 1941. Chem. Rev. 28, 179.

45. Palmer I.S., D.D. Fischer, A.W. Halverston, O.E. Olson. 1969, Biochim. Biophys. Acta. 177, 336.

46. Patterson K.Y., P. Kuitunen, E. Vuori, C. Veillon, O.A. Levander. 1987. Fed. Proceeding. 46, 879.

47. Patterson B.H., K. Helzoner, O.A. Levander, P.A. Mc Adam., C. Veillon, P.R. Taylor, L.A. Zerch. : TEMA 6 (in press).

48. Pope A.L., R. Moir, M. Somers, E.J. Underwood, C.L. White. 1979. J. Nutr. 109, 1448.

49. Reamer D.C., C. Veillon. 1983. J. Nutr. 113, 786.

50. Robinson J.R., M.F. Robinson, O.A. Levander, C.D. Thomson. 1985. Am. J. Clin. Nutr. 41, 1023.

51. Sandholm M. 1975. Acta Pharmacol. Toxicol. 36, 321.

52. Schwarz K., E. Sweeny. 1964. Fed. Proc. Fed. Am. Soc. Exp. Biol. 23, 421.

53. Schwarz K., C.M. Foltz. 1957. J. Amer. Chem. Soc. 79, 3292.

54. Schubert J.R., O. Muth., J.E. Oldfield, L.F. Remmert. 1961. Fed. Proc. Fed. Am. Soc. Exp. Biol. 20, 689.

55. Schroeder H.A., D.V. Frost, J.J. Balassa. 1970, J. Chronic Dis. 23, 227.

56. Shamberger R.J. 1983. Biochemistry of Selenium, Plenum Press N.Y.

57. Shuber A.P., E.C. Orr, M.A. Recwy, P.K. Schendel, H.D. May, N.L. Schauer, J.G. Ferry. 1986. J. Biol. Chem. 261, 12942.

58. Soda K., N. Esaki, T. Nakamura, H. Tanaka. 1981. In : Selenium in Biology and Medicine (M.M. Spallholz, J.L. Martin and H.E Ganther eds.). AVI Wesport, 427.

59. Stadtman T.C. 1987. FASEB J. 1, 375.

60. Stewart R.D.H., N.M. Griffiths, C.D. Thomson, M.F. Robinson. 1978. Br. J. Nutr. 40, 45.

61. Sunde R.A., G.E. Gutze, W.G. Hoekstra. 1981. J. Nutr. 111, 76.

62. Sunde R.A., J.K. Everson. 1985. In : Trace Elements in Man and Animals. TEMA 5. (C.F Mills, I. Bremner and J.K. Chester eds.). C.A.B., U. K., 480.

63. Sunde R.A., J.K. Everson. 1987. J. Biol. Chem. 262, 933.

64. Swanson C.A., D.C. Reamer, C. Veillon, J. King, O.A. Levander. 1983. Am. J. Clin. Nutr. 38, 169.

65. Thomson C.D., R.D.H. Stewart. 1974. Br. J. Nutr. 32, 45.

66. Thomson C.D., M.F. Robinson. 1986. Am. J. Clin. Nutr. 44, 659.

67. Underwood E.J. 1977. In : Trace Elements in Human and Animals., Acad Press, N.Y., 302.

68. Veillon C., C.J. Edmonds, M.F. Robinson, C.D. Thomson, V.C. Morris, O.A. Levander. 1985. In : Trace Elements in Man and Animals. TEMA 5 (C.F. Mills, I. Bremner and J.K Chesters eds.)., C.A.B., U. K., 495.

69. White C.L., M. Sommers. 1977. Austr. J. Biol. Sci. 30, 47.

70. White C.L. 1980. Austr. J. Biol. Sci. 33, 699.

INTESTINAL ABSORPTION OF SELENIUM IN RATS: EFFECTS OF SEGMENTS AND AGE.

S. Ciappellano, M. Porrini, F. Brighenti, G. Testolin
Dipartimento di Scienze e Tecnologie Alimentari e Microbiologiche - Sezione Nutrizione - Via Celoria n. 2 - 20133 Milano - ITALY

Introduction

Balance trials have ascertained that selenium salts are quite absorbed from the intestine, but information is very limited with regard to the absorption mechanisms of their inorganic forms. Furthermore nutrients absorption can be influenced by age, due to the changing of some digestive functions (1, 2). In the present work absorption of selenium as sodium selenite was studied in rats of varying age using everted intestinal sections with radiolabelled Se in the incubation medium.

Materials and Methods

Female Sprague-Dawley rats (30 days old) were divided into three groups of 20 animals. A commercial non purified rat diet containing about 180 µg Se/kg and water were provided ad libitum. At 3, 12, 30 months of age the animals were anaesthetized and killed by bleeding, the small intestine removed, rapidly cooled by flushing with 4°C Ringer-bicarbonate buffer (RBB) pH 7.0, weighed and the length measured in the relaxed state;

it was then divided into three equal sections. A tract of 10 cm in the middle of each section was resected, weighed, everted and cannulated on the perfusion apparatus. The sections of intestine, after being filled with serosal solution (5 mM glucose RBB), were plunged into 100 ml of mucosal solution (10 mM glucose and 50 µM Na-selenite RBB containing Na_2-$^{75}SeO_3$, Specific Activity: 74 Bq/nmol) kept at 37°C. The serosal solution was circulated by a peristaltic pump at about 1 ml/min. Gas (O_2:CO_2 - 95:5) was bubbled vigorously through the mucosal fluid to avoid the formation of unstirred water layers (UWL).
1 ml aliquots of the collected serosal solution were taken after 10, 20, 30, 40 min of incubation and Se determined. At the end of experiment the sections of intestine were removed, washed with RBB solution and used for Se determination.

Results

Selenium (sodium selenite) present in the mucosal buffer moves to the serosal side (table n. 1); this transmucosal movement is a function of incubation time in all proximal, medial and distal sections of intestine. In the proximal tract, after 40 min of incubation, the selenium transferred is significantly lower ($p < 0.05$) in young animals than in adult and old ones. In the medial tract old rats present higher values than young ones ($p<0.05$) but do not differ significantly from adult ones. In the distal tract, adult rats transfer the highest amount of selenite in the serosal side, while young rats show the lowest Se values. After 40 min, the total amount of Se transferred is not significantly different ($p>0.1$) for the proximal, medial and distal sections of young rats; for adult rats the amount transferred follows this order: distal > proximal > medial; for old rats: proximal > distal > medial. About Se absorbed and retained by the small intestine proximal section always retains the lowest amount of Se ($p<0.01$), while the distal tract of adult rats absorbs and retains the highest amount.

Table 1 Transferred Se, Retained Se and Trans/Ret Se Ratio in Sections of Small Intestine of Rats 3 - 12 - 30 Month Old

Age / Time	Young (13)		Adult (15)		Old (18)	
PROXIMAL GUT						
10 MIN	1.0	± 0.2	1.6	± 0.3	2.9	± 1.3
20 MIN	2.4	± 0.6	5.1	± 0.7	5.1	± 1.2
30 MIN	3.7	± 0.8	8.9	± 0.9	8.4	± 2.2
40 MIN	5.0	± 0.9a	13.1	± 1.5a	13.4	± 2.8a
Retained Se	8.4	± 1.0a	53.9	±15.7	39.2	± 4.7a
Trans/Ret Se	0.55		0.24		0.34	
MEDIAL GUT						
10 MIN	1.1	± 0.2	1.2	± 0.2	1.5	± 0.3
20 MIN	1.8	± 0.4	3.1	± 0.5	3.5	± 0.4
30 MIN	3.1	± 0.5	5.2	± 0.9	6.4	± 1.2
40 MIN	4.0	± 1.0a	7.5	± 1.3a	9.3	± 1.5a
Retained Se	24.5	± 4.5A	96.4	±15.5A	112.1	±11.5A
Trans/Ret Se	0.22		0.08		0.08	
DISTAL GUT						
10 MIN	0.9	± 0.2	1.9	± 0.5	1.5	± 0.4
20 MIN	1.8	± 0.3	5.5	± 1.2	3.4	± 0.8
30 MIN	2.5	± 0.6	11.7	± 2.6	7.0	± 1.3
40 MIN	5.4	± 0.9A	18.8	± 4.2A	10.6	± 2.1A
Retained Se	38.7	± 3.3A	137.8	±21.7A	117.6	± 9.0A
Trans/Ret Se						

Values (nmol Se/g wet wt) at different time are mean ± SE with number of experiment in parentheses. Values, within columns, bearing common superscript are significanly different (a: $p<0.05$ - A: $p<0.01$) according to Variance analysis.

Conclusion

Se transfer in young rats shows no differences in all proximal, medial and distal sections of small intestine. Adult and old rats transfer a significantly higher ($p < 0.05$) amount of Se in all sections than do young rats, but the trend is always towards an increase with time . Probably, aspecific absorption mechanisms increase during ageing of animals. From our results it is evident that, for all groups tested, the intestinal proximal tract is more effective in selenite transfer from the mucosal to serosal side; the transfer efficiency (TRANSFERRED/ RETAINED Se RATIO) is higher in the proximal than in the medial and distal tracts (table n.1). Furthermore, even though young rats show a lower transfer of Se than adults and old ones,it should be noted that the younger the animal the more efficient the transfer.

Acknowledgement

Research work supported by CNR, Italy. Special grant I.P.R.A. Sub-project 3. Paper n. 1762.

References

1. Thomson A.B.R., M.Keelan. 1986. Can.J.Physiol.Pharmacol. 64, 30.

2. Esposito G.,A.Faelli,M.Tosco,M.N.Orsenigo,R.Battistessa. 1985. Am.J.Physiol. 249, G328.

STIMULATION OF INTESTINAL ABSORPTION OF SELENITE BY CYSTEINE AND GLUTATHIONE

E. Scharrer, R. Würmli, S. Wolffram

Institute of Veterinary Physiology, University of Zürich,
Winterthurerstrasse 260, CH-8057 Zürich, Switzerland

Introduction

In areas where selenium deficiency occurs in man or domestic animals selenite is often administered orally. Unlike selenate, which is actively absorbed from the small intestine by the sulphate carriers (1), selenite in general appears to be absorbed by diffusion (2,3). Nevertheless bioavailability of Se from selenite is high (4). Since, unlike selenate, selenite is spontaneously reduced in the presence of thiols (5) and Anundi et al. have shown that glutathione enhances accumulation of Se from selenite in isolated rat intestinal epithelial cells (6), we have studied in vivo absorption of Se from ^{75}Se labelled selenite as affected by cysteine and glutathione. Absorption of Se from ^{75}Se labelled selenate was investigated for comparison. Cysteine and glutathione normally occur in the lumen of the small intestine. Cysteine is released from proteins during digestion and glutathione is ingested via food and in addition enters the small intestine through the bile (6).

Materials and Methods

Absorption of Se from ^{75}Se labelled selenite or selenate (concentration: 10 µmol/l) was studied in the absence and presence of thiols (concentration: 1 mmol/l) using ligated loops of midjejunum of anesthetized adult male rats (7). Phenolred was employed as unabsorbable marker. Luminal disappearance of Se and accumulation of Se in the jejunal wall was measured. The values were related to 100 mg intestinal dry weight. ^{75}Se activi-

Selenium in Medicine and Biology

ty in blood samples taken from the respective jejunal vein at the end of the experiment (duration of experiment: 30 min) was also determined. ^{75}Se activity was measured in a gamma counter.

Results

L-Cysteine and glutathione stimulated luminal disappearance of Se from selenite to about the same extent (Table 1). Accumulation of Se from selenite in the jejunal wall appeared to be also slightly increased by both thiols (Table 1). ^{75}Se activity in blood samples obtained from the respective jejunal vein at the end of the experiment was elevated 2.6- and 2.9-fold by L-cysteine and glutathione, respectively (Table 1). Absorption of Se from selenate was not affected by the thiols (Table 1).

Table 1 Effect of L-Cysteine (L-Cys) and Glutathione (Glu) (Concentration: 1 mmol/l) on Absorption of Se from ^{75}Se-labelled Selenite or Selenate (Concentration: 10 µmol/l) by Ligated Loops of Midjejunum (Duration of Experiment: 30 min)

	Luminal Disappearance nmol/100mg	Intestinal Accumulation nmol/100mg	^{75}Se Activity in Blood of Jejunal Vein cpm/ml
Selenite			
Control (9)	3.9 ± 0.4	2.6 ± 0.3	11672 ± 1666
+ L-Cys (9)	7.5 ± 0.7*	3.4 ± 0.2	30142 ± 5105*
+ Glu (9)	8.7 ± 0.7*	3.5 ± 0.2	33601 ± 4574*
Selenate			
Control (6)	11.5 ± 1.1	1.4 ± 0.1	39871 ± 2723
+ L-Cys (6)	10.3 ± 0.7	1.7 ± 0.1	42961 ± 7994
+ Glu (6)	13.2 ± 0.9	1.6 ± 0.1	51412 ± 8932

Values are means + SEM; number of rats in parentheses
* significantly different from control ($p < 0.05$)

In general Se from selenite was absorbed slowlier from the intestinal lumen than Se from selenate, in particular in the absence of thiols ($p < 0.01$). This is also reflected by the ^{75}Se activity of the blood samples (Table 1). On the other hand ^{75}Se accumulation in the intestinal wall was higher for selenite than for selenate ($p < 0.05$).
In an additional experiment D-cysteine was as effective as L-cysteine regarding stimulation of absorption of Se from selenite (luminal Se disapearance: control, 4.0 ± 0.7; presence of D-cysteine, 8.4 ± 0.9; presence of L-cysteine, 8.7 ± 1.4; intestinal Se accumulation: control, 1.9 ± 0.2, presence of D-cysteine, 3.7 ± 0.3; presence of L-cysteine, 4.1 ± 0.6; n=6).

Discussion

The results presented clearly show that cysteine and glutathione stimulate transfer of Se from selenite across the jejunal epithelium into the circulation. The effect apparently is not stereospecific, because the L- and D-form of cysteine were equally effective under our in vivo conditions. It is well known that, unlike selenate, selenite is non-enzymatically reduced in the presence of thiols whereby selenotrisulfides are formed, which may be reduced further to elemental selenium via the respective selenopersulfides and hydrogen selenide as intermediates (5).
Since elemental selenium is poorly absorbed (4), the stimulating effect of cysteine and glutathione on absorption of Se from selenite is most likely due to formation of selenotrisulfides and maybe also of selenopersulfides and hydrogen selenide. Whether the effect of thiols on absorption of Se from selenite is attributable to facilitated transport of these compounds across the mucus-rich praeepithelial unstirred layer (8) or to enhanced transport across the epithelium remains to be investigated. Since cysteine and glutathione are present in the intestinal lumen under physiological conditions the results presented might have physiological implications. In particular the high bioavailability of Se from selenite (4) might be related to the reaction of selenite with thiols in the lumen of the small intestine.

Acknowledgment

Supported by the Schweizerische Nationalfonds, Grant No. 3.807-0.86.

References

1. Wolffram, S., B. Grenacher, E. Scharrer. 1988. Q.J. Exp. Physiol. 73, 103.

2. McConnell, K.P., G.J. Cho. 1965. Am. J. Physiol. 208, 1191.

3. Ardüser, F., S. Wolffram, E. Scharrer. 1985. J. Nutr. 115, 1203.

4. Combs, G.F., S.B. Combs. 1986. The Role of Selenium in Nutrition. Academic Press, Inc.

5. Ganther, H.E. 1979. Adv. Nutr. Res. 2,107.

6. Anundi, I., J. Högberg, A. Ståhl. 1984. Acta Pharmacol. Toxicol. 54, 273.

7. Scharrer, E., M. Hosser. 1980. Pflügers Arch. 388, 165.

8. Winne, D., H. Görig, U. Müller. 1987. Naunyn-Schmiedeberg's Arch. Pharmacol. 335, 204.

TRANSPORT OF SELENOAMINO ACIDS ACROSS THE INTESTINAL BRUSH BORDER MEMBRANE (BBM)

S. Wolffram, B. Berger and E. Scharrer
Institute of Veterinary Physiology, University of Zürich,
Winterthurerstrasse 260, CH-8057 Zürich, Switzerland

Introduction

Selenoamino acids are considered to be the main selenium (Se) source in naturally occuring food and feedstuffs. Although there is some experimental evidence for a common intestinal transport site for Se-methionine (Se-Met) and the sulphur analogue methionine (Met) in hamster small intestine (1) the exact transport mechanism(s) for Se-Met or for other selenoamino acids has not been elucidated yet. We therefore have investigated transport of Se-Met and Met across the intestinal BBM. To characterize transport of selenocystine some additional experiments were performed.

Materials and Methods

For the experiments BBM vesicles (BBMV) isolated from pig jejunum by a Ca^{++} precipitation method were used. Aliquots of BBM were stored under liquid nitrogen until use. The final BBM fraction showed a 11 to 20-fold enrichment in the activity of the BBM marker enzyme alkaline phosphatase with respect to the original homogenate. Uptake of ^{75}Se-labeled Se-Met, ^{3}H-labeled Met and ^{3}H-labeled cystine was determined by a rapid filtration technique.

Selenium in Medicine and Biology

Results

Figure 1 shows the time course of Se-Met and Met uptake by the BBMV in the presence of an inwardly directed Na^+- or K^+-gradient (100 mmol/l). Se-Met as well as Met uptake at incubation periods of 15 s to 30 min was significantly higher under Na^+-gradient conditions compared to K^+-gradient conditions whereas equilibrium uptake (60 min) was similar under both conditions.

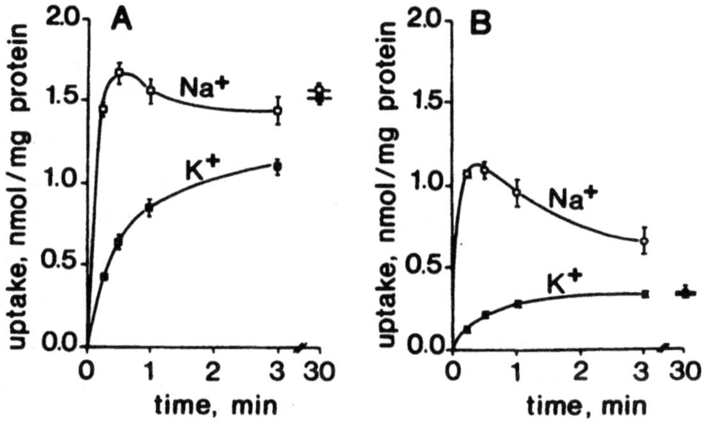

Figure 1: Uptake of Se-Met (A) and Met (B) by pig jejunal BBMV.
BBMV were preloaded with 300 mmol/l D-mannitol, 20 mmol/l HEPES/Tris, pH 7.4; incubation media contained (mmol/l): 100 D-mannitol, 100 NaCl(o) or KCl(•), 20 HEPES/Tris, pH 7.4 and 0.1 Se-Met or Met; values are means ± SEM of 6 determinations of 3 membrane preparations each determined in duplicate.

Equilibrium uptake of Se-Met as well as Met was inversely related to the osmolarity of the incubation medium indicating transport of both substrates into the vesicular lumen. Binding of Se-Met, however, exceeded binding of Met about 10-fold.

For calculation of the apparent kinetic constants (V_{max},K_m) of

Na^+-dependent Se-Met and Met transport initial uptake (incubation: 1.5 s) as a function of extravesicular substrate concentration was measured. Na^+-dependent uptake was obtained by subtracting uptake under K^+-gradient from uptake under Na^+-gradient conditions. Uptake curves for Se-Met as well as Met fitted a kinetic model with a single active component with similar kinetic constants for both substrates (V_{max}, nmol/mg protein x 1.5 s: 1.70 (Se-Met), 1.25 (Met); K_m, mmol/l: 0.24 (Se-Met), 0.23 (Met)).

In additional experiments the substrate specificity of the transport site(s) for Se-Met and Met was examined. Initial uptake of Se-Met and Met (0.1 mmol/l) under Na^+-gradient conditions was measured in the presence of various L-amino acids (10 mmol/l). Besides mutual inhibition between Se-Met and Met (90% inhibition of control uptake) uptake of both amino acids was inhibited to a similar extent by Cys, Ala, Leu, Phe, and Ser (50-90% inhibition of control uptake) whereas Lys and Glu showed no clear effect. As shown in Figure 2 preloading BBMV with 10 mmol/l Se-Met clearly stimulated uptake of Met (0.1 mmol/l) under Na^+-equilibrium conditions ($Na^+_{in} = Na^+_{out}$ = 100 mmol/l).

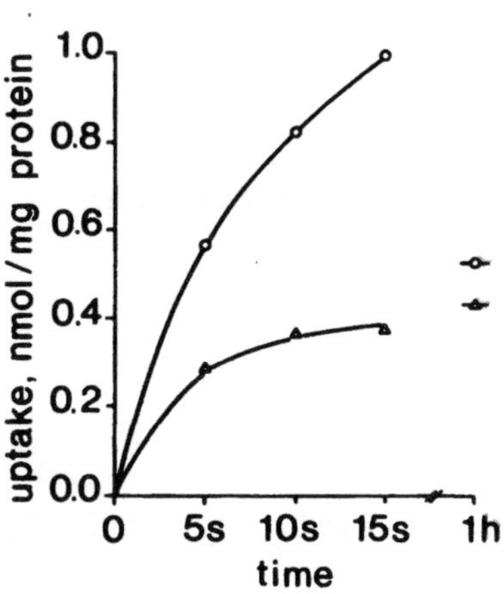

Figure 2: Stimulation of Met uptake by intravesicular Se-Met. BBMV were preloaded with (mmol/l): 100 D-mannitol, 100 NaCl, 20 HEPES/Tris, pH 7.4 (Δ) and 10 Se-Met (o); incubation medium contained (mmol/l): 100 NaCl, 110 D-mannitol, 0.25 Se-Met, 0.2 Met, 20 HEPES/Tris, pH 7.4; values are triplicate determinations of a single membrane preparation.

To get some insight in the mechanism(s) for Se-cystine transport across the intestinal BBM, uptake of L-cystine (0.01 mmol/l) was determined in the absence or presence of DL-Se-cystine (1.0 mmol/l) or of various L-amino acids (0.5 mmol/l). Whilst DL-Se-cystine and the cationic amino acids Arg and Lys caused a 30-50% inhibition of L-cystine uptake the neutral amino acid L-Met inhibited uptake of L-cystine only by about 20%. Thus, Se-cystine appears to inhibit mainly uptake of L-cystine by the carrier(s) for cationic amino acids.

Conclusions

It is concluded from the results that Se-Met and its sulphur analogue Met are transported across the intestinal BBM by a common Na^+-dependent mechanism for neutral amino acids. Replacement of sulphur in Met by Se appears to be without effect on the transport characteristic of the amino acid. Se-cystine mainly inhibits uptake of cystine by the carrier(s) for cationic amino acids suggesting transport of Se-cystine by the transport site(s) for cationic amino acids.

Acknowledgement

This research was supported by the Schweizerische National Fonds grant no. 3.807-0.86.

References

1. McConnell, K.P., G.J. Cho. 1967. Am.J.Physiol. 213, 150-156.

KINETICS OF SODIUM SELENITE BY INTRAVENOUS INJECTION IN RABBITS : STUDY OF SELENIUM AND GLUTATHIONE PEROXIDASE IN PLASMA.

O. GUILLARD, S. BOUQUET, C. TALLINEAU, A. PIRIOU and P. COURTOIS.

Institute for Medicines, Laboratory of Biochemistry and Pharmacology, University of Medicine, 86034 - POITIERS, FRANCE.

Introduction

Selenium (Se) was at first recognized as an indispensable trace element for animals and more recently for humans (1). Over the last few years, there has been increased interest in the dynamic of Se metabolism (2, 3). However, studies on Se kinetics in different animal species have revealed that the kinetic of this metal is a complex function, whatever the chemical form may be. Moreover, only a few publications contain kinetics studies and bioavailability on rabbits by intravenous route (IV) with sodium selenite. On the other hand, Se is an essential component of the enzyme glutathione peroxidase GSHPx in human or animal plasma. In rabbits, no study on the correlation of the plasma GSHPx activity to Se intravenously administered doses has been published.

The purpose of the present experiment is to obtain more complete information on the dynamic of sodium selenite in rabbits following the intravenous route, with the help of pharmacokinetic analysis ; this is accompanied by the study of plasma GSHPx activity.

Materials and methods

Animals and drug

Experiments were carried out using 11 male New Zealand rabbits weighing 3.6 ± 0.15 kg. They were placed in metabolism stainless steel cages. Room temperature was maintained at 22 ± 1°C. At the end of an 8-day (d) acclimation period, the 11 rabbits were divided into two groups. Group 1

Selenium in Medicine and Biology

(8 animals) received sodium selenite. Group 2 (3 animals) received sterile physiological saline so as to investigate exclusively the plasma GSHPx activity in relation to group 1. All the animals fast 24 hours (h) before the experiment with water *"ad libitum"* free Se such as the purified diet basal controlled by electrothermal atomic absorption spectrometry (EAAS). Sodium selenite (Na_2SeO_3, 5 H_2O, Merck) is dissolved in sterile physiological saline super pure making a 0.04 % (w/v) solution (as Se).

Time and techniques of collection, dosing

For Se determination in plasma, blood samples are taken at the following time intervals (h) : 0, 0.25, 0.50, 0.75, 1, 2, 4.50, 6, 12, 24, 72 and for the plasma GSHPx activity at 0, 6, 12, 24, 72, 120, 192, 240, 290 h. To avoid variations in the endogenous Se levels which may be due to circadian rhythms, samples were taken at the same time each day. Urine collection is effectuated over a 12 d period. From 0 to 24 h, blood samples are taken from jugular vein ; after having shaved the rabbits neck, subcutaneous anesthesia is effectuated (xylocaïn 1 %) and the jugular vein rendered accessible, a catheter is introduced (Ø : 1.2 mm) with heparin (25.000 U.I.) at 1 % in physiological saline (v/v) ; between 6 and 24 h , the catheter is filled with heparin at 10 %. The wound is enriched by benzylpenicilline sodique. From 72 to 290 h, blood samples are taken from the marginal ear vein. The dose injected into the marginal ear vein of each rabbit is 1.02 µmol/kg at a volume of 2 ml during 3 minutes.

Analytical procedure

Measuring Se in plasma and urine is carried out with a Perkin-Elmer model Zeeman 5000 equipped with an HGA 500, an AS 60 and a selenium electrodeless discharge lamp. Stabilized Temperature Platform Furnace (STPF) concept is used (4). The matrix modifiers used in this study were nickel nitrate hexahydrate for plasma and a mixed modifier magnesium nitrate/Ni for urine, prepared in nitric acid and triton X-100 (Merck). Argon and oxygen as purge and alternate gas, respectively. The method of standard additions and lyophilized human serum, Seronorm^R (Nycomed) batch 105 (plasma) and batch 108 (urine) were used. Plasma GSHPx activity was assayed by the method of Paglia and Valentine (5) using cumene hydroperoxide as substrate.

Data analysis

By ordinary least-squares analysis. Plasma and urine concentration data are fitted and plasmatic equation described a standard model of the form: $C_{p(t)} = Ae^{-\alpha t} + Be^{-\beta t} + Ce^{-\gamma t}$. The volume of distribution (Vd_{area}) was calculated by the equation : Vd_{area} = dose / $AUC._{0-\infty}$.Ke and total clearance was determined by Cl_t = dose / $AUC_{0-\infty}$. The elimination half-life which agrees with the terminal phase for Se excreted / time is $T_{1/2}$ elim. = 0.693/Ke. The results are expressed as means ± SEM. Student's t test is used for statistical analysis.

Results and discussion

After a 1.02 µmol/kg of Se by IV bolus, C_{max} was 15.6 ± 3.7 µmol/l at time 0.25 h.The evolutionn of plasma concentration-time has presented many phases. The first ($T_{1/2}$ α = 0.15 h) and second ($T_{1/2}\beta$ = 3 h) slopes, correspond to a rapid distribution into extravascular tissues.However, these pharmacokinetics parameters suggested that they would agree with an open three - compartment model ; indeed, in showing evolution of Se plasmatic concentration, a tri-exponential function can be considered of which the last slope may be in correlation with urine elimination. The semi-logarithmic concentration for 8 rabbits is shown in fig. 1 following equation : Cp(t) = 12.250 e-4.470t + 2.087 e-0.261t + 2.470 e-0.008t .The calculated theoretical points are in close agreement with the observed mean points.We are led to think that these values in plasma can not be directly compared with the studies of Hasegawa (6) and Thomson and Stewart (7) in rats or with Jereb et al. (8) in humans by IV as regards the whole body retention of ^{75}Se. Plasma clearance (CL_{tot}) and Vd area, were : 2.82 ± 0.63 ml/min and 0.23 ± 0.04 l/kg,

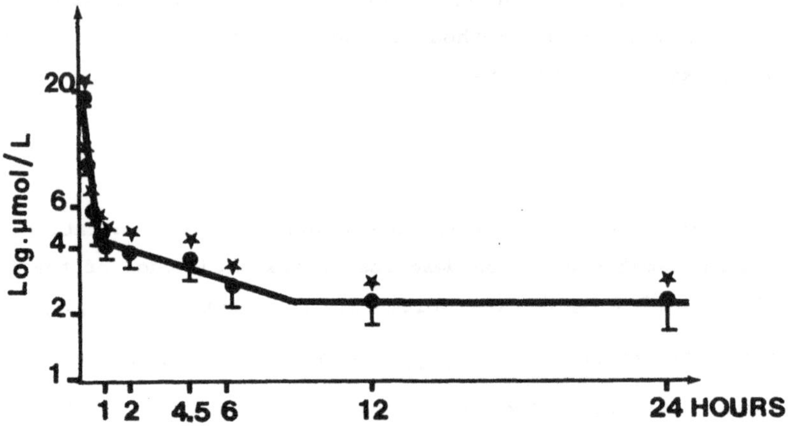

Fig. 1. Plasmatic evolution of Se in function of time in 8 rabbits after administration single dose by IV (1.02 µmol/kg.).★. Observed points ; ● Theoritical points (± S.E.M.) .

respectively. This last value is in agreement with a distribution, mainly in plasma and hightly vascular tissues.
The urinary parameters shown terminal half-life of elimination is : 82 ± 9 h (3.41 d) calculated over a 290 h time-collection. With the percentage of elimination, our results for Se levels in urinary excretion in the first 24 h are shorter (30 %) than those reported (7) in rats (16 %). An amount of 80 % is excreted during 12 d, and this elimination can be explained by existence of tissulary compartments such as the liver and kidney in agreement with many studies (9, 10). The elimination of Se via the kidney is an important feature of body Selenium homeostasis. As regards the study of plasma GSHPx activity, after injection of a rather large dose of Se (1.02 µmol/kg), we observe during 12 d, increases in enzymatic activity identical to group 2 ; the latter are insignificant (NS). To oral Se supplementation in humans several authors (11, 12) note a significant increase of plasma GSHPx activity but only in depleted subjects. All these investigations show that origin and nature of plasma GSHPx activity in rabbits or humans are not yet clear (12) ; in the former, results are at times contradictory (13, 14). In this pharmacokinetic analysis, our results take into account the existence an endogenous

Se pool closer physiological reality. These parameters furnish convincing arguments for future studies with others routes and others Se salts.

Acknowledgement

This research was supported by HERBAXT Lb. ZI Nord. Bât 5. BP 71. 77200 TORCY CEDEX France.

References

1. Mutanen, M. 1986. Ann. Clin. Res. 18, 48-54.

2. Janghorbani, M., L. Kasper and V. Young. 1984. Am. J. Clin. Nutr. 40, 208-218.

3. Young, V.R., A. Nahapetian and M. Janghorbani. 1982. Am. J. Clin. Nutr. 35, 1076-1088.

4. Slavin, W., D.C. Manning and G.R. Carnrick. 1981. At. Spectrosc. 2, 135-145.

5. Paglia, D.E. and W.N. Valentine. 1976. J. Lab. Clin. Med. 70, 158-169.

6. Hasegawa, A. 1972. J. Hyg. Chem. 18, 70-75.

7. Thomson, C.D. and R.D.H. Stewart. 1973. Br. J. Nutr. 30, 139-147.

8. Jereb, M., R. Falk, B. Jereb and C. Lindhe. 1975. J. Nucl. Med. 16, 846-850.

9. Burk, R.F. 1976. In : Trace Elements in Human Health and Disease, Vol. 2 (A.S. Prasad, ed.). Academic Press, New-York. London. pp. 105-133.

10. Nève, J., M. Henry, A. Peretz and J.P. Mareschi. 1987. Cah. Nutr. Diet. 22, 145-161.

11. Lane, H.W., S. Dutrick and D.C. Warren. 1981. Proc. Soc. Expl. Biol. 167, 383-390.

12. Steiner, G., H. Menzel, I. Lombeck, F.K. Ohnesorge and H.J. Bremer. 1982. Eur. J. Pediatr. 138, 138-140.

13. Lee, YH., D.K. Layman and R.R. Bell. 1979. Nutr. Rep. Int. 20, 573-577.

14. Tappel, M.R., J. Chaudiere and A.L. Tappel. 1982. Comp. Biochem. Physiol. 73 B, 945-949.

RETENTION AND BIOLOGICAL HALF-LIFE OF (^{75}Se)SELENOMETHIONINE IN FOUR PATIENTS WITH DUCHENNE MUSCULAR DYSTROPHY

T. Westermarck
Helsinki Central Institute for the Mentally Retarded
SF-02400 Kirkkonummi, Finland

T. Rahola and M. Puhakainen
Finnish Centre for Radiation and Nuclear Safety
P.O.Box 268, SF-00101 Helsinki, Finland

R. Lappalainen and T. Louhimo
Childrens Castle Hospital
SF-00250 Helsinki, Finland

Introduction

Duchenne muscular dystrophy is caused by a genetic defect affecting membrane structure and permeability of both skeletal and cardiac muscles and accompanied by an increased level of lipid peroxidation (Hunter et al. 1986). Various experiments have demonstrated that the disease is associated with an increased elimination of selenium (Nève et al. 1985). Gamstorp et al. (1986) recently reported an possible improvement of the muscular strenght of Duchenne patients after treatment with selenite and vitamin E. Previously we studied the long-term turnover of selenite-Se in patients with Duchenne disease (Westermarck et al. 1982) and found an increased elimination of Se. Now we have studied the organic selenium compound methionine-Se.

Selenium in Medicine and Biology

Material and methods

Retention and biological half-life of selenium were studied with whole-body counting and ^{75}Se-selenomethionine in four patients with the Duchenne type of muscular dystrophy (DMD). The four male patients were aged 5 to 8 years. The patients were orally given water solutions containing approximately 2 kBq ^{75}Se per kg body weight. No selenium carrier was added to the ^{75}Se-solutions (Amersham,(^{75}Se) Injection BP). The patients fasted overnight before administration of the tracer doses in the morning. They were also asked not to eat for 1-2 hours after the administration but their diets were not otherwise controlled. The whole-body counter measurements were done in an ironroom with a multi-detector linear scanner (Suomela 1983). The measurement performed 24 hours after the administration of the dose was taken to represent 100 per cent of the tracer. The activity excreted before the first measurement was disregarded and caused an individually varying error in the retention percentage, but did not affect the determination of the biological half-life. Whole-body counting measurements were done about once a month for about 200 days. Selenium supplementation was started after about 150 days except for one patient (HK). The statistical counting error was negligible during the first 150 days increasing thereafter to 20 per cent at day 300. All counting rates were corrected for physical decay of ^{75}Se.

Results

The average retention of the long term component of ^{75}Se-selenomethionine was found to be 94 ± 5 per cent and the average biological half-life 75±7 days. In Table I the retentions and biological half-lives of the studied patients are presented.

Table I. Administration, long-term retention and biological halflife of ^{75}Se-selenomethionine in children suffering from DMD.

Patient	Age Y	Sex	Weight kg	Administered ^{75}Se kBq	Retention %	^{75}Se $T_{1/2}$ (biol.) d
RA	7	M	19	36	100±1	75±2
HK	5	M	17	31	96±1	67±2
AL	8	M	20	41	89±1	75±2
SL	6	M	18	37	92±1	84±2
Average					94±5	75±7

In figure 1 the retention curve of one of the patients (SL) given selenium supplementation and in figure 2 the retention curve for a patient (HK) not given selenium supplementation are shown. The effect of the selenium supplementation can clearly be seen in figure 1.

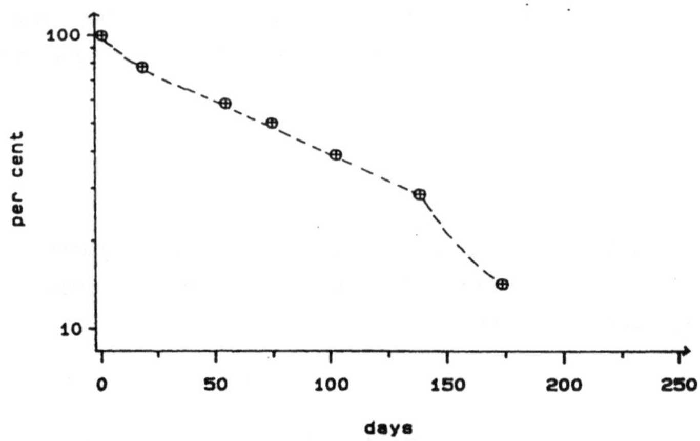

Fig.1. Retention curve of ^{75}Se-activity in the whole-body after the administration of ^{75}Se-selenomethionine to a DMD patient (SL) given selenium supplementation from day 140.

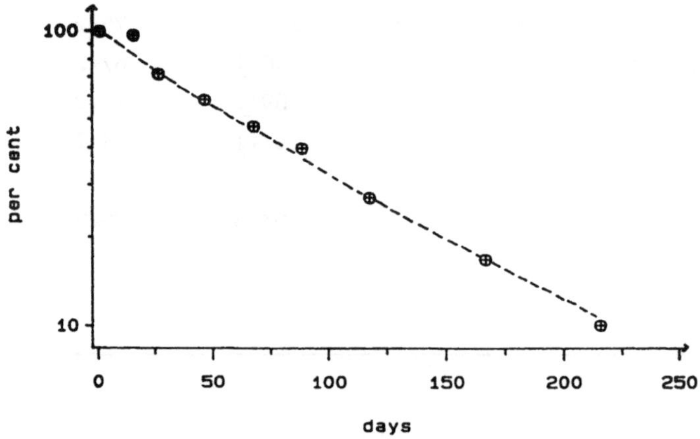

Fig.2. Retention curve of ^{75}Se-activity in the whole-body after the administration of ^{75}Se-selenomethionine to a DMD patient (HK).

Discussion

In an earlier investigation of DMD patients, who were administered selenium as ^{75}Se-sodiumselenite, the retention found varied between 30 and 80 per cent and the biological half-life was found to be 65 ± 6 days (Westermarck et al. 1981). In an investigation on four healthy women (Griffiths et al. 1976) using ^{75}Se-selenomethionine as a tracer the absorption was found to vary from 95.5 to 97.3 per cent and the biological half-lives from 207 to 290 days. The corresponding values for three of the same subjects as in the above mentioned investigation studied using ^{75}Se-sodiumselenite (Thomson et al. 1974) were 70, 64 and 44 per cent and 96 to 144 days. The biological half-lives found in the present investigation are about one third of the half-lives found by Griffiths.

References

1 Gamstorp, I., Gustavson, K.H., Hellström, O. and Nordgren, G., 1986. J. Child Neurology, 1, 211-214.

2 Griffiths, N.M., Stewart, R.D.H. and Robinson, M.F., 1975, 1976, Br. J. Nutr. 35, 373-382.

3 Hunter, M.I.S., and Mohamed, J.B., 1986, Clin. Chim. Acta. 155, 123-132.

4 Nève, J., Vertongen, F. and Molle, L. 1985, Clinics in Endocrinology and Metabolism, 14, 629-656.

5 Suomela, M., 1983, Whole-body counter studies in radiation protection and clinical research, Institute of Radiation Protection, Helsinki, STL-A-45.

6 Thomson, C D. and Stewart, R.D.H., 1974, Br. J. Nutr. 32, 47-57.

7 Westermarck, T., Rahola, T., Suomela, M. and Kallio, A-K., 1981, The Fourth International Symposium on Trace Element Metabolism in Man and Animals (TEMA-4), Perth Western Australia, 506-509.

SELENIUM IN THE ANTERIOR PITUITARY AFTER A SINGLE INJECTION OF 75-SE L-SELENOMETHIONINE

H. Grønbæk
Department of Neurobiology, Institute of Anatomy, University of Aarhus, 8000 Aarhus C, Denmark

O. Thorlacius-Ussing
Department of Surgery I, Aarhus County Hospital, 8000 Aarhus C, Denmark

Introduction

In 1976 the National Research Council concluded that growth retardation may be the best indicator of selenium toxicity (1). Recent studies have demonstrated decreased growth hormone and somatomedin C secretion parallel to growth retardation in rats exposed to sodium selenite (2,3). Schoental demonstrated selenium to accumulate in the anterior pituitary after subcutaneous injection of sodium selenite and we have lately confirmed this accumulation also after intraperitoneal injection (4). McAdam and Levander observed trends toward less growth inhibition in rats consuming organic selenium compared to those consuming inorganic compounds (5). No information is, however, available concerning selenium accumulation in the anterior pituitary after exposure to selenomethionine.
The aim of this very preliminary study is to estimate the selenium content in the anterior pituitary and other, especially endocrine, organs after intraperitoneal injection of 75-Se selenomethionine.

Selenium in Medicine and Biology

Material and Methods

For this preliminary study we used 9 male Wistar rats divided in 3 equal groups. At t=-5 min. the rats were lightly anesthetized with pento-barbital. At t=0 min. each rat was injected with a solution containing L-selenomethionine (1.7 mg Se/kg) and 75-Se L-selenomethionine (862 kBq/rat). Less than 10 min. after injection the rats were whole-body counted, and this value was used as 100% reference value. Two, 24, and 48 hours later the rats in one of the groups were recounted and afterwards exsanguinated by v. porta puncture. Following organs were removed: anterior pituitary, adrenals, thyroid gland, testis, liver, kidney, spleen and 1 ml of blood and the wet weight were recorded before the content of 75-Se was measured in a well counter. All estimates were corrected for background radiation and physical decay of the isotope.

Results

The anterior pituitary had the highest selenium concentration of the endocrine organs (7.6 ± 1.8 (SD) µg/g wet weight), this being double the concentration observed in the thyroid gland (3.3 ± 0.2 (SD) µg/g wet weight) and the adrenals (3.1 ± 0.2 (SD) µg/g wet weight) and 8 times that found in testis (0.9 ± 0.1 (SD) µg/g wet weight). In descending order of selenium concentration the individual organs appeared as follows: liver > anterior pituitary > kidney > thyroid gland > adrenals > spleen > blood > testis (Table 1).

Generally the selenium content in most organs decreased only slightly during the first 48 hours. In the liver, however, the concentration decreased rapidly, 62% had disappeared during 48 hours, whereas the concentration in testis practically remained constant.

The whole body content decreased nearly 50% during the experimental period of 48 hours.

Table 1. Selenium Concentration in Different Organs.

Time	Anterior Pituitary	Thyroid Gl.	Adrenals	Testis
2 h	7.6±1.8	3.3±0.2	3.1±0.2	0.9±0.1
24 h	6.0±1.0	2.7±0.2	2.3±0.03	0.8±0.1
48 h	6.9±1.2	2.9±0.6	1.8±0.1	0.8±0.1

	Liver	Kidney	Spleen	Blood	Whole body
2 h	11.1±0.4	5.2±0.7	2.0±0.9	1.2±0.01	1.7±0.1
24 h	5.1±1.5	5.4±0.5	1.7±0.2	0.8±0.1	1.2±0.1
48 h	4.2±2.5	3.9±0.3	1.3±0.04	0.8±0.04	0.9±0.04

The table shows the selenium content in µg/g wet weight ± SD in the particular organs investigated at 2, 24 and 48 hours after the injection.

Discussion

This present preliminary study demonstrates that selenium rapidly accumulates in the anterior pituitary after a single intraperitoneal injection of 75-Se L-selenomethionine. Compared to the other endocrine organs (thyroid gland, adrenals and testis) a very high selenium concentration was observed in the anterior pituitary. The observed distribution of selenium among the different organs investigated is in accordance with that found by others (for review see (6)). Compared to a previous study using the same amount of selenium as 75-Se sodium selenite (4), some distinct differences were observed. 1) The selenium concentration in the anterior pituitary after a single injection of 75-Se L-selenomethionine was 2.5 times the concentration after injection of 75-Se sodium selenite (7.6 µg/g versus 2.9 µg/g wet weight). 2) We only observed a slight decrease in selenium concentration in the anterior pituitary during the 48 hours experimental period, 2/3 of the selenite had disappeared during the same period.

The rather high amounts of selenium in the anterior pituitary is obscure. The only known biological function of selenium in mammals is in the enzyme glutathione peroxidase.

Although the actual concentration of this enzyme in the anterior pituitary is unknown there is no well established theory which may suggest a high concentration. In the secretory granules of the pituitary cells, however, high amounts of sulphur containing proteins are present (7). The metabolism of sulphur and selenium has much in common and the selenium may substitute sulphur in these proteins explaining the high amounts of selenium in the anterior pituitary compared to other endocrine organs. Since these proteins also seems important for the secretion of pituitary hormones (8) substitution of sulphur with selenium within the proteins may explain both the high concentration observed and the decreased growth hormone secretion.

Acknowledgement

This study was supported by grants from the Danish MRC (12-6581), NOVO Foundation and Jemo-Pharm Ltd., Denmark.

References

1. National Research Council. Selenium. Washington, D.C.: National Academy of Sciences. 1976.

2. Thorlacius-Ussing, O., A. Flyvbjerg, J. Esman. 1987. Endocrinology 120, 659.

3. Thorlacius-Ussing, O., A. Flyvbjerg, K. Damm Jørgensen, H. Ørskov. 1988. Acta Endocrinologica 117, 65.

4. Thorlacius-Ussing, O., F. Taagehøj Jensen. 1988. Biol. Trace Element Res. (in press).

5. McAdam, P.A., O.A. Levander. 1987. Nutrition Research 7, 601.

6. Bopp, B.A., R.C. Saunders, J.W. Kesterson. 1981. Drug Metabolism Reviews 13, 271.

7. Slaby, F., M.G. Farquhar. 1980. Mol. Cell. Endocrinol. 18, 33

8. Lorenson, M.Y., L.S. Jacobs. 1984. Endocrinology. 115, 1492.

SELENIUM METABOLISM AND AVAILABILITY IN RAINBOW TROUT

Mogens Gissel Nielsen
Institute of Zoology and Zoophysiology, University of Aarhus, Building 135, DK-8000 Aarhus C, Denmark

In many regions of the world fish products are one of the main sources of selenium in human nutrition. At the same time fish often contains heavy metals, and occasionally in unacceptable high concentrations. A considerable amount of human intake of e. g. mercury comes from the consumption of fish. Further, selenium and mercury can interact and detoxicate each other in animal organism, and especially in marine organism the concentration of both elements can reach extremely high values. Bryan reported (2) concentrations of mercury and selenium in liver from the black marlin (**Makaira indica**) on 41.6 and 21.6 ppm, respectively. Similar values for the muscle were 29.2 and 8.8 ppm. The highest concentrations are reported from marine mammals where the mercury concentration in the liver of sea lion and short finned pilot wale were 747 and 454 ppm, respectively; the values for selenium were 260 and 59.4 ppm.

The biological availability of selenium is strongly dependent on the concentration of heavy metals, because selenium in heavy metal compounds is much less available to the organism than most other inorganic or organic selenium-compounds. At the same time the toxic effect of the mercury is reduced. The dynamic of selenium in fish is therefore of great importance for the human nutrition.

In our first experiments with rainbow trouts (5), we were most concerned about the ecological effects of selenium, since it is known to be toxic for many organisms in concentrations higher than a few ppm. Experiments with chronic exposure to

Selenium in Medicine and Biology

selenite (Na_2SeO_3) showed that concentrations higher than 0.1 ppm Se were letal for the fish. Independent of the concentration of selenium in the water the fish reaches their maximum selenium concentration within four weeks, Fig 1.

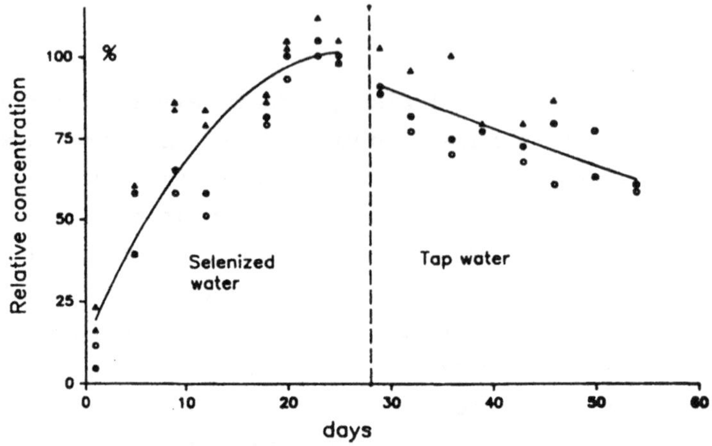

Fig 1. Uptake and elimination of selenium in rainbow trout. The concentration of selenium in the water ranged from 0.01 to 0.4 ppm Se, and after 28 days the fish were transferred to tap water.

The figure shows the uptake and elimination of selenium at different concentrations, - the values are expressed as percentage of maximum concentration, in order to compare fish exposed to different concentrations of selenium. The elimination expressed as the biological halflife was calculated to 29 days independent of the concentration of selenium in the fish, so the chances of bioaccumulation in trouts do not exist. These results agree with our results obtained with eels,(4) where the biological halflife was found to be 27 days.
Exposure to acute concentration of selenium showed that the trouts could survive up to 100 ppm selenium for half an hour, and it showed also that fish surviving the first 10 days after exposure had excreated so much selenium that later death was not likely to be an effect of the selenium treatment.
In food-chains and also in human nutrition fish with high con-

centrations of selenium are not necessarily toxic due to the low availability of selenium from marine products. According to (3,6) the biological availability of fish-products for humans ranges from 9 to 33 percent of that of sodium selenite, while these values for corn and wheat are 84 and 110 percent, respectively. In consequence of that, a high intake of selenium in fish does not necessarily fulfil the human requirement of this element.

The biological availability of selenium in trout meat cooked in different ways was investigated in feeding experiments with rats. During the cooking process the selenium content decreased significantly, e.g. the loss of selenium was 30 % after boiling 8 minutes, and 25 % after 10 minutes frying. The various concentrations of selenium in the five differently treated food-types were not reflected on the concentration of selenium in the organs in the rats.

The distribution of selenium in different organs of rainbow trouts was followed at exposure to selenium through the water and intraperitoneal injection.

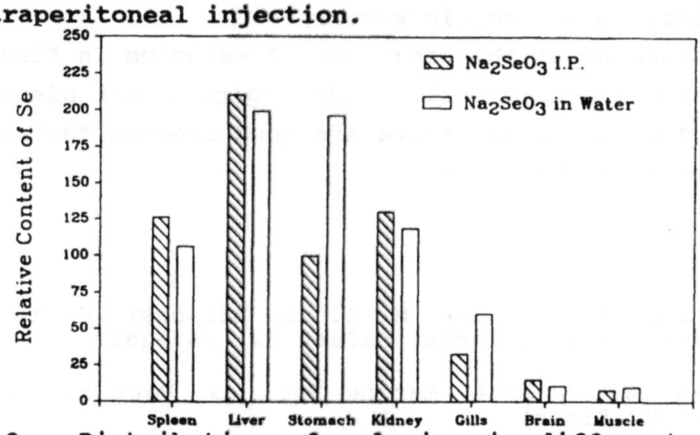

Figure 2. Distribution of selenium in different organs of rainbow trouts 14 days after intraperitoneal injection and after exposure to selenized water for two weeks. The selenium content is expressed as percentage of the mean concentration of the seven organs.

In one of the experiments the trouts were kept in water containing 0.1 ppm Se, and samples of fish were taken frequently for analyzing the selenium concentration in the different organs. In another experiment selenium was injected intraperitoneal and the trouts were kept in running water in different periods, before they were killed and the selenium concentration in the different organs was analyzed. Fig. 2 shows the distribution of selenium 14 days after the start of the treatment, and the concentrations are expressed as percentage of the mean concentration of the seven organs. As it can be seen from the figure the variation in the selenium concentration between the different organs is quite remarkable, whereas the difference between the two treatments is rather small, except for the stomach and gills.

The distribution of selenium in trout tissue as a function of time was very important for our later work with histochemical demonstrating mercury (1). It was demonstrated, that a selenium treatment of the fish two hours before killing and fixation increases the stainability of mercury, which may be due to visualization of organic mercury.

The understanding of the behaviour of selenium in fish is important from a nutritional, toxicological, and histological point of view, and thus, there are good reasons for an increased research in this field.

References

1. Baatrup, E., Nielsen, M. G. and Danscher, G. 1986. Ecotoxicol. and Environm. Safety 12, 267-282.

2. Bryan, G. W. 1984. Marine Ecol. 5, Part 3, 1289-1431. (Ed.: O. Kinne)

3. Cantor, A. H., Scott, M. L. and Noguchi, T. 1975. J. Nutr. 105, 96-105.

4. Gissel-Nielsen, G. and Nielsen M. G. 1973. Ambio 2,4, 114-117.

5. Nielsen, M. G. and Gissel-Nielsen, G. 1978. Agriculture and Environment 4, 85-91.

SELENIUM HOMEOSTASIS

Dietrich Behne
Hahn-Meitner-Institut
Research Group "Trace Elements in Health and Nutriton"
D-1000 Berlin 39, F.R.G.

Introduction

As stated about 100 years ago by C. Bernard, regulatory mechanisms are responsible for a "fixité du milieu intérieur" and are the means by which the organism achieves relative freedom from the changes in its environment (1). With essential elements the concentration in the body can vary only within a certain range without biological functions being disturbed, and it is therefore of great interest to investigate the ways in which the organisms are protected against an insufficient or excess dietary supply of these elements. I would like to discuss here to what extent this "fixité du milieu intérieur" applies to selenium. Ideal homeostasis would mean that the selenium in the body compartments is maintained at a certain level irrespective of the dietary intake. Without the effect of homeostatic control the selenium level in the compartments would increase steadily with increasing amounts of the element taken up by the organism.

Depending on its concentration selenium has essential or toxic effects. Although there are differences with regard to the species and also differences with regard to the chemical form of the selenium administered, one can generally say that the lower limit for optimum functioning is reached at a selenium level in the diet of about 0.1 mg/kg and toxic effects are found with diets with a selenium content of more than 5 mg/kg. Compared with other essential elements this range between

Selenium in Medicine and Biology

deficiency symptoms and signs of toxicity is rather small, which suggests that the organisms are relatively poorly protected against great variations in the dietary selenium supply.

On the other hand investigations of the selenium levels in some tissues show that the element content does not change steadily with increasing selenium intake. When, for instance, the results of several studies on the liver selenium content of male rats fed diets with selenite or selenate are being compared (2-5), two plateaux are apparent, one for low selenium intake up to a dietary level of about 0.05 mg/kg, the other in the range between 0.1 and 0.5 mg Se/k diet. These data suggest that within certain limits homeostatic mechanisms for the regulation of the selenium metabolism do exist.

In the discussion of these mechanisms I will first consider the aspects of absorption and excretion, that is the homeostasis with regard to the whole organism, then move on to the question of the maintenance of the selenium contents in the different tissues and finally look into homeostatic control at a molecular level. These points will be discussed with reference to the results of a few relevant experiments. More comprehensive references to the studies on the metabolic fate of selenium can be found in survey articles such as those of Bopp et al. (6) or Levander (7).

Absorption and excretion

In the regulation of the selenium status of the organisms absorption does not appear to be a decisive factor. This was shown, for instance, in a study on rats which had been fed diets with varying contents of selenium added as selenite (8). All of the rats absorbed 95 to 100 % of the element administered irrespective of the selenium content of the diets which ranged from deficient to nearly toxic levels. The increase in the dose administered by addition of the relatively large amount of 5 µg selenium did not have a significant effect on the absorption either. The same was found in other species including man. Absorption was high after oral intake of selenium in several chemical forms and did not appear to be dependent on the selenium status of the organism (6,7).

Regulation takes place, however, via changes in the urinary excretion rate and thus in the biological half-life of the element in the organism. This was shown in experiments on rats which were fed diets with different selenium levels. After injection of a small amount of the labeled element the half-excretion time for the tracer decreased from 20 days, when a diet with 0.1 mg Se/kg was fed, to only 1 day in the animals which were fed a diet with the high content of 1 mg/kg (9). A particularly interesting effect was found for rats with a very low selenium intake (10). Here no differences in the retention of the element were found between the animals fed the basal diet and those which received the same diet with 0.03 mg Se/kg added. This indicates the existence of a kind of threshold level of selenium in the diet. Above this level the excretion of the element increases with increasing selenium intake while below the threshold only a constant small percentage of the selenium is lost via the urine. This change in the urinary excretion behavior seems to be an important homeostatic mechanism for minimizing losses of the element during periods of insufficient selenium supply.

Tissue selenium levels

In the absorption and excretion studies information was obtained on the homeostatic control of the element in the whole organism. As the next step I would like to discuss the question to what extent regulation mechanisms are responsible for the levels and the metabolism of selenium in the different body compartments. The analysis of the selenium content in the tissues of rats with a normal selenium status showed that the element is unevenly distributed (11). Calculated per dry mass the highest levels in the testes and in the kidneys differed from the lowest in the muscle and the brain by about one order of magnitude. Great differences were also found with regard to the distribution of glutathione peroxidase. The enzyme-bound fraction of the element was particularly high in the erythrocytes and in the liver and particularly low in the testes. By comparing the total amount of selenium and glutathione peroxidase it was estimated that more than 60 % of the selenium in the body of the rat is not bound to the enzyme. This finding was the basis of our research on new selenoproteins which I will discuss later.

In order to find out whether homeostatic mechanisms are responsible for the maintenance of the selenium levels in the different tissues and to investigate to what extent they are affected by changes in the selenium supply, the distribution and retention of the element were determined during selenium depletion (12). In rats which had been fed a selenium-deficient diet for 10 weeks, the greatest decreases were found in the liver and in the erythrocytes, the smallest in the testes and adrenals. When the animals were then given a small amount of labeled selenium, the retention was highest not in the severely depleted tissues but in the very organs which had lost only little of the element. The relatively slight decrease in the selenium contents in the testes and adrenals could not be explained solely by a greatly reduced turnover, as in this case the selenium levels would already have been reduced to a much greater extent by the growth of the organs during the depletion period. It is probably here rather a case of reutilization of the metabolized element, which in selenium deficiency is not excreted but instead is transported back to certain priority target tissues.

As these results suggested a specific regulation of selenium a similar experiment was carried out in which a large number of tissues was investigated (5). In this study extremely selenium-depleted rats of the second generation were used. After these animals and selenium-adequate controls had been given a very small amount of labeled selenite, the deficient animals retained about 20 to 50 times more of the dose in the adrenals, brain, corpora lutea, ovaries, pituitary, testes and thyroid than the control animals. In tissues such as the heart, the muscle and the liver, however, the retention was only slightly different in the two groups. Regulatory mechanisms apparently exist which ensure that during periods of insufficient selenium intake the content of the element is kept up above all in the brain and in the reproductive and the endocrine organs. The maintenance of the selenium level in the liver, the heart and the skeletal muscle is last in this hierarchy. This difference in the distribution might explain the fact that in a low selenium state lesions are first found in the latter tissues. The occurrence of lesions during the first stages of selenium depletion therefore indicates that the tissues affected are of only secondary importance with regard to the maintenance of the selenium level.

Nothing is known so far about the mechanisms which are responsible for this regulation of the selenium levels. However, in the case of certain tissues there is some evidence that sexual hormones might be involved in the selenium metabolism. In an experiment on rats, in which the production of gonadotropic hormones was interrupted by hypophysectomy, the selenium content in the atrophic testicular tissue decreased to about a third of the level in the intact controls (13). After treatment with pregnant mare's serum gonadotropin, which has the same effects as the gonadotropic hormones, LH and FSH, the testes were regenerated and the testis selenium level increased again to almost the level of the untreated intact males. These changes were found in animals during selenium depletion.

This means that due to the effect of the gonadotropic hormones selenium is transported to the testes from other, already selenium-deficient tissues and shows again that the testis is one of the main priority target organs for selenium supply. In several studies an influence of sexual hormones on the selenium metabolism has also been observed in the liver. The liver glutathione peroxidase activity in the male rat is about half of that in the female. Castration or administration of estradiol to normal or castrated male rats produced a significant increase in the activity of this enzyme (14,15). However, at our present state of knowledge these findings only indicate relations between hormones and tissue selenium levels and the question of the direct involvement of sexual hormones in the regulation of the selenium metabolism still has to be clarified.

An important point to be considered in all the studies on selenium metabolism and its regulation is the chemical form of the element in the diet. In most of the studies the element was administered in the form of selenite. However, in plant material, a large proportion is present in the form of selenomethionine. Whereas at low dietary levels the uptake, retention and turnover of selenium administered either as selenite or as selenomethionine was found to be similar (16), several investigations have shown that at a high dietary intake there are pronounced increases in the tissue contents of the element when, instead of selenite or selenate, the organic form of the element is present in the diet (4,17). These differences are largest in the muscle. The findings suggest that at high dietary intake the homeostatic regulation of the tissue selenium levels is not so effective when the element is administered in the organic form, probably because the organism does not then differentiate so well between methionine and selenomethionine.

Selenoproteins

So far selenium homeostasis has been discussed in this survey without reference to the chemical forms of the element in the tissues. The only selenoprotein known as yet to have biological functions in mammals is glutathione peroxidase and most of the studies have concentrated on the investigation of the metabolism and the functions of this enzyme. However, as mentioned earlier, less than 40 % of the selenium in the rat was found to be bound to glutathione peroxidase. It was therefore of interest to find out to what extent other selenoproteins are present in the organism and should be included in the studies on selenium.

For the investigations of these selenoproteins and their homeostatic regulation rats were used which had been fed a selenium-deficient diet for several generations (5). After the administration of labeled selenite to these animals the proteins were separated by means of analytical SDS-polyacrylamide gel electrophoresis and the ^{75}Se-containing compounds identified by autoradiography. By investigating a large number of tissues a total of 13 selenium-containing proteins or protein subunits have been detected. Tests have shown that the appearance of these compounds is not due to inspecific binding of the element or only a gradual proteolytic fragmentation of a larger selenoprotein. Most of them were found in all of the tissues investigated.

After the detection of these selenoproteins the question arose of whether, as with the priority supply of selenium to certain tissues, there is also a priority supply of the element to certain proteins. To answer this question an investigation was carried out on the selenium incorporation into the proteins from a small dose of labeled selenium given either to selenium-deficient rats or to rats which were supplemented again with the element during this experiment (5).

In all tissues investigated a considerable difference in the distribution of the tracer was found. In the replete animals the main part of the selenium dose was incorporated into the glutathione peroxidase. In the deficient animals a much smaller part of the activity was found in the enzyme and instead other selenoproteins were preferentially supplied. The findings show that there is also a hierarchy in the selenium supply at a molecular level and that at a low dietary intake of the element the maintenance of the glutathione peroxidase level is of only secondary importance compared with that of other newly found selenoproteins.

Conclusions

It has been found that within certain limits the selenium metabolism is controlled by regulatory mechanisms. Absorption does not appear to be homeostatically controlled. Homeostasis is achieved, however, via changes in the excretion rate. Especially at a high selenium intake the regulation is dependent on the chemical form of the element in the diet. With insufficient selenium intake homeostatic mechanisms are responsible for the priority supply of the element to certain specific target tissues and within the tissues to certain selenoproteins, and here the maintenance of glutathione peroxidase seems to be of only secondary importance compared with that of other newly found selenoproteins.

References

1. Bernard, C.. 1878. Les Phenomenes de la Vie 1, 113, 121.
2. Cary, E.E., W.H. Allaway, M. Miller. 1973. J.Anim.Sci. 36, 285.
3. Behne, D., M. Duk, W. Elger. 1986. J.Nutr. 116, 1442.

4. Salbe, A.D., O.A. Levander. 1987. Fed.Proc. 46, 1153.

5. Behne, D., H. Hilmert, S. Scheid, H. Gessner, W. Elger: Biochem.Biophys.Acta (in press)

6. Bopp, B.A., R.C. Sonders, J.W. Kesterson. 1982. Drug Metabol.Rev. 13, 271.

7. Levander, O.A.. 1986. In: Trace Elements in Human and Animal Nutrition, vol.2 (W. Mertz, ed.). Academic Press, Orlando, p. 209.

8. Brown, D.G., R.F. Burk, R.J. Seely, K.W. Kiker. 1972. Int. J.Vitam.Nutr.Res. 42, 588.

9. Burk, R.F., D.G. Brown, R.J. Seely, C.C. Scaief III. 1972. J.Nutr. 102, 1049.

10. Burk, R.F., R.J. Seely, K.W. Kiker. 1973. Proc.Soc.Exp. Biol.Med. 142, 214.

11. Behne, D., W. Wolters. 1983. J.Nutr. 113, 456.

12. Behne, D., T. Höfer-Bosse. 1984. J.Nutr. 114, 1289.

13. Behne, D., T. Höfer, R. von Berswordt-Wallrabe, W. Elger. 1982. J.Nutr. 112, 1682.

14. Capel, I.D., A.E. Smallwood. 1983. Res.Comm.Chem.Path. Pharmacol. 40, 367.

15. Igarashi, T., T. Satoh, S. Ono, K. Iwashita, M. Hosokawa, K. Ueno, H. Kitagawa. 1984. Res.Comm.Chem.Path.Pharmacol. 45, 225.

16. Thomson, C.D., R.D.H. Stewart. 1973. Br.J.Nutr. 30, 139.

17. Whanger, P.D.. 1986. J.Amer.Coll.Toxicol. 5, 101.

EFFECTS OF CORTICOSTEROID TREATMENT ON SELENIUM STATUS IN MAN AND ANIMALS

A. Peretz[+], J. Nève[++], J. Fontaine[+++], F. Vertongen[++++], and J.P. Famaey[+]
Free University of Brussels, Saint-Pierre Hospital, [+]Department of Rheumatology, and [++++]Department of Medical Chemistry; rue Haute, 322, B-1000, Brussels; [++]Toxicology and Bioanalytical Chemistry Unit, and [+++]Department of Pharmacology; Campus Plaine, B-1050, Brussels, Belgium.

Introduction

Acute and chronic corticosteroid treatments have been reported to modify the metabolism of trace elements such as zinc and copper, with alterations in plasma, urine or tissue levels (1). These modifications were identified in humans suffering from inflammatory diseases, in healthy animals as well as in cultured cells (2). To our knowledge, no data have been published on a similar effect on selenium (Se) metabolism although low plasma Se levels have been characterized in patients suffering from diseases such as arthritis where corticosteroid treatment is frequent (3,4).

The first part of this paper is devoted to the modifications in Se status, as assessed by various biochemical indices, that we previously described in patients with rheumatoid arthritis treated with usual drugs including low to high doses of corticosteroids (4). Then, we analyze the effects on Se levels in several body fluids and tissues of acute and chronic corticosteroid administration to healthy rats.

Subjects and Methods

Human subjects were rheumatoid arthritis patients (RA), 38 females and 7 males , 23-82 years old (58±15). All had active disease, and 40 of them had active inflammation (ESR > 20 mm/h). Nineteen of these subjects were treated with corticosteroids (C⊕ group), prednisolone or methylprednisolone, at prednisolone equivalent doses of 7.5 mg daily (n=7); 20 mg (7); and 60 mg (5). All the other patients were treated with non steroidal antiinflammatory drugs (C⊖ group). Data were compared with those from age and sex matched controls originating from the same area. Fasting blood samples were collected from the subjects for the determination of several biochemical parameters including plasma and erythrocyte Se levels, as well as the activity of the selenodependent enzyme glutathione peroxidase (GSH-Px) in erythrocytes. Details about the methods and the quality controls have been previously published (5,6). Animals were a total group of 330 male Wistar rats, 45-50 days old, weighing 180-220g. Long-term corticosteroid treatment was investigated by providing to animals an available complete maintenance diet that was previously enriched with prednisolone, 8 mg/kg (Usine d'Alimentation Rationnelle, Villemoisson,F). An approximate dose of 1mg prednisolone/kg rat was provided daily. Groups of treated animals and respective controls receiving unsupplemented diets were killed by exsanguination 3 or 5 weeks after the beginning of the study. Blood was collected; liver and femurs were dissected, washed and lyophilized, and the Se content was determined (5-7). Short-term corticosteroid treatment was investigated by a single intraperitoneal injection of a saline suspension of prednisolone (0.1 ml/100 g rat), at a level of 3 mg prednisolone /kg rat. The animals were killed 7,16,24 or 72 hours after the injection. Controls were injected with the vehicle alone.

Results

Table 1 shows the values of parameters of Se status that were determined in different subgroups of RA patients and in controls ($\bar{X} \pm SD$):

Subjects		n	Plasma Se µg/L	Erythr. Se µg/g Hb	Erythr. GSH-Px U/g Hb
RA	all	45	77 ± 20	0.40 ± 0.09	6.8 ± 1.7
	C⊕ all	19	66 ± 21	0.36 ± 0.08	7.3 ± 3.3
	7.5 mg	7	93 ± 16	0.38 ± 0.09	7.4 ± 1.7
	20 mg	7	70 ± 16	0.37 ± 0.11	7.4 ± 2.9
	60 mg	5	44 ± 12	0.40 ± 0.08	8.0 ± 2.9
	C⊖	26	86 ± 15	0.42 ± 0.08	6.9 ± 1.5
	FC I-II	28	81 ± 20	0.41 ± 0.07	6.4 ± 1.4
	FC III-IV	17	72 ± 19	0.37 ± 0.08	7.2 ± 2.5
Controls		45	84 ± 15	0.40 ± 0.10	7.2 ± 1.8

C= corticosteroid treated at various doses; FC= Functional Class (Steinbrocker criteria);

Significantly lower plasma Se levels were observed in subjects treated with corticosteroid at all doses when compared to those treated with other drugs and with controls ($p < 0.001$, Student t test). Significant differences were also disclosed for the same parameter when subjects treated with high doses corticosteroid (20 and 60 mg) were compared to those receiving 7.5 mg ($p < 0.025$).No statistical differences were observed for all other parameters or groups. No relationship could be established between plasma Se levels and variables such as disease duration, presence of extraarticular features or of the rheumatoid factor, and several inflammatory parameters. Low plasma Se levels in subjects with severe functional impairment (FC III-IV) did not reach statistical significance.

Table 2 shows the Se status of rats during chronic or acute prednisolone treatment ($\bar{X} \pm SD,n$):

		Treated rats	Controls
Chronic oral administration			
Plasma Se, µg/L	3 weeks	441 ± 31 (15)	450 ± 39 (10)
	5 "	471 ± 41 (15)*	503 ± 33 (10)
Erythr. Se,µg/gHb	3 "	1.4 ± 0.3(15)	1.4 ± 0.8(10)
	5 "	1.2 ± 0.3(15)	1.0 ± 0.1(10)
Liver Se,µg/g DW	3 "	2.8 ± 0.3(14)	3.0 ± 0.2(10)
	5 "	1.8 ± 1.3(14)	2.0 ± 0.8(10)
Bone Se,µg/g DW	3 "	180 ± 45 (8)	155 ± 17 (10)
	5 "	117 ± 9 (13)	120 ± 5 (10)
Acute intraperitoneal administration			
Plasma Se, µg/L	7 hours	442 ± 21 (8)	422 ± 18 (7)
	16 "	451 ± 15 (8)	420 ± 24 (7)
	24 "	426 ± 26 (8)	424 ± 28 (7)
	72 "	463 ± 31 (8)	437 ± 13 (8)
Erythr. Se,µg/gHb	7 "	1.4 ± 0.3(8)	1.1 ± 0.3(7)
	16 "	1.4 ± 0.1(8)	1.3 ± 0.3(7)
	24 "	1.2 ± 0.4(8)	0.9 ± 0.3(7)
	72 "	1.3 ± 0.3(8)	1.4 ± 0.2(8)

* $p = 0.05$, Student t test ; DW= dry weight.

Chronic oral prednisolone administration resulted in a slight decrease in plasma Se levels after 5 weeks consumption of the prednisolone enriched diet but Se remained unchanged in other compartments. The amount of food consumed by animals as well as their body weight were not different from controls, excluding these factors that can affect Se status. Acute administration in adopted conditions had no influence on parameters of Se status.

Discussion

Low plasma Se levels have still been characterized in inflammatory rheumatic diseases (3,4) or during the progress of adjuvant arthritis of the rat (7). Interest for the element comes from the fact that it is part of the active site of GSH-Px, a protective enzyme against cell damages by peroxides and free radicals produced during the inflammatory process. High levels of peroxidation products have indeed been reported in serum and synovial fluid of RA patients (3,4). However, few attempts have been made to correlate these low plasma Se levels with clinical parameters of disease severity or biochemical indices of inflammation. In RA patients, TARP et al (3) showed some significant relations with disease duration and clinical arthritis activity, but were unable to disclose any effect of drug therapy. However, they only investigated two patients treated with corticosteroids.

In our RA patients, low plasma Se levels were associated both with severe functional impairement and high doses of corticosteroid administration, but no other influence could be detected. After a stratification of patients (4), the effect of treatment might be dissociated from that of functional impairement (In clinical practice, patients with severe disease usually receive corticosteroids), and a dose-dependent effect of corticoids was even evidenced (table 1). This is to our knowledge the first report of an alteration of Se status by these drugs.

Our attempt to demonstrate the effects of corticosteroid administration in healthy rats only disclosed a slight decrease in plasma Se levels after 5 weeks chronic treatment. Acute corticoid administration had no effect, unless in adopted conditions.

The interpretation of the corticoid effect on Se metabolism is rather hypothetical. An increased binding into several tissues, or a change in the renal handling of the element can be evoked on the basis of similar effects on zinc and copper metabolism (2,8).

This study is supported by grants from the Fonds National de la Recherche Scientifique de Belgique, and from Labcatal SA, France.

References

1. Aaseth,J., Munthe,E.1978. Scand.J.Rheumatol. 7, 237.

2. Cousins,R.J.1985. Physiological Rev. 65, 288.

3. Tarp,U., Overvad,K., Hansen,J.et al.1985. Scand.J.Rheumatol. 14, 97.

4. Peretz,A., Nève,J., Vertongen,F.et al.1987. J.Rheumatol,14, 1104.

5. Nève,J., Chamart,S.,Molle,L.1987. In: Trace Element Analytical Chemistry in Medicine and Biology (P.Brätter,P.Schramel,eds)W.de Gruyter,Berlin, 4,239.

6. Nève,J., Hanocq,M., Molle,L.,et al.1983. The Analyst 107, 934.

7. Nève,J., Fontaine,J., Peretz,A,et al.1988,Agents and Actions,in press.

8. Peretz, A.,Nève,J.,Famaey,J.P.1988, submitted.

Section II

Biological functions of selenium

BIOLOGICAL FUNCTIONS OF SELENIUM

J. Nève
Free University of Brussels, Toxicology and Bioanalytical Chemistry Unit, Campus Plaine 205/1, B-1050 Brussels, Belgium.

Introduction

Since the demonstration in 1957 by SCHWARZ of the essentiality of selenium (Se) for animals, considerable work has been devoted to this trace element. Its important function in the active site of the peroxide destroying enzyme glutathione peroxidase (GPx) was established by ROTRUCK in 1971, and evidence for human Se requirement was reported nearly simultaneously in 1979 by Chinese scientists in the Keshan Disease, and by VAN RIJ et al in a patient on prolonged total parenteral nutrition (1-4). Se impact on health has become a rapidly growing field of scientific research : several human Se deficiency states are now identified and the element is implicated in the prevention of some important diseases such as cancer, cardiovascular diseases and degenerative affections. Several reviews have covered quite well all these aspects (1-4), that will not be extensively rediscussed in the present paper. A special attention will be drawn on most recent and sometimes controversial data, and an effort will be made for a global view on Se in biology and biochemistry.

SELENIUM IN GLUTATIONE PEROXIDASE

GPx is the only known selenoenzyme in humans. It has a key physiological role in the metabolism of oxygen and its dysfunction has been implicated in the development of many clinical symptoms associated with Se deficiency.

Protein structure and mechanism of action

GPx has a molecular weight of between 76.000 and 95.000 Da,

depending on species and tissue. It consists of 4 identical subunits containing 4 g-atom Se/mole, 1 per subunit. Se has been identified in the active site of the molecule as seleno-cysteine, most likely in its selenolate form, at physiological pH and concentration of glutathione. LADENSTEIN and Colleagues (5) established the three dimensional structure of the bovine enzyme, characterizing the active site in flat depressions on the molecular surface. GUNZLER and CoWorkers (6) showed that the monomer of bovine GPx consists of 198 aminoacids. The active site was located at position 45 from the N-terminal end. Four cysteine residues, at positions 74, 91, 111 and 152, may be associated with structural integrity of the active site or even contribute to the catalytic function of the enzyme. Arginine residues at positions 50 and 177 are associated with the Se-CYS residue, and probably provide positive charges to guarantee proper orientation of the carboxylic functions of glutathione at the substrate binding site, and thereby directing the thiol group towards the selenolate ion of the catalytic center (6).

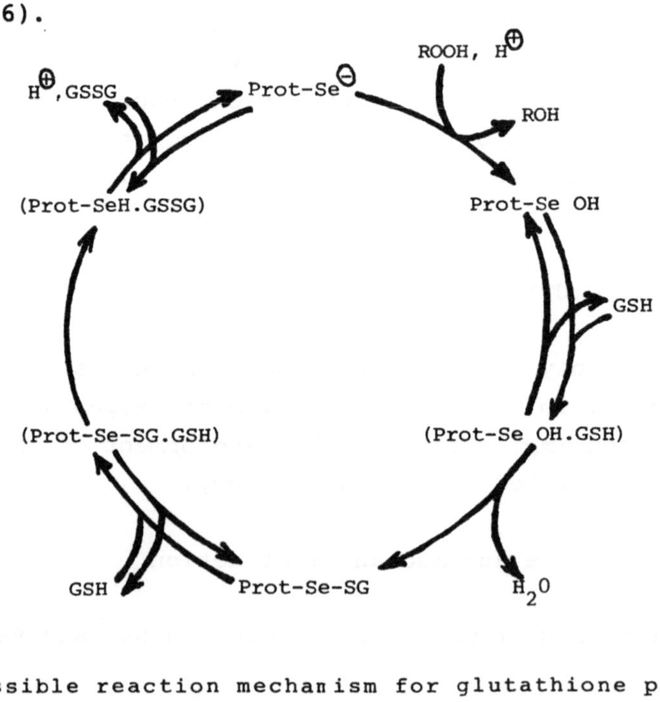

Figure 1: possible reaction mechanism for glutathione peroxidase.

The mechanism leading to Se incorporation into GPx has not been fully elucidated. A direct incorporation of Se-CYS into the peptide backbone using a specific t-RNA has been suggested, but a post translational incorporation of a Se compound such as selenide or selenoglutathione (G-S-Se-H) with an amino acid residue such as dehydroalanine, serine or cysteine, already present in the pre-GPx molecule, seems more likely and has been found possible in vivo (7). Recently, SUNDE and EVENSON (8) demonstrated with labelled amino acids that serine rather than cysteine is the precursor for Se-CYS in rat liver GPx. They suggested a co-translational mechanism for incorporation involving the replacement of the phosphate of phosphoserine with -SeH, either while phosphoserine is esterified to the seryl-t-RNA, or after phosphoserine incorporation into the polypeptide chain, leading to the synthesis of Se-CYS.

One of the supposed mechanism of action is summarized in Figure 1 (6). The first step of the catalytic action would be a simple bimolecular reaction of the substrate (peroxide) with Se-CYS (selenolate form, $R\text{-}Se^{\ominus}$), yielding to alcohol or water as reaction product and to an oxidized enzyme form (selenenic acid, R-Se-OH). This last form could thereafter complex glutathione through ionic interaction, and undergo an intramolecular reaction yielding to a selenosulfide derivative (R-Se-S-G). A second molecule of glutathione then reacts generating oxidized glutathione, and regenerating the selenolate form of the enzyme, now being ready for another cycle. Regeneration of glutathione by reduction of G-S-S-G is mediated by the NADPH dependent G-S-S-G reductase. There is a good agreement on the existence of a selenolate as the resting form of the enzyme; however, the oxidized forms of the active site are more controversial (9). GPx, unlike other peroxidases, has high specificity for its donor substrate glutathione, but a very low specificity for acceptor substrates which may be a wide variety of peroxides such as H_2O_2, lipid, steroid, nucleic acid, thymine and prostaglandines... hydroperoxides.

Tissue distribution and activity. Relations with selenium status.

GPx is generally present in the cytosol and mitochondrial matrix of cells although activity has been measured in several body fluids. It has been identified in many organs, but in different amounts in various tissues and with substantial differences between animal species (10). In the rat, high activity is found in the liver and erythrocytes; intermediate activity in kidney, spleen, adrenals and lung; and low activity in plasma, thymus, skeletal muscle, brain and testis (11). In humans, high activity is measured in liver and blood cells. In human erythrocytes, only 10 % of Se is associated to the enzyme, contrary to approximatively 80 % in various animal species. Recent data suggest that human plasma GPx, which represents only a few % of the plasma Se content, is different from the known cellular enzyme. Structure as well as catalytic activity are similar, but the plasma enzyme is a glycoprotein, immunologically different from the cellular form (12).

Relations between Se status and GPx activity are rather complex and depend on many factors such as species and tissue and on nutritional influences (10). GPx activity correlates well with Se content in blood or organs in individuals with marginal or deficient Se status, but not in adequate individuals or at higher levels of Se intake (1,2,10). Se deficiency, even marginal, always results in a decrease in cellular and plasma GPx activity. This reduction has been clearly related in many animal species and even in man to the development of clinical signs of Se deficiency (1-4). Experiments in the rat demonstrated that both GPx activity and GPx protein levels as assessed by radioimmunoassay decrease exponentially during progressive dietary Se deficiency indicating that Se status regulates the level of GPx protein as well as of activity. The loss of GPx protein occured at lower rates than the loss of activity, suggesting the presence of an inactive GPx polypeptide during short-term Se deficiency (13,14). Se supplementation of deficient individuals is accompanied by a rise in GPx activity (with different rates according to the body cells or fluid considered) which ends by a plateau of GPx activity when Se requirement for the enzyme is satisfied (1-3). Measurement of GPx activity actually consists in the most convenient technique available for assessing Se status. The enzyme activity not only represents biologically active Se, but such as indicated before, it appears to achieve saturation in some tissues at Se intakes that could be defined as satisfying or adequate nutritional requirements (15).

Metabolic functions of glutathione peroxidase.

Oxygen derived highly reactive species such as peroxides and free radicals are continuously produced in aerobic cells during the normal operation of a wide spectrum of biological processes (16). Moreover, they can be generated by the action of exogenous factors such as redox cycling substances and drug oxidation, cigarette smoke, radiation and various stimuli able to activate phagocytes. These derivatives either participate in biochemical reactions, or attack unspecifically a variety of biological structures causing oxidative damages. Figure 2 gives a simplified scheme for the production of these active species in different subcellular localisations, and the regulation of the steady state concentration by several enzymes including GPx (17).

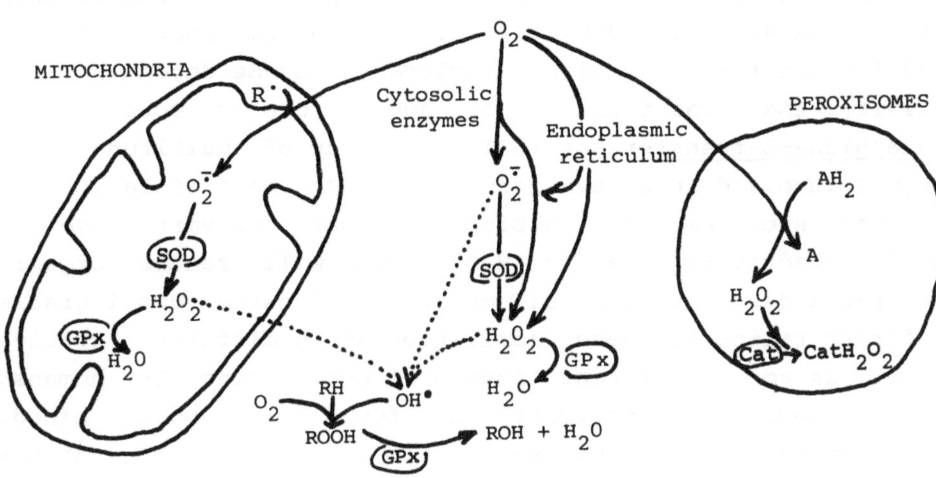

Many events commence with the production of superoxide radicals ($O_2^{\bar{\cdot}}$) by both enzymatic and non enzymatic means. The disproportion of $O_2^{\bar{\cdot}}$ is catalysed by cytosolic and mitochondrial superoxide dismutases (SOD) that return half of the oxygen and reduce the balance to H_2O_2. This last species can also be produced directly by a variety of enzymes, not only within the peroxisomes, but also in the microsomal, mitochondrial and soluble fractions of cells. The presence of H_2O_2 and $O_2^{\bar{\cdot}}$ in cells favours the formation of additional oxygen species such as the hydroxile radical ($OH^{\cdot}$) and singlet oxygen. Removing H_2O_2 from the cell is catalysed by two enzymes : catalase, which is confined almost exclusively in peroxisomes, and glutathione peroxidase, which is mostly present in cytosol and mitochondrial matrix. GPx is however the sole enzyme having the capacity to reduce organic hydroperoxides. This last property also includes lipid hydroperoxides in membranes, such as recently demonstrated by URSINI and coworkers (18). They postulated the existence of a phospholipid hydroperoxide GPx, specialized for lipophilic interfacial substrates. Other low molecular weight compounds or enzymes also take part in the lowering of the steady state concentrations of oxidizing species. "Free radical scavengers" include water soluble compounds such as glutathione, ascorbate and urate as well as lipid soluble derivatives such as α-tocopherol (vit. E). The function of the vitamin is complementary to GPx, particularly in biomembranes where it protects lipophilic constituents (16,17). LAWRENCE and BURK (19) discovered some years ago that Se-deficient rat liver also disclosed a non Se-dependent GPx activity, that was found to be due to one or more of the glutathione-S-transferases (GST), a group of multifunctional enzymes involved in biotranformation and/or detoxification of a broad spectrum of xenobiotics. These enzymes have no catalytic activity against H_2O_2, but will reduce organic hydroperoxides. Differential measures indicated considerable different ratios of these activities among different species as well as among different tissues of one species. In humans, only Se-dependent activity is found is whole blood, erythrocytes, platelets and biopsy skeletal muscles. A small non Se dependent GPx activity is measured in plasma, and a larger activity in biopsy liver supernatant and homogenates (20). In conditions of decreasing or increasing Se status in animals, inverse relations were established between GPx and GST activities, suggesting that GST can function as a substitute for GPx in some organs and experimental conditions.

Consequences of peroxidation in biological systems.

Production of reactive oxygen species is essential to biochemical processes such as hydroxylation, biosynthesis of thyroxin, oxidation of fatty acids, oxidation of alcools... and for phagocytosis (16). However, peroxides and free radical can also become cytotoxic, particularly when the antioxidant capacity of the cell or the organelle is deficient. Free radicals can damage many biomolecules which are essential to cellular integrity by addition to unsaturated bounds. Table 1 summarizes the pathological consequences due to perturbation of some of these molecules.

Table 1 : Some cellular free radical targets and consequences of chemical modification (16).

Small molecules	
.Unsaturated and thiol containing amino-acids	.Protein denaturation and cross-linking,enzyme inhibition
.Nucleic acid bases	.Cell cycle change,mutations
.Carbohydrates	.Cell surface receptor changes
.Unsaturated lipids	.Cholesterol and fatty acid oxidation, lipid cross-linking. Organelle and cell permeability changes
.Cofactors,enzymes	.Decreased availability and activity

.Neurotransmittors	.Decreased availability and activity
.Antioxydants	.Decreased availability
Macromolecules	
.Protein	.Peptide chain scission, denaturation
.DNA	.Strand scission, base modification. Mutations.
.Hyaluronic acid	.Change in synovial fluid viscosity

One of the most studied consequence of oxidative conditions (or "stress") is lipid peroxidation caused by reaction on cholesterol and fatty acids. This process can become autocatalytic after initiation and be accompanied by the production of carbonyl byproducts (e.g. malondialdehyde), or epoxides. Lipid peroxidation destroys plasma and intracellular membranes leading to changes in fluidity and permeability, and inactivation of a lot of membranal enzymes and protein receptors. Products of peroxidation can also cause damages to molecules that are distant from the lipid phase and deplete protective mechanisms (21).

Lipid peroxidation is however not exclusively a destructive process. This activity is indeed associated with the metabolism of arachidonic acid, and the production of oxygenated biomolecules. GPx and Se have both been involved in several steps of this metabolism, and appear to have important regulatory functions.

As a general rule, lipid peroxides (peroxide "tone") regulate the production of oxygenated metabolites of arachidonic acid (HPETE and derivatives). Alterations in peroxide levels due to GPx dysfunction results in modifications in the amount and type of derivatives that are produced. In lipoxygenase pathway, the reduction of HPETE to HETE depends on a peroxidase. BRYANT et al. showed in rat platelets that this activity is coupled through GPx to glucose metabolism and disclosed a decreased conversion 12 HPETE to 12 HETE in Se deficient animals (22). In similar conditions, GAIROLA and TAI found a selective inhibition of LTB_4 biosynthesis in pulmonary alveolar macrophages of Se deficient animals (23). They suggested that this effect on leukotrienes synthesis is not mediated by GPx, but is due to alterations in LTA_4 synthetase or LTB_4 hydrase. Modifications concerning 5-HETE and LTB_4 derivatives have important consequences on chemotactic and immunoregulatory properties of involved cells. In the cyclooxygenase pathway, alterations in the production of different end products were demonstrated by modifications of Se intake in animals. The balance between the proaggregatory and vasoconstrictory metabolite, thromboxane A_2, TxA_2, measured in platelets, and the antiaggregatory and vasodilatotory derivative prostacyclin, PGI_2, assessed in endothelial cells of aortic tissues, was evaluated. Increased production of TxA_2 and decreased PGI_2 synthesis were disclosed in Se deficiency states. Opposite effects on the PG synthetases were suggested as possible mechanisms. On the contrary, Se administration caused decreased production of TxA_2 in platelets, and, in some organs, increased PGI_2 production (24,25). GUIDI et al. recently proved that GPx was regulating some of the aspects of TxA_2 formation (26). These data suggest that Se modulates platelet aggregation. Finally, an effect of Se was also found in prostaglandins synthesis (PGE_2).

Human diseases related to glutathione peroxidase.

Clinical conditions involving oxygen species will be reviewed eslewhere in this book. The involvement of Se deficiency in human peroxidative diseases has been demonstrated in KESHAN (cardiomyopathy) and KASHIN BECK diseases (osteoarthropathy); in membrane abnormalities affecting erythrocytes, platelets and muscles during several diseases; in liver, brain, and several organ damages accompagying the exposition to exogenous compounds or several diseases such as neuronal ceroid lipofuscinosis, aging, cirrhosis... (1-4). The preventive effect of Se against cardiovascular diseases has been interpreted by a limitation of peroxide induced damages on lipid structure and metabolic integrity, particularly in arachidonic acid and cholesterol metabolism. The Se dependent protection has also been implicated against cancer where mutagenecity and tumor promoter activity has been demonstrated for lipid peroxidation products (28). Finally, some of the effects of Se as a regulator of immunity have been explained by the maintenance of membrane integrity of immunocompetent cells (28).

SELENIUM DEPENDENT METABOLIC MODULATIONS

A number of Se-dependent metabolic alterations have been identified in man and animals and a lot of them do not apparently involve GPx (29).

Modifications in metabolism of several drugs and foreign compounds was studied at different levels of Se intake, and changes in chemical toxicity were disclosed. An enhanced toxicity due to Se deficiency was demonstrated for derivatives which undergo a redox cycling in the organism, forming free radicals and oxidative species : the 5-nitrofurane derivatives (nitrofurantoine and furazolidone), the bipyridinium herbicides (paraquat and diquat), CCl_4 and other chlorinated hydrocarbons, and finally sodium nitrite. The correction of nutritional Se deficiency, was able to protect animals against induced toxic damages. However, for another class of compounds, Se deficiency decreased the toxicity. This was demonstrated for hepatotoxic compounds such as acetaminophen, diethylmaleate, iodipamide and the mycotoxin aflatoxin B_1. These compounds are usually detoxified by glutathione and/or binding involving glutathione S-transferase enzymes, which are increased during Se deficiency. A last class of derivatives includes drugs for which the toxicity can be prevented in animals by Se supplementation to diets that are apparently adequate for Se : doxorubicin, bromobenzene, tetracyclin, gentamycin and the anticancer cis-diamine dichlorplatinum drug (29).

Considerable work has also been devoted on the protection by Se against chemically induced carcinogenesis in animals.

Under defined dietary conditions Se has been shown to counteract to various degrees the tumorigenesis of organs or tissues induced by a wide range of chemical carcinogens such as 3-methylcholantrene, α-pyrene, 2-acetylaminofluorene, 7-12 dimethylbenzanthracene ... Modulation of the activities of carcinogen metabolizing enzymes has been evidenced. Se was able to inhibit activation and accelerates detoxification of carcinogenic compounds (27,30,31).

Protection by Se through GPx against peroxidative damages induced by foreign compounds has been evidenced in some cases. However, it seems that Se has an independent more general influence on intermediary metabolism, including hepatic drug metabolism. REITER and WENDEL (32) conducted investigations on the effects of Se deficiency in mice on the activity of GPx and several drug metabolizing enzymes. They noted three kind of effects that occured at different moments after initiation of deficiency. The loss of GPx activity in liver was the earlier marker of Se deficiency whereas changes affecting other enzymes appeared in a later phase when liver GPx activity had completely disappeared. A small number of enzyme activities however remained unchanged. The activity of a second group of enzymes at least doubled in Se deficiency. Glutathione S-transferase activity was the first to increase, followed by changes in many other enzymes such as glutathione reductase, ethoxycoumarine deethylase, cytochrome P-450 hydroperoxidase, heme oxygenase, UDP glycuronyltransferase and glutathione thiol transferase... A third group of enzymes was characterized by enzymatic rates that dropped to about one half or less : NADPH cytochrome P-450 reductase, flavin dependent monooxygenase and sulfotransferase. During repletion of deficient animals, different time constants for the recovery of enzyme activities were observed, ranging from half life of 5.7 h to 29 h for GPx and 44 h for glutathione-S-transferase. These experiments confirmed previous ones showing the influence of Se, not mediated by GPx, on the metabolism of sulfhydrils group. In deficient animals, glutathione S-transferase activity and glutathione synthesis are both increased. Interorgan metabolism of glutathione is affected with an increased release of glutathione from the liver and resulting higher plasma and urine levels (33). The hepatic capacity of glutathione conjugaison is increased, with consequences on the detoxification of some xenobiotics. Supplementation with Se was also claimed to increase the hepatocellular level of glutathione in response to an enhanced consumption of this species and production of reduced glutathione, with a possible stimulant effect on immunocompetent cells, what afford to explain the immunostimulant capacity for Se that was demonstrated in animals (28). Finally, WENDEL and OTTER (34) indicated that modulations of metabolic pathways in the liver also involve enzymes of intermediary metabolism. They demonstrated an increased glucose turnover and tolerance in Se deficient animals, with increased 6-phosphogluconate dehydrogenase and succinate dihydrogenase activities, and suggested possible Se-dependent interactions with endocrine glands. The same group also showed during Se deficiency a stimulation of DNA replication rate in liver cells where proliferation was induced, and underlined the role of Se in various regulation systems in the organism (34).

SELENIUM INTERACTION WITH MINERALS

Many studies in animals and man proved that Se is able to interact in the body with a wide range of toxic minerals such as As, Cd, Hg, Cu, Ag, Pb or Pt..., and can modifiy their toxicity to various degrees. The element is highly effective in the prevention of testicular damage, high dose mortality, teratogenecity, cardiotoxicity and hypotensive effect induced by Cd; of renal and brain damages due to organic and inorganic Hg compounds; and of renal and lethal effects of Pt compounds used as anticancer drugs. Se has mild protective effects against liver necrosis and exsudative diathesis induced by Ag salts and toxic effects of Pb on red blood cells (1,2).

Different mechanisms have been proposed to explain these interactions. First, a direct reaction of Se as selenide with the metal leading to the formation of biologically inactive forms. This has been demonstrated for Cd, Pt, Pb, Ag and Hg. Accumulation of granules of metal selenides has been showed in some organs. Metals seem to be redistributed in the body and not necessarily excreted in greater amounts. Secondly, a diversion by Se of binding of metals such as Hg and Cd from critical proteins to less critical ones has been evidenced. Se would react with thiol groups of some molecules, forming selenotrisulfides centers that have a strong affinity for metals. Metabolic interference such as the one for As and Se, which have common excretory patterns, has also been evoked as a possible mechanism. Finally, Se effects have been linked to the prevention of oxidant stress induced by metals such as Cd and Ag and to modulation of liver detoxification functions (35,36).

Most of these studies were led with Se and the toxic element administred parenterally at rather high doses. Recent investigations however confirm that the nutritional level of Se present in ordinary commercial diets for animals is sufficient to alleviate some metal toxicities. Several studies also proved that dietary Se supplementation has an influence on normal levels of these elements in some biological compartments. Different Se compounds have however not the same ability to interact with other minerals (37).

OTHER BIOLOGICAL ROLES FOR SELENIUM

Se is largely associated with a wide variety of proteins in nearly all organs, but the physiopathological activity of these compounds remain poorly known. A 10.000 Da selenoprotein has been identified in human muscle, and could be related to muscle disease associated with Se deficiency. Keratinoid selenoproteins, 15-20.000 Da, have been characterized in spermatozoa and are responsible for flagella integrity. Other proteins characterized in animals or bacteria generally contain active Se as selenocysteine (1-4).

Extensive investigations that were led on the protective effects of Se against cancer have suggested that biological Se forms can act on cell proliferation (tumor cells) particularly at high intake levels. Such as indicated before, Se could increase the hepatic concentration of both reduced (GSH) and oxidized (GSSG) forms of glutathione. LE BOEUF and HOEKSTRA showed that increase in GSSG results in the activation of a protein kinase inactivator of eukaryotic initiation factor 2 (el F-2). Other authors demonstrate that the element can modulate cell proliferation and/or viability in culture at some concentrations by altering rates of synthesis of RNA, DNA or proteins... (38,39). Finally, POIRIER and MILNER or VERNIE (27) proposed that high Se intake could result in the production in greater amounts of a reductive metabolite of Se, selenodiglutathione (G-S-Se-S-G), which is a potent inhibitor of amino acid incorporation into proteins by a specific inhibition of eukaryotic elongation factor 2.

Acknowledgement : The author gratefully acknowledges the Fonds de la Recherche Scientifique et Médicale de Belgique for its various supports.

References.

1. Nève,J., Vertongen,F., Molle,L., 1985, Clin. Endocrinol. Metab. 14, 629.
2. Nève,J., Henry, M., Peretz,A., Mareschi, J.P. 1987. Cah. Nutr. Diet. XXII, 145.
3. Levander,O.A. 1986. In : Trace element in human and animal nutrition, vol. 5 (W. Mertz, ed). Academic Press, Orlando, pp. 208.
4. Combs,G.F., Combs,S.B., 1984. Ann. Rev. Nutr. 4, 257.
5. Epp,O., Ladenstein, R., Wendel,A. 1983. Eur. J. Biochem. 133, 51.
6. Gunzler,W.A., Steffens,G.J., Grossmann,A. et al. 1984. Z. Physiol. Chem. 365, 195.
7. Sunde,R.A., Hoekstra, W.G., 1980. Biochem. Biophys. Res. Com. 93, 1181.
8. Sunde,R.A., Evenson,J.K. 1987. J. Biol. Chem. 262, 933.
9. Chaudière,J., Wilhelmsen,E., Tappel,A. 1984. J. Biol. Chem. 259, 1043.
10. Nève,J., Vertongen,F., Molle,L. 1984. In : Industrial uses of selenium and tellurium (Se-Te Develop. Assoc. ed), STDA press, pp 537.
11. Behne,D., Wolters,W. 1983. J. Nutr. 113, 456.

12. Takahashi,K., Avissar,N., Whitin,J., Cohen,H. 1987. Arch. Biochem. Biophys. 256, 677.
13. Takahasi,K., Newburger,P., Cohen,H.J. 1986. J. Clin. Invest. 77, 1402.
14. Knight,S.A., Sunde,R.A. 1987. J. Nutr. 117, 732.
15. Levander,O.A., 1986. J. Amer. Coll. Nutr. 5, 37.
16. Freeman,B.A., Crapo,J.D. 1984. In : Advances in the Biology of Disease. vol. 1. (E. Rubin, I. Danyanov, eds). Williams et Wilkins, London, pp. 26.
17. Chance,B., Sies , H., Boveris , A. 1979. Physiol. Rev. 59, 527.
18. Ursini,F., Maiouno,M., Gregolin,C. 1985. Biochem. Biophys. Acta, 839, 62.
19. Lawrence,R.A., Burk,R.F. 1976. Biochem. Biophys. Res. Com. 71, 952.
20. Thomson,C.D. 1985. Biochem. Int. 10, 673.
21. Slater,T.F., Cheeseman,K.H., Davies,M. et al. 1987. Proc. Nutr. Soc. 46, 1.
22. Bryant,R.W., Simon,T.C., Bailey,J.M. 1982. J. Biol. Chem. 257, 14937.
23. Gairola,G., Tai,H. 1985. Biochem. Biophys. Res. Com. 132, 397.
24. Toivanen,J.L.1987. Prostag. Leukotrienes and Med. 26, 265.
25. Doni,M.G., Bonaccorso,G., Piva,E. 1983. Haemostasis, 13, 248.
26. Guidi,G.S., Schiavon,R., Biasioli,A. et al. 1984. J. Lab. Clin. Med. 104, 574.
27. Vernie,L.N. 1984. Biochem. Biophys. Acta 738, 203.
28. Kirimidjian,L., Stotzky,G. 1987. Environ. Res. 42, 277.
29. Combs,G.F., Combs,S.B., 1987. Pharmac. Ther., 33, 303.
30. Whanger,P.D. 1983. Fund. Appl. Toxicol. 3, 424.
31. Shamberger,R.J. 1985. Mut. Res. 154, 29.
32. Reiter,R., Wendel,A., 1983; 4,5, Biochem. Pharmacol. 32; 33; 34, 3063; 1923; 2287.
33. Hill,K.E., Burk ,R.F. 1985. Arch. Biochem. Biophys. 240, 166.
34. Wendel,A., Otter,R. 1987. Biochem. Biophys. Acta, 925, 94.
35. Diplock,A.T., Watkins,W.J., Heurson,M., 1986, Ann. Clin. Res. 18, 55.
36. Nordberg,G.F., 1978. Environ. Hlth. Perspect. 25,3.
37. Magos,L., Clarson,T, Sparrow,S. et al. 1987. Arch. Toxicol. 60, 422.
38. Medina,D. 1986. J. Am. Coll. Nutr., 5, 21.
39. Batist,G., Katki , A.G., Ferraw , V.J. et al. 1986. J. Am. Coll. Nutr. 5, 87.

STUDY OF SELENIUM ANTIOXIDANT PROPERTIES ON LIPID PEROXIDATION: USE OF NMR AND ESR SPECTROSCOPIES

F. Nepveu, M. Massol
Unité INSERM 305, Laboratoire de Chimie Bioinorganique, Université Paul Sabatier, 38, rue des 36-Ponts, F-31400 Toulouse

C. Capul, P. Derache
Unité INSERM 87, 2, rue F. Magendie, F-31400 Toulouse.

Introduction

Selenium, an essential trace element in animal and human diets, is present at the active site of Glutathione-peroxidase (GSH-Px) (1). This enzyme destroys lipid peroxides by reduced glutathione according to the reaction:

$$2GSH + ROOH \longrightarrow GSSG + ROH$$

Thus, this protective system decreases the formation of hydroxy radicals, preventing a free radical chain reaction and subsequent lipid peroxidation of cell membranes (2, 3, 4, 5). In this work magnetic resonance spectroscopies are used, in addition to optical ones, to study the effects of manganese-selenite and seleno-DL-methionine on linoleic acid peroxidation induced by the enzyme lipoxidase, in presence or in absence of GSH-Px or Vit E.

Methods

All reagents were purchased from Sigma, Prolabo and Merck. Manganese selenite dihydrate was obtained by precipitation of concentrated solutions of manganese chloride and sodium selenite.

Selenium in Medicine and Biology

UV-assay: the peroxidation system consists of 0.428mM linoleic acid in a borate buffer (0.2M, pH 9.0). The reaction was started by addition of 166 U/ml lipoxidase and then, conjugated diene formation has been followed with a UV-visible spectrophotometer at 234 nm.
ESR and NMR assays: In the same buffer, a mixture of linoleic acid 0.428 mM and of the spin trap DMPO 62.5 mM was incubated at room temperature. The reaction was started by addition of lipoxidase 2500 U/ml and ran over a 20 mn period, then the reaction mixture was transferred to a quartz flat cell for measurement on a Brucker ER200D ESR spectrometer. In this ESR study, GSH (8 nM), GSH-Px (16 mU) or Vitamin E (1 mM) were added to the mixture. For NMR-assay, proton relaxation time was measured by a pulsed NMR 20 MHz spectrometer at 37°C.

Results:

Under our conditions, no significant variation between the different systems was found by UV assays (Table 1). This means that lipoxidase activity is not affected by manganese selenite or seleno-DL-methionine.
T1, T2 proton relaxation time measurements of a medium reflect the paramagnetic relaxation enhancement due to the concentration of paramagnetic organic radicals or metallic species. In our experimental conditions, variation in T1 and T2 did not significantly differ from controls (Table 1): the very low paramagnetic species concentration and the rapid chain radical reaction show that this NMR assay cannot be applied to the study of the lipoperoxidation in such biochemical models.
The ESR spin trap method is a cumulative method for detecting free radicals. For investigation of lipid peroxidation, DMPO has often been used as spin trap. By this method, it has been

shown that DMPO gives identical signals with those observed in Fe^{2+}-catalysed reaction, showing a spectrum of DMPO-alkyloxide adduct (6, 7, 8). In our work the spin trapping technique has been used sucessfully for detection of a variety of free radicals (hydroxyl, superoxide, alkoxy). Vitamin E protects against a radical chain reaction propagation (Fig 1). ESR studies give some information about the effects of seleno-compounds. At high concentration, manganese selenite promotes free radical production (Fig 1). This observation could suggest that manganese selenite catalyses hydroperoxide breakdown, following the reaction:

$$LOOH + Mn^{2+} \longrightarrow LO^{\cdot} + Mn^{3+} + OH^{-}$$

Such results have been observed with other manganese salts (9). In contrast to this last observation, seleno-DL-methionine protects against a lipid peroxidation promoted by a free radical chain reaction (Fig 1).

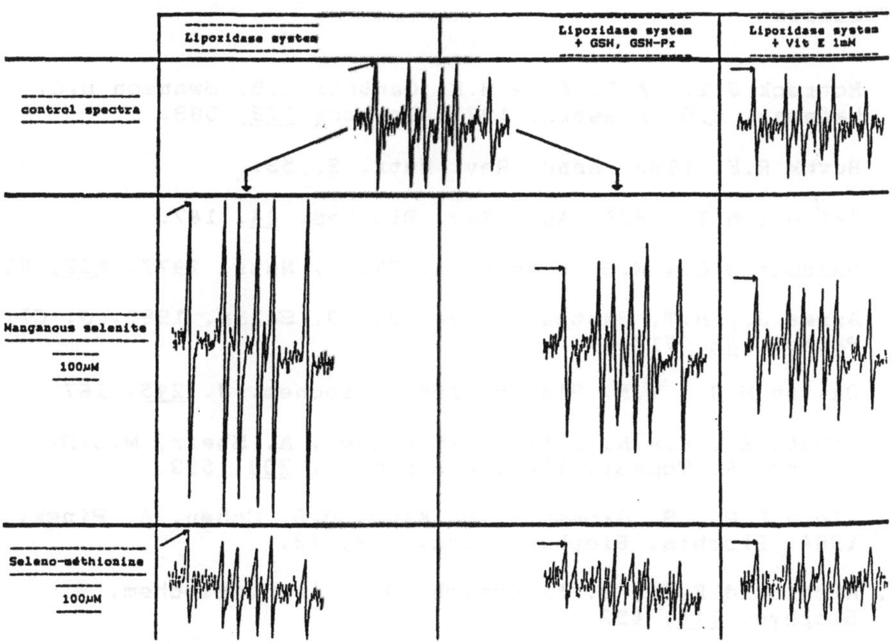

Fig 1: EFFECT OF SELENO COMPOUNDS ON FREE RADICAL PRODUCTION (ESR-Study g = 2.0155).

	Concentration µM	O.D.U / Min	T1 msec	T2 msec
Control	0	0.462 ± 0.047	3960 ± 298	2696 ± 20
Seleno-methionine	10	0.445 ± 0.018	3917 ± 139	3017 ± 353
	100	0.395 ± 0.019	3793 ± 101	2645 ± 5
Manganese selenite	10	0.442 ± 0.022	4075 ± 137	3027 ± 220
	100	0.385 ± 0.016	4074 ± 180	2854 ± 17

TABLE I: SELENIUM COMPOUND EFFECTS ON CONJUGATED DIENE FORMATION (UV-234nm Study) AND ON PROTON RELAXATION TIME (NMR 20 MHz Study)

Acknowlegments:

We are very grateful to F. Anglade and C. Faure for their technical assistance and INSERM for financial support.

References:

1. Rotruck J.T., A.L. Pope H.E. Ganther A.B. Swanson D.G. Hafeman, W.G. Hoestra. 1973. Science 179, 588.

2. Burck R.F. 1983. Annu. Rev. Nutr. 3, 53.

3. Fridovich I. 1975. Ann. Rev. Biochem. 44, 147.

4. Hafeman D.G., W.G. Hoestra. 1977. J. Nutr. 1977. 107, 656.

5. Aznar J., M.T. Santos, J. Valles, J. Salla. 1983. J. Clin. Pathol. 36, 712.

7. Davies M.J., T.F. Slater. 1987. Biochem. J. 245, 167.

8. Albano E., K.A.K. Lott, T.F. Slater, A. Steir, M.C.R. Symons, A. Tomasi. 1982. Biochem. J. 204, 593.

9. Klein B.P., S. Grossman, D. King, B.S. Cohen, A. Pinsky. 1984. Biochim. Biophys. Acta. 793, 72.

10. Archibald F.S., I. Fridovich 1982. Arch. Biochem. Biophys. 214, 452.

DIETARY FATS AND HEART LIPID PEROXIDATION STATUS

G. NALBONE, E. TERMINE, J. LEONARDI, P. LECHENE, M. CHAUTAN and H. LAFONT.

INSERM U 130, 18 Avenue MOZART, 13009 MARSEILLE (FRANCE)

Introduction

Although fish oils, are clearly an efficient way to prevent atherosclerosis (1), they may induce an increase in sensitivity to peroxidation because of their high unsaturation degree. Indeed, n-3 PUFA constituting fish oils are efficiently incorporated in membrane phospholipids of various organs. Among these organs, the heart has been shown to respond to dietary n-3 PUFA as supplied by fish oils (2, 3). In this work we investigated the effect of salmon oil on the selenium status of rat heart. To this end, we compared the effect of salmon oil to those of corn oil, saturated and low fat diets. Heart lipid peroxidation status was studied by investigating GSH-PX activities and ultramicroscopic aspect of myocardium.

Materials and Methods

Four groups of 6 male Wistar rats (IFFA-CREDO, L'Arbresle, France) weighing 190-210g were used. One group was fed a low fat diet (4.4 %, w/w) consisting of a lard and corn oil mixture with a polyunsaturated / saturated (P/S) ratio of 1.2. A second group received a corn oil diet (17 %, w/w) with a P/S ratio of 5. The third group received a salmon oil diet (12.5% w/w) supplemented with 4.5% corn oil, with a P/S ratio of 2 and the last group was fed a 15 % lard diet supplemented with 2% corn oil with a P/S ratio of 0.4 . Composition of the diets is given elsewhere (3). All the diets were supplemented with 1% cholesterol (w/w). Corn oil, as supplied, contained 45 mg of alpha-tocopherol per 100g of oil. Salmon oil was supplemented with

100 mg of alpha-tocopherol/100g of oil. Hence, the total amounts of alpha-tocopherol supplied by the control and lard diets were each 171mg/kg and were 246 mg and 315 mg for corn oil and salmon oil diet respectively. The rats were fed for 8 weeks. Tissue homogenization and lipid extraction were done as previously described (3). Choline and ethanolamine containing phospholipids were separated by HPLC as previously described (4). Fatty methyl esters (FAME) of phosphatidylcholines (PC) and phosphatidylethanolamines (PE) were separated as previously described (3). Heart selenium-dependent GSH-PX activity (EC. 1.11.1.9) was assayed according to Levander et al. (5) by using t-butylhydroperoxide to initiate the reaction. Heart glutathione S-transferase (EC. 2.5.1.18) activity was assayed according to Habig et al. (6) by using 1-chloro-2,4-dinitrobenzene as substrate.

Results and Discussion

Over the 8-week feeding period no significant difference ($p > 0.05$) was observed in body weight gains although rats fed high fat diets tended to have higher body weights than those of control group. Heart weights were not significantly different. Protein, total phospholipid, cholesterol and triglyceride concentrations in heart were not significantly altered ($p > 0.05$) by the dietary manipulations. Only triglycerides tended to increase in the hyperlipidic groups. The proportion of choline and ethanolamine containing phospholipids was not modified. Marked variations were observed in the fatty acid composition of PC and PE. The n-6/n-3 ratio of cardiac PC and PE of the salmon oil group was decreased as a consequence of n-3 PUFA incorporation. The highest increase in the unsaturation index of PL was seen with the salmon oil group (+ 26% in PC and + 13% in PE). Ultramicroscopic examination of ventricle revealed that few lipid droplets were occasionally present in cardiomyocytes of rats fed control diet, corn oil and lard diets. The lipid droplets were tightly associated to the mitochondria. These lipidic structures were more frequently observed in hearts of rats fed salmon oil.

In addition, the most striking feature of this group was the significant accumulation of dense granules structurally identified as lipofuscin. As shown in Table 1, heart selenium-dependent GSH-PX activity tended to be higher in the salmon oil group than in the other groups. This increase was significant ($p < 0.05$) when the values were compared to those of the lard diet. They were +36% and +54% for the total and specific activities respectively. Heart glutathione S-transferase activities as well as liver selenium-dependent GSH-PX activity were not significantly modified among the four groups.

TABLE 1: ACTIVITIES OF GSH-PX AND GLUTATHION S-TRANSFERASES IN HEART AND LIVER OF RATS FED VARIOUS DIETS.

	LOW FAT	CORN OIL	SALMON OIL	LARD
Heart GSH-PX Units/g heart	48.9±6.6	47.0±4.6	60.5±10.9	44.1±3.1
GSH S-Tr. Units/g heart	2.8±0.4	2.6±0.3	2.3±0.5	2.2±0.3
Liver GSH-PX Units/mg prot.	1.3±0.3	1.2±0.1	1.5±0.1	1.4±0.3

As the purpose of our study was to investigate the effect of dietary PUFA on the basal selenium status, no supplementation in this element was performed. Our basal diet contained approximately 60 µg of selenium/kg which is near the limit recommended. The specific activities of the GSH-PX measured in heart and liver of rats gave values in the range to those reported by Levander et al. (5), thus suggesting that our basal diets did not induce selenium deficiency. However, fish oil diet might have induced some selenium deficiency since lipofuscin accumulated in heart. This was not the case since heart selenium-dependent GSH-PX was not lowered in the salmon oil group, and was even 50% higher than that measured in heart of rats fed the saturated lard diet. Moreover, liver GSH-PX activity, which is used as an index of selenium status in animals, was kept at the same level in all the experimental groups thus making selenium deficiency in the salmon oil group improbable. In addition, if GSH-PX activity had been insufficient in the salmon oil group, we should have observed some activation of the glutathione S-transferases, since these enzymes have been demonstrated to compensate for inadequate activity of the selenium-dependent enzyme. In fact, the glutathione S-transferases remained unchanged, which again argues against subnormal selenium status in the salmon oil group.

In the light of a previous study (3) vitamin E status was probably lowered in salmon oil group. This alteration was probably sufficient to lower membrane-bound vitamin E level, thus rendering inadequate the protection of PUFA against free-radicals and leading to a higher level of lipid peroxidation. In conclusion, a balance has to be defined between the amount of fat and the n-6/n-3 ratio of the diet, in such a way that the beneficial effects of n-3 PUFA on the cardiovascular system are preserved but with a limited incidence on the lipid peroxidation status.

References

1. Dyerberg, J. Bang, H.O., Stofferson, E., Moncada, S. and Vane, J.R. 1978, Lancet, 2, 117.

2. Swanson, J.E. and Kinsella, J.E. 1986, J. Nutr. 116, 514.

3. Nalbone, G. Termine, E., Léonardi, J. Portugal, H., Lechene, P. , Calaf, R., Lafont, R. and Lafont H. J. Nutr. (in press).

4. Termine, E., Léonardi, J., Lafont, H., and Nalbone, G. 1987, Biochimie, 69, 245.

5. Levander, A.O., De Loach, D.P., Morris, V.C. and Moser, P.B. 1983. J. Nutr. 113, 55.

6. Habig, W.H., Pabst, M.J. and Jakoby, W.B. 1974, J. Biol. Chem. 249, 7103.

EFFECT OF SELENIUM SUPPLEMENTATION IN RAT DIET ON BLOOD PLASMA LIPID PARAMETERS

A. M. Crespo

Departamento de Zoologia e Antropologia - Faculdade de Ciências, Universidade de Lisboa, Bloco C_2 Campo Grande, 1700 Lisboa, PORTUGAL

R. E. Pinto

Departamento de Química - Faculdade de Ciências, Universidade de Lisboa Bloco C_1 Campo Grande, 1700 Lisboa, PORTUGAL

Introduction

Selenium is an essential element in nutrition. Its deficiency is responsible for several pathological conditions. However very rich selenium diets may produce toxic effects (1). Changes in lipid concentrations were observed in several tissues of rat when drinking water contained 54 ppm selenium (2,3). A positive correlation of unsaturated fatty acids in human serum and selenium levels was observed (4).
The aim of the present work was to investigate the effect of light selenium supplementation to the rat diet on lipid metabolism in blood plasma.
Wistar male rats three months old were treated for one,three and six months with sodium selenite given in drinking water containing 2 p.p.m. selenium. Rats drank about 120µg/kg rat body weight.
Protein, phospholipids, acylglycerols, total and free cholesterol were measured in plasma. High density lipoprotein (HDL) and low density lipoprotein (LDL) cholesterol and phospholipids were also determined. Chromatography (GLC) of fatty acids was carried on both total lipids and cholesterol esters in plasma.

Selenium in Medicine and Biology

Results

Differences between treated and non treated rats after one month were not significant .

Levels of protein (6.1±0.6 g/l), acylglycerols (0.80±0.13mM), total phospholipids (1.1±0.15mM) and HDL phospholipids (0.8±0.08mM) did not change also in rats treated for three and six months.

Plasma cholesterol did not change after one and three months of treatment (Fig. 1). Cholesterol increased in rats that drank water containing selenium for six months. Free cholesterol changed from 0.38 ± 0.06mM to 0.62±0.1mM. Cholesterol esters (1.2±0.12mM) increased about 45% and total cholesterol (1.5±0.15mM) increased about 60%. However cholesterol esters/ /free cholesterol ratio does not change significantly.

Fig.1

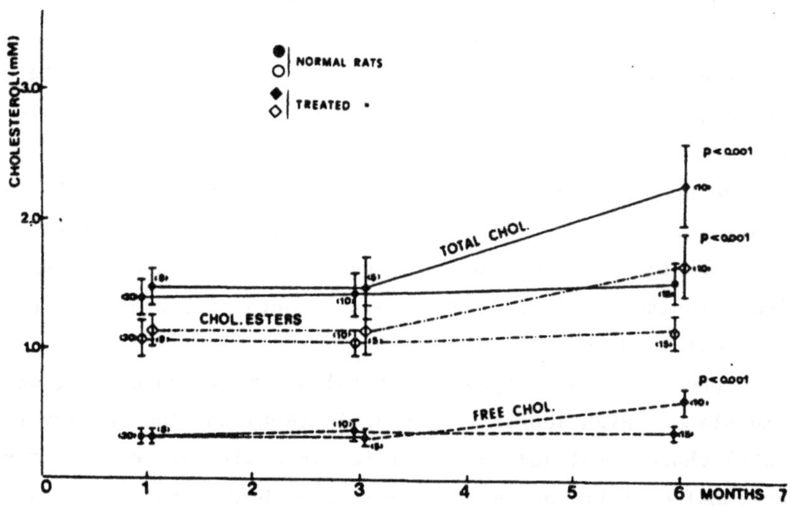

Total cholesterol both in HDL and in LDL follow the increase (Fig. 2) observed in plasma total cholesterol (Fig. 1).Total cholesterol in HDL (0.91±0.12mM) increased about 35% and in LDL (0.45±011mM) increased 75%.

Fig. 2

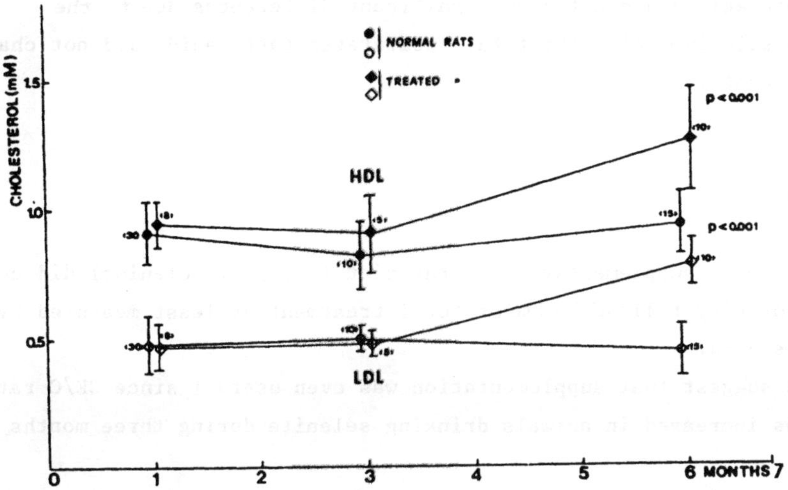

Cholesterol esters/free cholesterol ratio (CE/C) did not change in LDL (Fig. 3) but CE/C in HDL had an increament of near 40% since increased from 4.5±0.9 non treated rats to 6.2±0.8 in treated ones.

Fig. 3

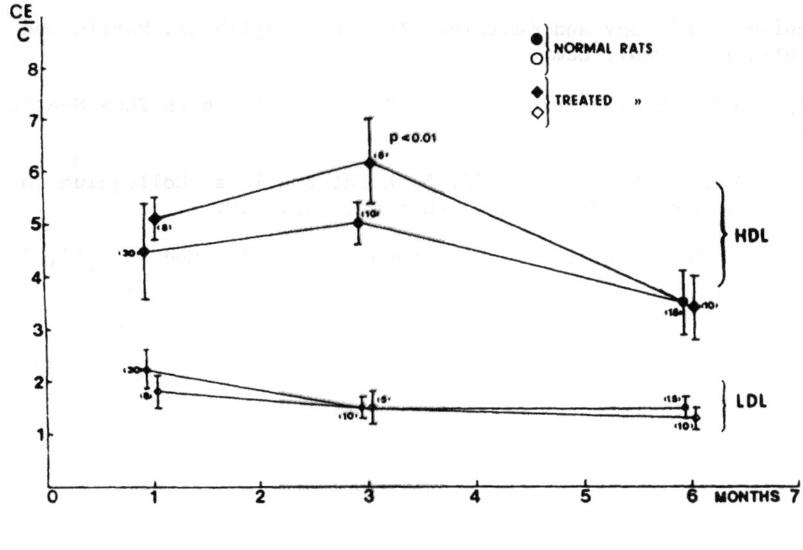

Preliminary measurements of both fatty acids from plasma and cholesterol esters (lauric,miristic,palmitic,palmitoleic,stearic,oleic,linoleic and arachidonic acids) did not show significant differences due to the diet with selenium. Also the total unsaturated fatty acids did not change significantly.

Discussion

Sodium selenite supplementation to rat diet (2 p.p.m. selenium) did not produce toxicity till half time of total treatment at least measured by the parameters used.

This work suggest that supplementation was even usefull since CE/C ratio in HDL was increased in animals drinking selenite during three months (Fig. 3).

However long time treatment (six months) produce increase in plasma and lipoproteins cholesterol (Figs. 1 and 2) what suggest a possible toxicity.

References

1. Selenium in Biology and medicine. 1980.Ed. Spallholz, Martin and Ganther. Avi. Publ. Comp.

2. Crespo, A.M., M.L. Henriques, R.E. Pinto. 1984. 16 th FEBS Meeting, Abst II - 174.

3. Crespo, A.M., R.E. Pinto. 1987. 5 th International Colloquium on lipid Metabolism and its Pathology. Atheroma, Sup.7 20.

4. Miettinen, TA, G. Alfthan, Jk Huttunen. 1983. Br. Med. J. 287, 517.

EFFECT OF SELENIUM INTAKE ON LIPID PEROXIDATION AND CARDIAC FUNCTIONS IN RATS

N. Koukay, M.J. Richard, A. Favier

Laboratoire de Biochimie C, Hôpital A. MICHALLON, BP 217X, 38043 Grenoble Cedex, France.

S. Mouhieddine, J. De Leiris

Laboratoire de Physiologie Animale, URA C.N.R.S. 632, Université J. Fourier, 38000 Grenoble, France

Introduction

Interest of selenium in human cardio-vascular diseases, originates from observations of cardiomyopathy, blood pressure changes and sudden death in selenium-deficient animals (1). A link between marginal selenium deficiency and coronary heart disease has been originally proposed on the basis of cross-sectional population comparisons (2). Much recent attention has been focused on the reactive oxygen species role (3,4) in post-ischemic tissue injury.

In this paper, we report the influence of selenium (Se) intake on antioxidant protective systems and cardiac functions during an in vitro myocardial ischemia in rats.

Materials and Methods

Animals : Forty male Wistar rats (180-200 g) were housed in stainless steel cages with free access to food and bidistilled water. 20 rats had been fed, for six weeks, a Se-deficient diet containing 0.06 ppm Se whereas control rats had been fed the same diet supplemented with 1 µg/g selenium as Na_2SeO_3. After an overnight fasting, they were killed by

Selenium in Medicine and Biology

exsanguination and an in vitro myocardial ischemia was performed according to the technique of Langendorff (1895) on the half of each group.

Methods : Cardiac frequence and coronary flow were recorded during perfusion and reperfusion phases of ischemia.

Then, isolated hearts were homogenised in 9 vol. of ice-cold 50 mM TRIS-HCL buffer, pH 7.4 containing 1 mM DTPA. After centrifugation (1000 g ; 20 min) supernatants were stored -20°C until analysis.

Selenium deficiency was verified by plasma Se determination (electrothermal AAS) and by measurement of glutathione peroxidase (GPX) activity (5) in red blood cells (RBC), livers and hearts. Plasma and tissue samples were also performed for lipid peroxidation products : malondialdehyde (MDA) (6), conjugated dienes (CD) (7) and organic hydroperoxides (HPO) (8). Other anti-oxidant systems were determined : glutathione transferase (GTX) catalase and superoxide dismutase (SOD). Alpha tocopherol and xanthine oxidase were also measured.

Results

Selenium status, glutathione peroxidase activity, lipid peroxidation products, and cardiac functions are presented in tables 1, 2 and 3.

Se deficiency does not depress growth of rats.

Se-GPX activity in RBC, liver and heart confirm Se deficiency in the appropriate experimental group. Concerning GTX, SOD, catalase, and XO activity no significant differences are observed.

We find a significantly greater lipid peroxide production in Se-deficient animals than in control animals specially in heart muscle. MDA production is significantly higher ($p < 0.010$) in these Se deficient hearts.

Vitamin E was also affected. It increased significantly in

deficient rats (data no shown).
Peroxidation markers and enzymatic protective systems are not correlated with ischemia.

Table 1 : Selenium status and glutathione peroxidase activities.

Se Concentration in Diet	Plasma Se µM/l	Plasma GPX UI/l	RBC GPX UI/g Hb	Liver GPX UI/g prot	Heart GPX UI/g prot
	***	***	***	***	***
0.060 ppm	2.364 ± 1.130	1951 ± 350.0	281.2 ± 41.05	136.6 ± 41.95	208.4 ± 35.32
1.000 ppm	4.416 ± 1.156	3026 ± 621.7	642.8 ± 119.2	523.4 ± 77.48	256.0 ± 22.80

Table 2 : Lipid peroxidation products (MDA)

Se Concentration in Diet	Plasma MDA uM/l	Liver MDA uM/g prot.	Heart MDA uM/g prot.
		*	**
0.060 ppm	6.448 ± 2.004	0.659 ± 0.101	0.702 ± 0.059
1.000 ppm	5.787 ± 2.714	0.592 ± 0.080	0.601 ± 0.059

Table 3 : Effects of selenium deficiency on cardiac function in rats.

	PRE ISCHEMIA		POST-ISCHEMIA	
Se Concentration in Diet	Coronary Debit ml/min	Cardiac Frequency Pouls/min.	Coronary Debit pouls/min.	Cardiac Frequency pouls/min.
0.060 ppm	8.3 ± 0.63	305 ± 14.2	4.5 ± 1.26	136 ± 147
1.000 ppm	7.7 ± 0.93	300 ± 34.8	5.8 ± 2.53	112 ± 146

T Test Student *** $p < 0.001$ ** $p < 0.010$ * $p < 0.050$

Discussion

After six weeks, rats fed 0.06 ppm Se, show a significant decrease of their plasma selenium. On the other hand, this study confirms that tissue levels of GPX are greatly dependent of dietary Se intake. Whole liver homogenate of deficient rats contain 26 % of the original GPX activity. The Se effect is less marked on heart GPX. The results presented here are in agreement with other studies which demons

trate the protective effect of GPX on the oxidative stress on cell membranes (9,10). However, if serum selenium deficiency induces an increase of peroxidation injury, cardiac functions are not affected in ours experimental conditions.
We just note that the lipid peroxide production which arises, leads to an increased vulnerability of animals. But it would be necessary to potentialize oxidant stress for modify cardiac physiological activities. Imura (11) shows an interaction between selenium and cis diamminedichloroplatinum (cis DDP) side effects ; he demonstrated that the appropriate utilization of trace elements such as selenium in cancer chemotherapy, efficiently, reduces the side effects of anticancer agents without compromising their antitumor activities.
In order to make better evaluation of the Se effects on the anticancer agents which function by generating free radicals others experiments continue in our laboratory.

References

1. Huttunen, J.K. 1986. Acta.Physiolo.Toxicol. 29, suppl VII, 311.
2. Salonen, J.T. 1982. Lancet, ii : 175.
3. Mc Cord, L.M. 1985. N.Engl.J.Med. 312, 159.
4. Diplock, A.T. 1985. Nutr.Res. suppl 1, 175.
5. Gunzler, W.A., H. Kramers, L. Flohe. 1974. Z.Klin.-Chem.Klin.Biochem. 12, 444.
6. Yagi, K. 1976. Biochem.Med. 15, 212.
7. Wickens, D., M.H. Wilkins, J. Lunec, G. Ball, T.L. Dormandy. 1981. Ann.Clin.Biochem. 18, 158.
8. Heath, R.L., A.L. Tappel. 1979. Annal.Biochem. 76, 184.
9. Hill, K.E., R.F. Burk, J.M. Lane. 1987. J.Nutr. 117, 99.
10. Hopkins, J., G.R. Tudhope. 1973. Br.J.Haematol. 25, 563
11. Imura, N., A. Naganuma, M. Satoh, Y. Koyama. 1987. Sangyo Ika. Daignju Zassi. 9, 223.

PROSTANOIDS MAY MEDIATE THE RADIOPROTECTIVE EFFECTS OF SELENIUM AND WR2721.

P. Bienvenu, F. Herodin, M. Fatome, J.F. Kergonou
Centre de Recherches du Service de Santé des Armées, 38000 Grenoble, France.

Introduction

Ionizing radiations are thought to act on biological targets primilary by water radiolysis, generating deleterious free radicals, such as the very reactive hydroxyl radicals ·OH which are able to peroxidize hydrogen and organic molecules. Less reactive substances such as the superoxide anion $O_2^{\cdot -}$ and hydrogen peroxide H_2O_2 are also produced if molecular oxygen, $O_2^{\cdot -}$, is present. Such active oxygen species may also be generated, but on a lesser extent, during normal physiologic processes and could be increased during illnesses such as infections or inflammations. These oxygen derivatives are also considered as stimulators of lipid peroxidation and prostanoid synthesis.
In view of counteracting the deleterious effects of ionizing radiations, powerful radioprotective chemicals, such as aminothiols e.g. WR-2721 (=WR)= S-2-(3-aminopropylamino) ethyl phosphorothioic acid, (Dose Reduction Factor = DRF: 2.7 for 500mg/kg doses in mice) have been discovered. Furthermore, it has been shown that an increased tolerance to WR-2721 could be exerted by some selenium salts, which were also radioprotective (1).
The synthetic analog of an arachidonic acid metabolite, dimethyl-Prostaglandin E_2 has also revealed an interesting radioprotective capacity (DRF: 1.7 for 1.6mg/kg dose in mice) (2).
Furthermore arachidonic acid metabolites may be modified either by WR2721, which decreases their urinary excretion (3) or by selenium derivatives (4). Therefore, in the present work, after checking the radioprotective effects of WR2721 and selenium, we tried to modulate them by a simultaneous treat-

Selenium in Medicine and Biology

ment with cyclooxygenase inhibitors such as acetylsalicylic acid.

Material and Methods.

Swiss CD1 male mice, Cobs(R), weighing 28-30g, obtained from Charles Rivers France, were placed in plastic containers and irradiated in a unilateral ^{60}Co γ-ray field at a dose of 0.6Gy/minute.
Analytical grade "Sigma" sodium selenite (=Se), was dissolved in saline 16mgSe/100ml), i.e. 1.6mg/kg) and injected intraperitoneally either 24 hours before 7.8Gy irradiation, or 30min. before 8 or 12Gy irradiations.
Acetylsalicylic acid (=ASA)= Aspirine du Rhône 500(R), provided by RP-Labo, was dissolved in drinking water (0.50g per liter) either during a 25 days period beginning 5 days pre-irradiation for 7.8Gy irradiation or during a week beginning 4 days before the 8 and 12Gy irradiations.
WR2721 was provided by Direction des Recherches Etudes et Techniques and dissolved in saline, and then injected by intraperitoneal route, about 20 minutes before irradiation. Depending on the animal groups tested, three different doses were given, as indicated in Table 1. Either an Indomethacin solution, 25mg/ml, Indocid(R)(=Indo) from Merck Sharp and Dohme, filling a sterile Alzet(R) type 2001 (200μl) osmotic pump, or a 2ml sterile PBS suspension of Biogel P300(R) (=Biogel) from BioRad Labs were placed 3 days before irradiation under the skin on the back of the animals, after slight anesthesia. Statistical evaluations were done by using confidence limits on percentages, as given by MAINLAND, HERRERA and SUTCLIFFE, cited by D. SCHWARTZ (5).

Results and discussion.

An approximate LD_{50} value close to 7.8Gy was observed for groups of 30 mice. The respective survival rates at 30th day were: 6.7% for 9Gy; 26.7% for 8.5Gy; 40% for 8Gy; 83.3% for 7.5Gy and 96.7% for 7Gy. The survival rates in Table 1 were also evaluated after a 30 days survey period following

irradiation. These results clearly show the protective effect of selenium for the lowest doses. Even for 9Gy, a nearly letal dose, an increased 10% survival, not reported in the table, was observed.
Selenium may also act synergistically with small doses of WR-2721.

Table 1: 30 Days Survival Rates % after Irradiation.

TREATMENT	RADIATION DOSE		
	7.8Gy	8Gy	12Gy
Controls	40.0	15.0	0
Se	75.0**	25.0	0
WR	[c]55.0	ND	[a]72.2** [b]20.0
Se + ASA	94.7**	60.0**	ND
Se + WR	ND	ND	[a]65.0** [b]50.0**
WR + ASA	[c]68.4**	ND	[b]0
WR + Indo or Biogel	ND	ND	[b]10.0

**significant differences, p 0.01 with controls; groups of 20 mice; ND= not done; a: WR=125.1mg/kg; b: WR=67.3mg/kg; c: WR=6.25mg/kg.

Furthermore, a synergistic effect was also obtained even for doses greater than the 7.8Gy LD_{50} by simultaneous oral administration of an acetylsalicylic acid solution. For 12Gy radiation doses, largely exceeding LD_{100}, ASA could also conversely exert antagonistic effects.
As radiation damage is supposed to involve primarily oxidative or peroxidative effects on organic biomolecules, selenium radioprotective efficiency might be based on its peroxide-scavenging capacity. Glutathione might not be a limiting factor in that process, because of its modest radioprotective capability.
No clearcut biochemical basis has been hypothesized for aminothiol radioprotection; we therefore tentatively suggest that WR-2721, as well as ASA might act directly by chelating metals involved in oxygen metabolism, e.g.

iron in cytochrome P-450, or copper, a superoxide dismutase (SOD) component, or providing SOD-like activity. As some prostanoids, such as PgE_1 and PgE_2 have been shown to diminish $O_2^{\cdot -}$ generation (6,7), they may also be implicated in such protective effects.

In conclusion, it appears that a selenium salt, as well as the aminothiol WR-2721, significantly protected mice against γ-irradiation. Furthermore, prostanoids might be involved in these effects as far as some cyclooxygenase inhibitors could alter the radioprotective efficiency of WR-2721 and selenium, which was thus increased for radiation doses close to the 7.8Gy LD_{50}.

Acknowledgement

We thank I. Cavero (Rhône-Poulenc Santé, Vitry-sur-seine for providing osmotic pumps, and J.D. Laval, for technical assistance.

References

1. Weiss, J.F., R.L. Hoover, K.S. Kumar. 1987. Free Rad.Res.Comms. 3, 33.
2. Walden, J.R., T.L.,Patchen M., Snyder S.L. 1987. Radiat.Res. 109, 440.
3. Donlon, M., L. Steel, E.A. Helgeson, W.W. Wolfe, G.N. Catravas. 1985. Int.J.Radiat.Biol. 47, 205.
4. Parnham, M.J., E. Graf. 1987. Biochem.Pharm., 36, 3095.
5. Schwartz, D. 1963: Méthodes statistiques. Flammarion, p. 284.
6. Weissmann, G: 1987. Drugs 33 (S1), 28.
7. Simpson, P.J., S.E. Mitsos, A. Ventura, K.P. Gallagher, J.C. Fantone, G.D. Abrams, M.A. Schork, B.R. Lucchesi. 1987. Amer.Heart.J. 113, 129.

SOME NEW ASPECTS OF THE PHYSIOLOGICAL ROLE OF SELENIUM : REGULATION OF GLUTATHIONE AND HYDROPEROXIDE METABOLISM

Vasanthy Narayanaswami* and K. Lalitha Norden

Department of Chemistry, Indian Institute of Technology, Madras-600 036, India

* Present address : Clayton Foundation Biochemical Institute, The University of Texas, Austin, TX78712, Texas.

Introduction

Selenium functions as selenocysteine at the active site of glutathione peroxidase, GSHPx(1). This one plays a critical role in mammalian hydroperoxide metabolism by taking part in the removal of hydrogen peroxide and other hydroperoxides that are associated with aerobic life processes (2). The contribution of glutathione, GSH, to hydroperoxide metabolism need not be overemphasised as this singularly abundant tripeptide maintains a potentially reducing environment with an impressive functional versatility while also regulating certain key metabolic events by thiol-disulfide exchange mediated alterations (3, 4). The putative role of selenium, Se, in regulation of glutathione metabolism and redox-linked processes as also its contribution to mitochondrial functional and structural integrity has been evaluated in this investigation in Japanese quails, <u>Coturnix coturnix japonica</u>.

Results

Selenium appeared to increase the mitochondrial functional efficiency by enhancing the oxygen uptake with succinate and pyruvate/malate as substrates to an extent of about 65 % at a dietary level of 2.0 ppm when compared to that in basal low selenium group (0.05 ppm Se)

Selenium in Medicine and Biology

in both liver and heart of Japanese quails. An increase in the succinate dehydrogenase activity in liver mitochondria by 86 % (Table 1) was accompanied by a significant influx of Ca^{2+} ions to a similar extent. Dietary Se also enhanced the activity of mitochondrial superoxide dismutase by about two fold in both liver and heart and GSHPx activity towards hydrogen peroxide by about 50 %. A concomitant decline in the level of mitochondrial thiobarbituric acid-reactive peroxidative endproducts was characteristic of Se supplementation. The GSHPx activities in the soluble fractions of the hepatic and cardiac tissues also increased by about 15 % and 50 % respectively with Se. However, the soluble glutathione S-transferase, GST, activity was high in Se deficiency and declined significantly to about one-half with a dietary Se level of 2.0 ppm. The high activity of GST was thus probably responsible for depleting the levels of cytosolic GSH to about 0.5 mM in liver and 1.9 mM in heart of Se deficient quails from the normal level of about 2.75 mM. A trend of decrease was also observed in the glutathione content on the mitochondria.
Changes in the thiol content were progressively reflected in the cellular redox poise as observed in the response of malate dehydrogenase, MDH, and aspartate aminotransferase, AAT, enzymes involved in transfer of reducing equivalents. The liver mitochondrial MDH activity was elevated by a five-fold extent in Se deficient status. A corresponding decline in the liver soluble AAT activity was characteristic of Se deficiency; the low level of activity increased by about 70 % with dietary Se.

Discussion

Newer aspects of the roles of selenium especially with regard to glutathione and hydroperoxide metabolism have emerged, thereby underscoring its significance in some of the intermediary metabolic processes. Regulation of Ca^{2+} flux across the mitochondrial membrane by Se appears to be an attractive possibility with a resultant control of some of intramitochondrial Ca^{2+} mediated processes as calcium ions have been demonstrated to regulate some of the dehydrogenases and thereby channelise reducing equivalents into the electron transport chain (5). The dual role of Se in enhancing mitochondrial functions including oxygen

Table 1 : Influence of dietary selenium on some metabolic parameters in the hepatic and cardiac tissues of quails

Parameters		Liver		Heart	
		Basal	2.0 ppm Se	Basal	2.0 ppm Se
1. Glutathione content μ mol/g	Soluble	0.5 ± 0.05	2.74 ± 0.16	1.96 ± 0.18	2.75 ± 0.2
	mitonchon-drial	0.39 ± 0.01	0.56 ± 0.02	0.47 ± 0.01	0.67 ± 0.01
2. Glutathione peroxidase nmol NADPH oxidised/min/mg protein	Soluble	322.4 ± 11.2	368.7 ± 30.5	66.9 ± 2.1^{a}	100.9 ± 1.7^{c}
	mitochon-drial	213.3 ± 6.7^{a}	325.5 ± 4.2^{c}	186.8 ± 4.8^{a}	248.7 ± 6.1^{c}
3. Glutathione reductase nmol NADPH oxidised/min/mg protein	(soluble)	8.3 ± 0.2^{a}	15.9 ± 0.5^{c}	7.6 ± 0.4^{a}	11.6 ± 0.5^{c}
4. Glutathione S-transferase (soluble) nmol chloro 2.4 dinitro benzene conjugated/min/mg protein		493.4 ± 4.3^{a}	224.1 ± 12.6^{c}	108.1 ± 4.7^{a}	56.2 ± 2.8^{c}
5. γ-Glutamyl-transpeptidase nmol p-nitraniline released/min/mg protein		5.2 ± 0.7	8.8 ± 1.4	-	-
6. Malate dehydrogenase nmol NADH oxidised/min/mg protein	Soluble	17.9 ± 1.3	14.7 ± 1.7	25.2 ± 2.1	22.3 ± 2.8
	mitochon-drial	447.8 ± 3.6^{a}	69.9 ± 1.8^{c}	230.8 ± 7.1^{a}	163.3 ± 4.4^{c}
7. Aspartate aminotransferase μmol pyruvate released/10 min/mg protein	Soluble	2.59 ± 0.08^{a}	4.44 ± 0.2^{c}	2.86 ± 0.4^{a}	6.9 ± 0.5^{c}
	mitochon-drial	1.22 ± 0.06^{a}	1.37 ± 0.1^{c}	1.0 ± 0.04^{a}	1.38 ± 0.03^{c}
8. *Superoxide dismutase U/mg protein	Soluble	15.2 ± 0.5^{a}	22.9 ± 0.9^{c}	3.6 ± 0.1^{a}	6.0 ± 0.1^{c}
	mitochon-drial	7.3 ± 0.4	17.4 ± 1.3	13.3 ± 0.3^{a}	29.9 ± 0.4^{c}
9. Mitochondrial lipid peroxidation nmol Malondialdehyde/mg protein		5.93 ± 0.22^{a}	1.11 ± 0.06^{c}	14.3 ± 0.7^{a}	1.66 ± 0.03^{c}
10. Succinate dehydrogenase decrease in A 600 nm/min/mg protein		0.38 ± 0.04	0.71 ± 0.03	-	-

* SOD activity : one unit is the amount of enzyme capable of inhibiting by 50 % the rate of NADH oxidation observed in control in 8 minutes.

In a horizontal row, means bearing different superscripts vary significantly. Between a and c $p < 0.001$.

uptake on the one hand, and its contribution to removal of hydroperoxides through GSHPx is an interesting facet of this trace element. New considerations for the involvement of Se in maintenance of mitochondrial structural and functional integrity have emerged especially in view of the considerable membrane associations of this element (6).
The subdued energy metabolism and related mitochondrial efficiency corresponds to the low profile of reducing equivalents, in turn affecting the thiol pool. The predominance of pro-oxidative events in Se deficiency possibly regulates some of the thiol-disulfide exchange mediated reactions. The increased content of monomeric collagen in skin of quails in Se deficiency could thus be related to activation of latent collagenase by glutathione disulfide.
The mutually compensatory activities of GSHPx and GST as a variance of the Se status is an adaptive response to stringent conditions of oxidative stress. Detoxication mechanisms appear to be partially compensated to ensure removal of metabolites resulting from, or leading to oxidative injury. The delicate balance of oxygen metabolism thus appears to be maintained through the reducing potential of the system, generated and functionalised by the thiol pool, with Se critically poised as a mediator of these events. These responses to the status of Se highlight its metabolic significance in maintaining the cellular redox poise and attest to its indispensable role in intermediary metabolism.

References

1. Flohe, L. 1982. In : Free Radicals in Biology, Vol. 5 (W.A. Pryor, ed.) Academic Press, New York, p. 223-249.
2. Chance, B., H. Sies, A. Boveris. 1979. Physiol. Rev. 59, 527-605.
3. Meister, A., M.E. Anderson. 1983. Ann. Rev. Biochem. 52, 711-760.
4. Ziegler, D.M. 1985. Ann. Rev. Biochem. 54, 305.
5. Denton, R.M., J.G. Mc Cormack. 1980. FEBS Lett. 119, 1-8.
6. Vasanthy, N., K. Lalitha. 1987. Biol. Trace Elem. Res. 14, 87-89.

Section III

Assessment of selenium status

BIOLOGICAL PARAMETERS FOR ASSESSING SELENIUM STATUS

J. Nève
Free University of Brussels, Toxicology and Bioanalytical Chemistry Unit, Campus Plaine 205/1, B-1050 Brussels, Belgium.

Introduction

Assessment of selenium (Se) status of man is usually performed by the determination of Se concentrations and/or of activities of the selenodependent enzyme glutathione peroxidase (GPx) in various biological materials either common such as plasma and urine, or more difficult to handle with, such as erythrocytes, leucocytes, platelets and hair... Beside methodological problems involved with such determinations, interpretation of the results is often rather complicated. Indeed, little is known about the homeostatic regulation of Se levels in the biological compartments, and, on the other hand, it is clear that differences exist in the sensitivity of these parameters to events such as Se depletion and repletion. Moreover, some of them very poorly reflect Se intake or body levels.

DETERMINATION OF SELENIUM LEVELS

The use of dietary data for assessing Se status has been largely discussed elsewhere (1-5). Such measures involve a lot of difficulties due for example to time to time variations in Se intake and to differences in bioavailability... They have been found useful only in extreme cases of deficiency or excess, but they never give a correct representation of individual Se intake (4,5).

Plasma selenium

Plasma (sometimes whole blood) Se remains nowadays the most commonly used parameter of Se status of man. Under a wide range of Se exposure conditions, the trace element level in

Selenium in Medicine and Biology

plasma has indeed been correlated with Se intake (1,2). High variations in usual Se dietary intake around the world,due to differences in Se content of soils and food as well as in bioavailability,have led to characterize three kind of Se areas with low (some parts of China, New Zealand, Finland, Danemark...), intermediate (Germany, Belgium, France, Poland...) or high Se status (parts of Canada, United States of America, China...), based on differences in plasma Se levels of representative groups. Differences ranging from 20-50 µg Se/L plasma in deficient areas to 60-80 in intermediate and higher than 100-120 in Se-rich countries have been reported (1,2). A high variability of this parameter according to time has been evidenced as a consequence of several factors such as modifications in the origin of imported cereals (6) or systematic supplementation of animal and even human food that was decided in some countries (1,2,7,8). Decreased plasma Se levels have also been observed in several population groups and attributed to changes in bioavailability caused by acid rains, intensive soil fertilization or heavy metal pollution ... (2). Some food habbits (vegans and lactovegans...); consumption of regimens poor in proteins (metabolic diseases such as phenylketonuria...); parenteral or enteral nutrition with no added Se; protein calorie malnutrition; presence of several diseases such as inflammatory states, cancer or digestive disorders...; and treatment with drugs like corticosteroids have all a documented negative influence on plasma Se levels. Finally, age (prematurity, below 20 years or above 60), pregnancy and lactation are physiological factors associated with low plasma Se levels (1,2,6-17). True Se deficiency symptoms have been observed for plasma Se levels less than 30 µg Se/L while toxicity manifestations may appear for values higher than 400 µg/L. Optimal level between these two extremes is not known. The definition of such a value will in fact depends on many factors that for the moment are poorly characterized (usual exposure conditions, intake levels of several heavy metals and other environmental factors, type of food...).

Selenium depletion of human subjects, that was investigated by a controled low dietary Se intake, was associated with a rapid and marked drop in plasma Se levels. NEVE et al reported a significant decrease as soon as after 10 days in patients on total parenteral nutrition receiving less than 30 µg Se/day (13); the response occured more rapidly than for other investigated parameters of Se status, demonstrating the value of plasma Se in the assessment of short-term changes in Se status. This was confirmed by the repletion of subjects with Se levels lower than normal, where a normalization was generally obtained after 3-4 weeks supplementation; the delay depending mainly on the extent of depletion and the dose of the selenoderivative dispensed (14,17-20). The chemical form of the derivative had also an influence on the rate of normalization:most studies demonstrated more rapid responses with Se-rich-yeast or-wheat and selenomethionine as compared to sodium selenite or selenate (14,18-23). The persistence of such a difference was observed even when inorganic Se was given at 1.5 to 3 times higher doses than the organic form (14). The most striking effects in the response of plasma Se to supplementation with different Se forms were however observed after normalization of Se levels in deficient subjects or during supplementation of subjects with initial higher Se levels (intermediate or high Se areas) : indeed, a stabilization of plasma Se generally occured with selenite and selenate (14,18,19,23,25) while a steady increase with no tendency to plateau, even after 300 days supplementation, was noted in subjects receiving selenomethionine (14,19,21). This was interpreted as a non specific incorporation of the amino acid in proteins (21). Natural organic Se-forms such as Se-rich yeast and wheat produced more controversial results; some authors being able to disclose a plateau (17,20,22-24), while others not, even after 11 weeks supplementation (18). Differences occuring in the shape of this response may be due to the usual Se status of the subjects : plateau were indeed more rapidly disclosed in subjects with moderate Se status

(21,22) as compared to those with rather low plasma Se levels (18-20). In a consecutive administration of Se supplements at two levels (100 and 200 µg Se/day as Se yeast), NEVE et al (22) observed two successive plateaus of plasma Se levels occuring approximatively 15 days after the begining of supplementation, but the effect of the two-fold increase in Se intake was rather moderate (only a ± 10 % increase in plasma Se levels).

Homeostatic regulation of plasma Se seems therefore to depend on factors such as the chemical form of dispensed Se and the preexisting Se status. Such as suggested by several authors, it is possible that plasma Se is not an adequate monitor of Se exposure or body burden at moderate levels, below chronic toxicity. Toxic Se intake however results in high Se levels (1,2).

Erythrocyte selenium

Contrary to many animal species, only a small part of erythrocyte Se is bound to the selenodependent enzyme GPx (10 % in the German adult) (26). This explains the poor correlation generally observed with erythrocyte GPx activity, except in pronounced Se deficiency. Nature and functions of the remaining part of erythrocyte Se are poorly known, but several recent studies disclosed a great lability of erythrocyte Se. Moreover, under a wide range of Se exposure conditions, satisfactory correlations with plasma Se were calculated (1,2,12-14, 21,22).

A limited number of authors having some authority in the field have considered for a long time that erythrocyte Se was only a predictor of long-term Se status, i.e. usual intake (4). They based this assumption on the relatively long turnover of red blood cells in the circulation. More recently however, rapid and significant decreases in these Se levels were reported in subjects receiving low Se intake, and the results were also confirmed in animals (9,13). Investigations that have been multiplied on the effects of Se supplementation on erythrocyte Se levels in subjects with variable Se status have all demonstrated significant increases in subjects with low Se status either at rates comparable to plasma Se (18,20, 24) or at slower rates (14,19). Differences in relation to the chemical form of dispensed Se or to the preexisting Se status were also disclosed as well as the occurence of plateaus (19). Supplementation of subjects with intermediate Se status led either to late increase (21) or to no significant modification (22), unless during the observation period. The value of erythrocyte Se levels in the assessment of changing Se status therefore deserves to be reconsidered.

Urine selenium

Under a wide range of Se intake, urine Se represents the major part - 50 - 60 % - of total Se losses. Excretion by faeces is the following most important part, while sweat and breath are neglible excretory routes, except if high or toxic Se intake occurs. Se urine excretion has been sometimes related to other parameters of Se status under moderate exposure conditions. Few authors demonstrated significant relations with plasma Se or Se intake (1,2,12,27,28), giving a poor evidence for the use of urine Se levels in the assessment of Se intake in these conditions. Other authors were unable to demonstrate such a relation in several dietary conditions (27). The highly significant correlations noted by several investigators between urine Se and creatinine levels suggest that this parameter is a function of muscular mass of individuals. This was confirmed by the relations disclosed between urine Se levels and the calculated body mass index and lean body mass of individuals (29,30). Such relations deserve to be reexamined as skeletal muscle appears to contain the largest body pool of selenium.

Different renal plasma clearances of selenium have been evidenced between subjects with low and higher Se status, indicating excretion of a smaller proportion of Se presented to the kidneys in subjects with low Se status due to poor dietary intake (31). This observation confirmed previous ones suggesting the presence of regulation mechanisms of urinary Se levels to maintain Se homeostasis in the body, and demonstrating the use of urine Se levels as an early indicator of low Se status (31,32).

Single doses Se supplements resulted in variable responses in urine Se levels that were related to different renal handling of dispensed Se forms (31,32). Rapid excretion was noted for free inorganic compounds although part of the dispensed dose was retained in various body compartments (32). Selenomethionine contrary to sodium selenite appeared to be reabsorbed from the renal tubules after glomerular filtration (31). Very few data were unfortunately published on the effects of long-term Se supplementation. Significant increases with stabilization after 30 days were noted in two studies (21,22), suggesting the use of urine Se levels as a short-term indicator of increasing Se status (100 µg supplemental Se/day). However, NEVE et al noted that a two fold increase in Se intake had no effect on Se urinary excretion (22), indicating that other excretory routes are prefered under these conditions. Again more data are necessary to demonstrate the usefulness of urine Se levels in the assessment of moderate Se exposure.

Other selenium levels

There is few evidence nowadays proving the interest of hair selenium in the assessment of Se status of man, except in pronounced Se deficiency states (4). Sensitivity of the indice to changing Se status is very poor (1,2). The fragmented results that were published on platelet and leucocyte Se levels are insufficient to prove their value in diagnosis of Se status.

DETERMINATION OF GLUTATHIONE PEROXIDASE ACTIVITIES

Once methodological problems solved, measurement of GPx activity represents a convenient technique for assessing Se status. Such as suggested by LEVANDER (4), the enzyme activity indeed corresponds to biologically active Se, and appears to achieve saturation in some tissues at Se intake that could be defined as satisfying nutritional requirement. Several recent investigations suggest that adequacy of Se status in some population groups can be assessed by the response to supplementation of GPx activity in cells with short turn-over.

Erythrocyte glutathione peroxidase

Relations with Se status have been extensively studied in animals and significant correlations with blood, plasma or dietary Se have been established many times (34). In humans, strong correlations have only been observed in subjects with low Se status while poor or not significant relations were demonstrated in other cases (marginal Se status or higher plasma Se levels) (34). Several authors working on results from quite large population groups in different areas indicated that the relation holds true only when Se level in blood or plasma is below a threshold value, and that the relation exhibits a plateau above this concentration (13,16,34,35). This suggested that the Se requirement for the enzyme was met, and that above that level, Se status was no more governing GPx activity. Published whole blood Se concentrations were : 100 µg Se/L in New Zealand (34); 60 µg Se/L in Belgium (1,13,33); 130 µg Se/L in USA (16), and 160 µg Se/L in Finland (35). The quite large variability in these supposed optimal levels is due to the different Se requirements of investigated group (depending on many factors).

Experimental Se depletion in normal subjects indicated that GPx activity in erythrocytes is not a sensitive indicator of changing Se status. Significant modifications are only detected after several weeks (13,33), the delay being related to the long life span of red blood cells. This time, it is indeed clear that Se incorporation into pre- GPx occurs only during erythropoiesis and cell maturation and therefore that a modification in Se status will only influence newly synthetized cells (1,2,33). Similarly, supplementation of deficient subjects only results in a slow normalization of erythrocyte GPx activity, reaching a plateau near usual values after 80-120 days or more supplementation, depending on the preexisting Se status and the intake level of the selenoderivative (14,17,19,20). Using density gradients, COHEN et al (25) very well demonstrated that the increase in erythrocyte GPx activity following supplementation of Se-deficient subjects was due to the youngest cells. Responses in erythrocyte GPx activity to Se supplementation could only been demonstrated in subjects from low Se areas (Finland, New Zealand, China) or in patients with low Se status due to reduced Se intake (total parenteral nutrition, phenylketonuria) (14,17,19,20). Subjects with higher usual Se status do not apparently respond, unless during the observation periods (21).

Platelet glutathione peroxidase

This indice has been used several times in the assessment of Se status or adequacy since the pioneer work of LEVANDER et al, both in animals and in Se-deficient subjects (18). Very rapid responses in this parameter with a stabilization of activity after 15 to 30 days supplementation were indeed reported in several groups of subjects with low Se status from Finland (18), Germany (36) or New Zealand (20). Differences were disclosed according to the chemical form of the dispensed derivative (18). The plateau of activity was reported for plasma Se levels in the range 125 to 140 µg Se/L (18,20), which therefore can represent optimal levels for platelet GPx activity. The velocity of changes is here clearly related to the short life span of platelets in the circulation. NEVE et al recently led two investigations on subjects with higher usual Se status from Belgium (21,22). Se supplementation was again accompanied by rapid and significant (but more moderate) increases in platelet GPx activity, occuring for plasma Se levels in the range 110-120 µg Se/L. Finally, in a study performed in the USA with subjects with plasma Se levels around 90 µg Se/L, STEAD et al (37) were unable to disclose any response in platelet GPx activity. In the only one investigation that was devoted to Se depletion of dietetically treated patients, MENZEL et al (36) showed a rapid decrease for platelet GPx activity.

These studies suggest that platelet GPx is not only a sensitive index of Se status (significant correlations have been reported with plasma Se levels: 21,22) but that its response to supplementation can give information on the adequacy of Se status in some population groups.

Plasma glutathione peroxidase

Very rapid responses to supplementation were disclosed by three groups of investigators (19,20,38) in subjects with low Se status or pronounced Se deficiency, with a stabilization of activity after 10 days (38) or 6-8 weeks (19,20). Other authors were however unable to disclose such a response in subjects with low (18) or intermediate Se status (21,22). The use of this indice in the assessment of short-term changes in Se status should be confined to patients with pronounced Se deficiency (38).

OTHER INDICES OF SELENIUM STATUS

Several functional physiological measurements such as in vivo platelet aggregability; killing capacity of neutrophils and determination of breath hydrocarbons have been explored, but their use is not frequent enough to draw definitive conclusions for their use in assessment of Se status (3).

Acknowledgements : the author thanks all investigators who collabored in studies on the assessment of Se status : Profs L. MOLLE, M. HANOCQ, F. VERTONGEN and Y.A. CARPENTIER; Drs A. PERETZ, S. VAN ERUM, N. THONNART and D. GNAT; Mr and Mrs Y. GREGOIRE, S. CHAMART, C. SOMMEREYNS and A. HARMEGNIES. Thanks are also expressed to the Fonds National de la Recherche Scientifique et Médicale de Belgique and the Ministères de l'Emploi, du Travail et du Budget de Belgique for their supports.

References

1. Nève, J., Vertongen, F., Molle, L. 1985. Clin. Endocrinol. Metab. 14, 629.
2. Nève, J., Henry, M., Peretz, A., Mareschi, J.P. 1987. Cah. Nutr. Diet. XXII, 145.
3. Levander, O.A. 1986. In : Trace elements in human and animal nutrition, vol. 5 (W. Mertz, ed). Academic Press, Orlando, pp. 208.
4. Levander, O.A. 1986. J. Am. Coll. Toxicol. 5, 37.
5. Levander, O.A. 1985. Fed. Proc. 44, 2579.
6. Alfthan, G. 1987. Personal communication.
7. Koivistoinen, P., Huttunen, J.K. 1986. Ann. Clin. Res. 18, 13.
8. Gissel-Nielsen, G. 1986. Ann. Clin. Res. 18, 61.
9. Van Rij, A.M., Thomson, C.D., Mc Kenzie, J.M., Robinson, M.F., 1979. Am. J. Clin. Nutr. 32, 2076.
10. Van Caillie-Bertrand, M., Degenhart, H.J., Fernandes, J. 1986. Pediatr. Res. 20, 574.
11. Lombeck, I., Kasperek, K., Harbisch, H.D., et al. 1977. Eur. J. Pediatr. 125, 81.
12. Wasowicz, W., Zachara, B.A. 1987. J. Clin. Chem. Clin. Biochem. 25, 400.
13. Nève, J., Vertongen, F., Thonnart, N., Carpentier, Y.A. 1984. In : Trace element analytical chemistry in medicine and biology, vol. 3 (P. Bratter, P. Schramel, eds). W. de Gruyter, Berlin - New York, pp. 139.
14. Nève, J., Vertongen, F., Thonnart, N., et al. 1986. Acta Pharmacol. Toxicol. 59, 166.
15. Lane, H.W., Warren, D.C., Taylor, B.J., Stool, E. 1983. Proc. Soc. Exp. Biol. Med. 173, 87.
16. Lane, H.W., Dudrick, S., Warren, D.C. 1981, ibid ref. 15, 167, 383.
17. Lombeck, I., Kasperek, K., Bachman, D. et al. 1980. Eur. J. Pediatr. 134, 65.
18. Levander, O.A., Alfthan, G., Arvilomni, H. et al. 1983. Am. J. Clin. Nutr. 37, 887.
19. Luo, X., Wei, H., Yang, C. et al. 1985. Am. J. Clin. Nutr. 42, 439.
20. Thomson, C.D., Ong, L.K., Robinson, M.F. 1985. Am. J. Clin. Nutr., 41, 1015.
21. Nève, J., Vertongen, F., Capel, P. 1988. Am. J. Clin. Nutr., 48, 139.
22. Nève, J., Van Erum, S., Dramaix, M., et al. 1988, This book.

23. Kumpulainen, J., Salmenpera, L., Siimes, A. et al. 1985. Am. J. Clin. Nutr. 42, 829.
24. Tarp, U., Hansen, J.C., Overvad, K., Thornling, E.B. 1986. Nutr. Res. 6, 853.
25. Cohen, H.J., Chovaniec, M.E., Mistretta, D., Baker, S.S. 1985. Am. J. Clin. Nutr. 41, 735.
26. Behne, D., Wolters, W. 1979. J. Clin. Chem. Clin. Biochem., 17, 133.
27. Robberecht, H.J., Deelstra, H. 1984. Clin. Chim. Acta, 136, 107.
28. Valentine, J.L., Kang, H.K., Spivey, G., 1978. Environ. Res. 17, 347.
29. Nève, J., Peretz, A. 1988. This book.
30. Hojo, Y. 1982. Bull. Environ. Contam. Toxicol. 29, 37.
31. Robinson, J.R., Robinson, M.F., Levander, O.A., Thomson, C.D. 1985. Am. J. Clin. Nutr. 41, 1023.
32. Young, V.R., Nahapetian, A., Janghorbani, M. 1982. Am. J. Clin. Nutr. 35, 1076.
33. Nève, J., Vertongen, F., Molle, L. 1984. In : Industrial uses of selenium and tellurium (Selenium - Tellurium Development Association, ed.). Darien, CT, 1984.
34. Thomson, C.D., Rea, H.M., Doesburg, V., Robinson, M.F. 1977. Brit. J. Nutr. 37, 457.
35. Tolonen, M., Westermarck, T., Halme, M., Sarna, S. 1986. Vita. Min. Spur. 1, 21.
36. Menzel, H., Steiner, G., Lombeck, I., Ohnesorge, F.K. 1983. Eur. J. Pediatr. 140, 244.
37. Stead, N.W., Leonard, S., Carroll, R. 1985. Am. J. Med. Sci. 290, 228.
38. Steiner, G., Menzel, H., Lombeck, I. et al. 1982. Eur. J. Pediatr. 138, 138.

SELENIUM CONCENTRATIONS IN HUMAN TISSUES

J.Aaseth
Hedmark Central Hospital N-2400 Elverum, Norway

Y.Thomassen
National Institute of Occupational Health P.O.Box 8149 DEP
N-0033 Oslo 1, Norway

Introduction

This contribution presents a survey of the distribution of selenium in man and factors which influence the selenium levels found in tissues. The main environmental pathway of selenium exposure in the general population is through food. Due to large variations in the amount of selenium in the soil of different geographical areas, the daily intake of a local population may be above or below the recommended selenium intake of 50-200µg/day(U.S.National Research Council,1980). The extreme estimated dietary intakes of selenium by adults range from 10 in the Keshan disease area to 5000µg/day in a selenosis area in another part of China (1). Dietary intakes between 30-200 µg/day can, however, be considered as more typical for the different population groups in the world. Cereals, meat and fish products are ostensibly the main dietary sources. The bioavailability of selenium is higher for foods of plant origin than for those of animal and fish origin (2,3). A wide variation in selenium concentrations is found in different types of food products. Vegetables and fruits are general low in selenium (<0.01 µg/g), while in meat products the concentrations vary from 0.03 to 4 µg/g. Seafood usually contains large amounts of selenium(about 0.5 µg/g). The selenium content in grain and

Selenium in Medicine and Biology

cereal products varies between 0.01 and 0.7 μg/g, depending on the type and where it is grown.

Reliability of published data

Accurate and reproduceable analytical data are essential for any retrospective or prospective studies relating selenium concentrations to health and disease and for establishing appropriate selenium intake and/or supplementation guidelines. Numerous selenium projects have been published in recent years by both the analytical and and medical communities without elaborate quality control measures. In sharp contrast to most trace elements the sampling, storage and preparation steps in the determination of selenium in human samples are free of contamination problems. The formation of intermediate volatile species of selenium should be considered during sample preparation. A broad range of analytical methods are capable of determining selenium in human tissues and food samples. The existence of suitable reference materials with recommended values for selenium has now faciliated method verification and quality assurance programs(4).

Selenium concentrations in human tissues

Whole blood The concentration of selenium in whole blood are different in different populations depending on the dietary intake. Figure 1 shows the influence of the dietary intake in some selected countries on the selenium concentration in whole blood. Factors which can influence this relationship are bioavailability and retention of dietary forms,

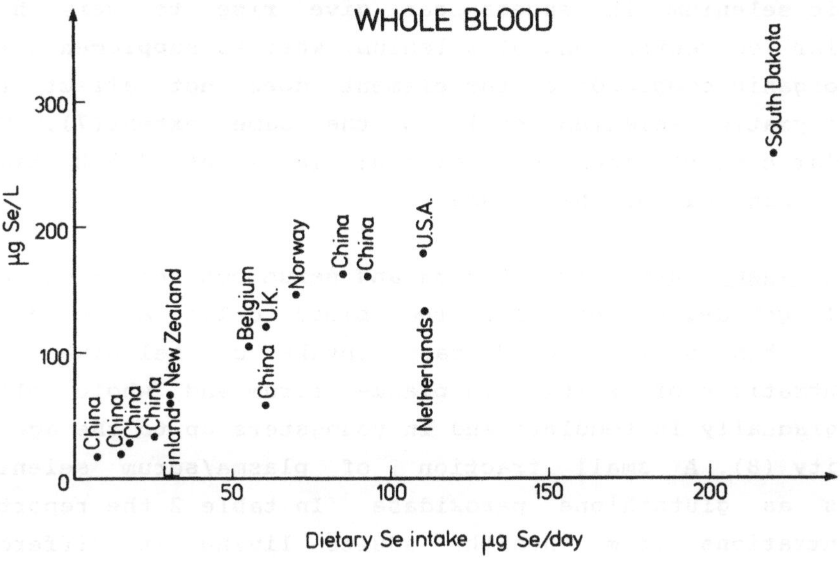

Figure 1: The relationship between the dietary selenium intake and the concentration found in whole blood.

physiological adaptions in metabolism, genetic variations and other mineral and vitamin interactions. In reports from China the upper and lower limits of safety are considered to be 440 and 27 µg Se/L, respectively(5). Table 1 gives an overview of the typical concentrations of selenium in healthy adults in different countries. In general blood selenium concentration, is lower in children than in adults. Furthermore, it has been reported that it decreases again after 60 years of age (6), presumably due to dietary changes.

Erythrocytes Changes in the intake of selenium are also reflected in the concentration of selenium in red blood cells after a period of approximately 4 weeks, slower than in plasma, since incorporation of ingested selenium into red blood cell proteins occurs only during erythropoiesis. Consequently erythrocyte selenium is considered to provide a long-term indication of general selenium status. Intake of organic selenium in excess can give rise to very high cellular concentrations of selenium, whereas supplementation by inorganic compounds of the element does not affect the non-enzymatic selenium pool to the same extent(7). The cellular concentration of selenium is about 1.5-2 times higher than that of the plasma.

Blood plasma and serum Plasma and serum concentrations are to a larger degree than for red blood cells affected by recent changes in the dietary intake of selenium. The concentrations of selenium in plasma, serum and whole blood rise gradually in toddlers and in youngsters up to the age of maturity (8). A small fraction of plasma/serum selenium occurs as glutathione peroxidase. In table 2 the reported concentrations from healthy adults living in different countries are listed. The whole blood and serum

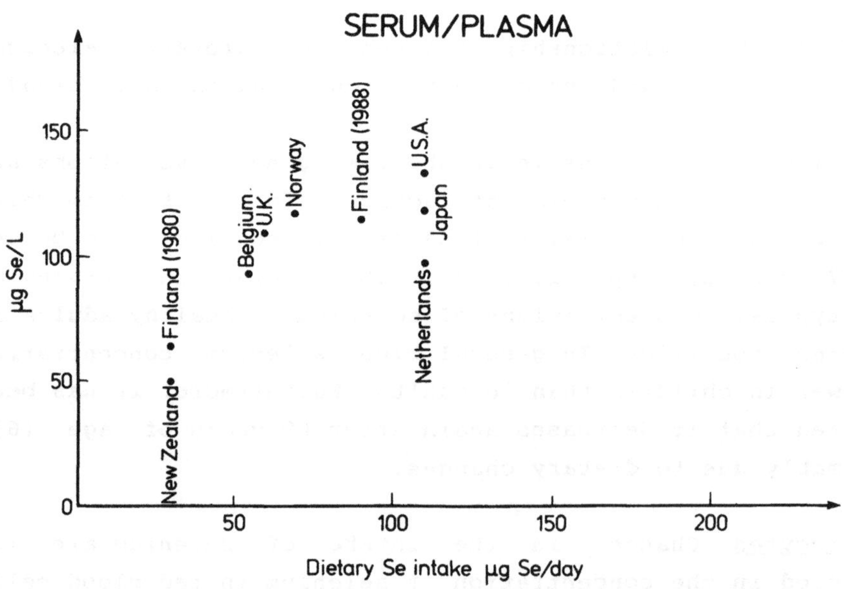

Figure 2: The relationship between the dietary selenium intake and the concentration found in serum/plasma.

concentrations varies from one geographical area to another due to differences in the daily intake of selenium. Figure 1 and figure 2 show the influence of the average daily intake of selenium on the selenium concentrations in both whole blood and serum. It is interesting to note that the relationship between the selenium intake and the body fluid concentrations seems to be more linear for whole blood than for serum. It should be pointed out that the figures are based on selected data which have been evaluated by the present authors with respect to analytical reliability.
Several factors may effect the selenium serum concentrations and hence the serum status of individuals. The most important factors are the dietary intake of selenium, occupational exposure, the alcohol intake, heavy metal exposure, selenium supplementation, age, physical activity and the health status.

<u>Urine</u> The amount of selenium excreated in the urine is used to estimate the dietary intake of selenium. Based on studies by Thompson (9) the average daily intake of selenium may be assessed roughly as double the 24 hour dialy urinary excreation. This relationship may not be valid for regions in which the daily intake is high or average. The normal selenium concentration in urine is below 100 µg/L with most reported values between 20-50 µg/L.

<u>Other organs</u> Distribution of selenium among internal organs has been studied by Schroeder, Frost and Balassa(10). They found that the relative concentrations in the tissues were as follows;
kidney > liver > spleen > testes > heart muscle > lung > brain

In a resent study, however, it is found that the thyroid gland contains 1.6 times more selenium than the liver (11).

The relative selenium concentrations in the various human tissues seems to be fairly constant, irrespective of the geographical origion of the analyzed material.

Table 1: Typical selenium concentrations reported in whole blood from healthy adults living in different countries (12).

Country	Year	µg Se/L	Comments
Europe			
Belgium	1983	125	
Denmark	1980	90	
Greenland	1984	110-180	Depending on number of meals of seals pr. week
East-Germany	1980	200	No quality assurance
Finland	1977	75	
Italy	1977	90	
Netherlands	1981	130	
Poland	1987	100	(13)
Sweden	1966	120	
United Kingdom	1984	145	
North-America			
Canada	1967	180	Ontario
USA	1968	100-270	Large geographical variations
South-America			
Venezuela	1972	360-800	Healthy school children

Africa			
Egypt	1972	70	
Asia			
China	1983	20-3200	Large geographical variations
India	1981	85	Data from Chandigard.
Japan	1980	200	
Taiwan	1983	100	
Thailand	1976	200	
Oceania			
Australia	1978	150	
New Zealand	1983	60	

Table 2: Typical selenium concentrations reported in serum or plasma from healthy adults living in different countries (12).

Country	Year	µg Se/L	Comments
Europe			
Belgium	1983	100	
Denmark	1980	80	
Finland	1977	75	
Finland	1988	110	(14)
France	1986	80	(15)
Greece	1986	60	(15)

Italy	1977	80	
Netherlands	1986	100	(15)
Norway	1987	125	(16)
Poland	1987	80	(13)
Portugal	1986	100	(15)
Spain	1986	85	(15)
Sweden	1987	85	(17)
United Kingdom	1986	115	(15)
West-Germany	1986	75	(15)
Yugoslavia	1987	70	(18)
<u>North-America</u>			
Canada	1982	140	Montreal
USA	1980	From 90 and upwards	Large geographical variations
<u>Africa</u>			
Zaire	1987	60-90	Large individual variations reported(19).
<u>Asia</u>			
China	1983	From 10 and upwards	Large geographical variations
Japan	1983	150	Healthy lactating mothers
<u>Oceania</u>			
Australia	1983	85	Healthy mothers
New Zealand	1983	50	

As almost no information on the concentrations of selenium in human tissues from developing countries is presently available, obtaining such data is important to identify and ultimately prevent any endemic selenium-associated diseases.

References

1. Levander, O.A..1982.In:Clinical Biochemical and Nutrional Aspects of Trace Elements(A.Prasad,ed.).Alan R.Liss,New York, p.345.

2. Mutanen, M.. 1986. Annals Clin. Res. 18 ,48.

3. Cantor, A.H.,M.L.Scott,T.Noguchi.1975. J.Nutr. 105 , 96.

4. Christensen, J.M., M.Ihnat, M.Stoeppler, Y.Thomassen, C. Veillon, M.Wolynetz. 1987.Frezenius Z.Anal.Chem. 326, 639.

5. Yang, G., S.Wang, R. Zhou, S.Sun. 1983. Am.J.Clin.Nutr. 37, 872.

6. Thomson, C.D., H.M.Rea, M.F.Robinson, O.W.Chapman. 1977. Proc. Univ. Otago Med. Sch. 55, 18.

7. Levander, O.A., G.Alfthan, H.Arvilommi, C.G.Gref, J.K.Huttunen, M.Kataja, P.Koivistoinen, J.Pikkarainen. 1983. Am.J.Clin.Nutr. 37, 887.

8. Verlinden, M., M. van Sprundel, J.C.Van der Auwera, W.J.Eylenbosch. 1983. Biol.Trace Elem.Res. 5(2), 103.

9. Thompson, C.D.. 1972. Proc.Univ.Otago Med. Sch. 50, 31.

10. Schroeder, H.A., D.V.Frost, J.J.Balassa. 1970. J.Chron.Dis. 23, 227.

11.Aaseth, J., H.Frey, E.Glattre, G.Nordheim, J.Ringstad, Y.Thomassen. To be published.

12.Aaseth, J., Y.Thomassen. In: Occurance and Distribution of Selenium. (M.Ihnat,ed.) CRC Press, Inc., Boca Raton. In press.

13.Wasowicz, W., B.A.Zachara.1987.J.Clin.Chem.Clin.Biochem. 25, 409.

14.Alftan, G., Personal Communication, National Institute of Public Health, Helsinki, Finland.

15 Thorling, E.B., K.Overvad, J.Geboers.1986. Ann.Clin.Res. 18, 3.

16.Ringstad, J., B.K.Jacobsen, Y.Thomassen. 1987. J.Trace Elem.Electrolytes Health Dis. 1, 27.

17.Barregård,L., Y.Thomassen, A.Schütz, G.Sällsten. 1987. In: Trace Elements in Human Health and Disease. Extended Abstracts from the second Nordic Symposium. Environmental Health 20, World Health Organization, Copenhagen, p.152.

18.Mikac-Devic, M. Institut for Clinical Chemistry, Hospital "Dr. M. Stojanovic", Zagreb, Yugoslavia. Personal communication.

19.Goyens, P., J.Golstein, B.Nsombola, H.Vis, J.E.Dumont. 1987. Acta Endocr. 114, 497.

PROBLEMS AND NEW TRENDS IN SELENIUM DETERMINATION IN BIOLOGICAL MATERIALS

P. Chappuis, F. Paolaggi, F. Rousselet
Laboratoire Central de Biochimie, Hôpital Lariboisière, 2, rue Ambroise Paré, 75010 Paris

C. Faugeron
Société PHIRA, 15, rue Champagne, 77400 Bussy Saint Martin

INTRODUCTION

The importance of selenium has been steadily increasing in biology and human medicine for several years, especially since the discovery that selenium is part of the prosthetic group of glutathione peroxidase. Thus, the motivation for its estimation, which was formerly only toxicological, is nowadays tending towards the determination of trace amounts of the element. There are indeed a wide range of different techniques available for selenium estimation, more than for any other trace element.

MINERALIZATION

Most techniques necessitate prior mineralization of the biological material (electrochemical, fluorimetric techniques, often atomic absorption). At the present time this step is relatively well codified. Welz and Coworkers showed (1). The insufficiency of nitric digestion alone (even when performed under pressure) to mineralize resistant organoselenium coupounds, such as trimethylselenium. The nitrosulfoperchloric digestion is shorter and more efficient than the nitroperchloric one. A part of the selenium is oxidized to state VI. The estimations made thereafter on

Selenium in Medicine and Biology

the mineralizate need, however, the element to be in the tetravalent state. The nitrosulfoperchloric digestion performed at 260°C is thus followed by a reduction step of the Se VI compounds to Se IV ones, generally using hydrochloric acid.

THE ELECTROCHEMICAL METHODS

The only pratically used method is the cathodic stripping voltammetric determination because it is sensitive, relatively specific and rapid. The procedure involves two steps : the first one concentrates selenium on an electrode ; the second yields a signal proportional to the selenium concentration. It is usually performed by pulse cathodic stripping voltammetry (2). The electrode used is either a hanging mercury drop electrode (the most frequent one) or a carbon fiber electrode coated with a mercury film.

THE FLUORIMETRIC METHODS (ADF)

These methods are among the oldest ones and depend on the fluorescence properties of piazselenols (combinations of Se IV and an aromatic diamine, generally diaminonaphthalene) which are extractible in an organic solvent (generally cyclohexane). This type of method, in contrast to human biology, is widely used by veterinarians, probably for reasons of sensitivity and applicability (biological media, fodder, etc.) : an intercomparison made recently (3) showed that 61 % of the participants were using this method.

The prior digestion must lead to Se IV (see above) ; it may give some parasitic fluorescence if nitric acid is not eliminated (4).

But the method has a considerable drawback :its mediocre reproducibility. The latter can be improved : - by standardizing the extraction pH of piazselenol at 1.8 (5) - by gassing the reaction medium with nitrogen in order to eliminate the oxygen responsable for the fluorescence quenching (6) - by avoiding the interference of iron in the formation

of piazselenol (4, 7). In order to improve the sensitivity two successive extractions of piazselenol have been proposed (7), so as to get rid of interfering fluorescent impurities. The procedure is delicate and necessitates a trained personnel, since it is relatively laborious, but its range of application is very wide : blood, urine, tissues, etc.
This method, known to be very sensitive, is curiously lacking in the literature comparative arguments with other selenium estimation methods (table 1).

HYDRIDE GENERATION INDUCTIVELY COUPLED PLASMA ATOMIC EMISSION SPECTROMETRY

Inductively coupled plasma emission spectrometry is not sensitive enough to allow the direct determination of selenium in biological fluids (blood, urine). However, when coupled to a hydride generation system functioning on the same principle as that described below (8), it becomes usable in biology. The detection limits are still not quite comparable with those of classical methods (table 1) but its linearity is wide.

ACID DECOMPOSITION HYDRIDE GENERATION ATOMIC ABSORPTION SPECTROMETRY (ADHAAS)

This method has for a long time suffered from the insufficient digestion of biological samples (see mineralization), but it is definitely interesting for many reasons. The setup is relatively cheap (an electrothermal atomization process, in particular, is not necessary). The range of application is wide (blood, plasma, urine, tissues). The sensitivity is great (see table 1) ; quite remarkably it has been shown to yield homogeneous results in different laboratories on urine selenium and weak concentrations of serum selenium (9). There are no noteworthy spectral interferences.
The drawbacks of ADHAAS are also well known. The digestion step necessitates a good experience. Other metal elements

can depress the response, probably through capture and decomposition of the hydride by metallic particles arising from the precipitate formed on the addition of sodium borohydride to the acid medium (10).

Table 1 : Comparative Sensitivities of the Usual Selenium Estimation Techniques

Ref.	Methods	Serum, total blood, or solutions ($nmol.l^{-1}$)	Sample after treatment (pg)	Urine ($mmol.l^{-1}$)	Tissues (pmol)
2	E.C.	1.3			
7		0.063			
22	ADF			4.9	5.7 (3sd)
4					5.7 (3sd)
32	EAAS	139 (1% Abs.)			
33		31 (3sd)			
23				126	
11		150			
18		63 (2sd)			
19		0.168	0.2		
16		103 (3sd)	2.7		
37		19 (2sd)			
1	ADHAAS	27 (1% Abs.)			
25		63 (2sd)			250
8	Hydrides/ ICP	10			

In parentheses is shown the system (sd or 1 % absorbance), when indicated, chosen by the various authors to express the sensitivity.
EC : electrochemisty ; ADF : fluorimetry ; EAAS : direct electrothermal atomic absorption spectrometry ; ADHAAS : hydride/atomic absorption atomique spectrometry.

ELECTROTHERMAL ATOMIC ABSORPTION SPECTROMETRY (EAAS)

This method, although looking attractive especially if used directly, raises some technical and setup difficulties that are still incompletely resolved.
The techniques including an extraction step necessitate a mineralization followed by chelation of selenium by an o-diamine and extraction with an organic solvent, which is then injected into the graphite furnace. The only advantage of this complicate method lies in the large reduction of the background which is damaging to the selenium analysis. The techniques after deproteinization are neither

used because of their complexity.
In contrast, the direct methods (EAAS) are widely used. The controversy about the radiation source is still open : for a long time EDL lamps were considered as more emissive and stable than hollow cathode lamps (11), but this notion is called in question since the arrival on the market of more efficient hollow cathode lamps (12).
The location, within the spectrum, of the selenium resonance line (196 nm) raises serious spectral interferences : e.g. iron displays about 40 absorption lines in this UV range and phosphorus generates molecular absorptions causing an unstructured background. This results in an overcompensation for those instruments equipped with a deuterium arc corrector system.
All these perturbations, which made some authors to reject the method, have been, however, corrected by numerous procedures.
The oldest used modifier, nickel, allows not only to limit the preatomic losses of selenium (probable formation of resistant nickel selenide) but also to lessen and to delay the phophorus molecular formations by transforming them into gaseous phosphorus (13), under the condition that more than 20 μg of it be used (14). Unfortunately the selenium signal decreases strongly when the nickel concentrations are very elevated (11, 15).
The use of copper alone allows to increase the sensitivity of the measurement (16) but it does not stabilize the organic forms of selenium. A copper/magnesium modifier was tried by Welz and Coworkers (15) who showed that the various valence states of selenium undergo different preatomic losses in the presence of nickel alone. The use of a mixed copper-magnesium modifier homogenizes this behavior, which is most important insofar as the calibration adopted by most authors (Se IV) is used to determine selenium in a biological matrix, where this element is most often present as Se II.

Saeed (13) uses, together with a deuterium arc corrector, a mixed platinum-nickel modifier for total blood, serum and urine (which are iron and phosphate-rich matrices). The use of platinum allows to transform PO^+ and P_2, which are partly responsible for the fine absorption lines, into gaseous phosphorus.

Another matrix modifier, palladium, is also used (17), sometimes in the presence of ascorbic acid, but whithout any noteworthy improvement of sensitivity (14).

Finally, some authors (16) recommend the use of a triple silver-copper-magnesium modifier that allows a unique calibration in an albumin-containing medium. Their method is remarkably precise : 1.8 to 2.9 %.

The evolution of the nature and the concentration of the matrix modifiers is still going on (table 2) : the current endeavors tend towards a universal modifier, adaptable to all biological matrices. This modifier has not been found yet, but is seems that the use in the diluting agent of magnesium nitrate and/or nitric acid is widely admitted (elimination of halogenated derivatives responsible for the strong absorption background at 196 nm). As to the calibration, the method of standard additions is widely used, but numerous authors have tried to simplify it by using a reference serum (see table 2).

Selenium estimation favors the use of Zeeman effect instruments equipped with an alternating current electromagnet, which correct the background at the exact analysis wavelength. On the other hand, there are very few articles about the correction efficiency by the Smieth-Hieftje system. However, the adequate use of matrix modifiers and elaborate atomization procedures (e.g. platform tubes), the exploitation of absorbance curves versus time of selenium and perturbing elements, may offer an alternative way to the correction by the Zeeman effect, as the latter is not used by most owners of atomic absorption instruments (12, 13, 14).

Table 2 : Main Modifiers Used in EAAS

			Matrix Modifiers (μg)					
Ref.	Matrix	m_0 (pg)	Ni	Mg	Cu	Pt	Others	Calibration
23	U	30	60	25				
11	S		25					HSA solution
4	S		25					Reference serum
34	S	34	12					Standard additions
35	B.P.S.			3	15			
24	U		20			150		Standard additions
14	S	15 à 20	20 + 4					Aqueous solutions
12	S			10	5			Standard additions
			15			120		Standard additions
13		60	15			60		Reference serum
19	B		34.7					Standard additions
36	S	25 à 29		11.2		5.6		Standard additions
		30 à 38	28					Standard additions
		18 à 26		11.2			1.46 (Pd)	Standard additions
16	S			1.47		7.9	9.51 (Ag)	BSA solution

U = Urines ; S = Serum ; B = Total Blood ; P = Plasma ; m_0 = characteristic mass

The platform pyrolytic tubes used according to the STPF (stabilized temperature platform furnace) concept are particularly well adapted to the estimation of a volatile element such as selenium : the platform minimizes the analyte and matrix condensation in vapor phase by delaying the atomization until the furnace internal temperature is in equilibrium with that of the walls. In that case measuring by peak surface is theoretically more precise. However, the progressive carbonaceous residue (built up) necessitates frequent platform cleaning (13). The latter is progressively wearing out in the angles and its reinsertion into its initial location becomes delicate. This problem is avoided by using the Auto Probe procedure (Philips), in which the platform is totally isolated from the walls of the graphite tube. Moreover, in this system, the rise in temperature that is responsible for analyte atomization is carried out only by a radiative process, contrarily to classical platforms. In contrast, wall atomizations give a more sensitive signal,

provided a pyrolytic tube is used. The carbonaceous residue can be greatly reduced by destroying the organic matrix with nitric acid in the diluent and carrying out part of the mineralization under oxygen (18).
The intervention of the tube in the atomization process should not be neglected : tubes coated with zirconium, lanthanum, tantalum or molybdenum (19) or with albumin.
In conclusion, the remarkable IUPAC investigations (9, 20, 21) allow to compare usefully the current methods of serum selenium determination. The exactness of the ADF, ADHAAS and EAAS methods is acceptable. As for the precision, all methods show the preponderant influence of each laboratory's own experience, particularly in the case of the EAAS method.

APPLICATIONS TO VARIOUS BIOLOGICAL MEDIA

The estimation of selenium in urine is more difficult than in serum, blood cells or total blood, because of the usual low levels which are about 10-fold less than in serum (22), the level of interfering phosphates, the presence of trimethylselenium whose resistance to mineralization is well known. For all these reasons the EAAS method is relatively little used, as it needs careful devising (23, 24). Anyway, few users, although experimented in that technique, have so far reported any results on urine (21). The ADF and ADHAAS methods seem, in spite of their complexity, the most well adapted techniques to that fluid.
The hydride-atomic absorption spectrometric method applied to biological fluids is the most used method to estimate tissue selenium (25). However, this method requires a large amount of biological material, of the order of 1 g, which is incompatible with usual biopsies. The mineralization-electrothermal atomic absorption analysis, although theoretically possible, offers no superior praticability. In contrast, some authors (26) have proposed a direct analysis on the solid sample : the weighed biopsy is deposited on an adapted atomizer, followed by a matrix modifier (Cu/Mg)

injection. By using this particularly attractive method (simplified procedure, limitation of contaminations, improvement of sensitivity since the sample is not diluted by reagents) the authors show that tissue selenorganic compounds undergo no preatomization loss up to a temperature of 1300°C. At this temperature the organic matrix is destroyed, especially with a supplementary step of mineralization under oxygen and a background correction by Zeeman effect, whose efficacy at high absorbances is well known. This method has been applied to liver biopsies (a sample of a few mg) using a solid reference sample (liver provided by N.B.S.).
As it has been implicated in male fertility, selenium has also been studied in human sperm by EAAS analysis (27, 28).

SELENORGANIC COMPOUNDS AND THE DIFFERENT VALENCE STATES OF SELENIUM

Some special cases necessitate the determination of some selenorganic compounds or some oxidation states of selenium, as the toxicity of selenium depends on its chemical form ; some fundamental studies deal with the different metabolic pathways of selenium after ingestion ; e.g. the follow-up of the urinary detoxication pathway may necessitate the estimation of trimethylselenium.
The different ionic and organic selenium derivatives can be separated by ion exchange chromatography or high pressure liquid chromatography (HPLC) which are generally applied to aqueous medium or urine. Selenium determination in the fractions can be made either by fluorimetry (29), by ICP spectrometry coupled with mass spectrometry (30) or by neutron activation analysis (31). Gas phase chromatography (GPC) methods have been applied to selenium estimation in blood and urine. Prior chemical treatment, comparable with that used for fluorimetric methods, makes them too tedious for a simple estimation of total selenium.

CONCLUSION

Some methods (e.g. X-ray fluorescence, neutron activation, isotopic dilution - mass spectrometry) necessitate a heavy and specialized equipment. They are interesting when they are utilized for comparison between the various techniques (20, 21).

The profusion of methods and equipments used in selenium estimation in biological media is largely due to the analytical difficulties in atomic absorption spectrometry, as mentioned above.

Numerous physicochemical procedures are being explored in order to reduce these obstacles. The interest and role of selenium in human medicine largely justify these efforts.

REFERENCES

1. Welz, B., M. Melcher, J. Neve. 1984. Anal. Chim. Acta. 165, 131-140.

2. Huang B.X., H.C. Zhang, G.G. Pu, F. Yin, S.C. Zheng, H. Yang, 1985. Anal. Letters. 18, 279-285.

3. Koh. T.S., J. Assoc. Off. Anal. Chem. 1987, 70 (4), 664-667.

4. Alfthan G. 1984. Anal. Chim. Acta. 165, 187-194.

5. Bayfield, R.F., L.F. Romalis. 1985. Anal. Biochem. 144, 569-576.

6. Watkinson, J.H. 1982. Anal. Chim. Acta. 134, 417-419.

7. Tamari, Y., S. Ohmoris, K. Hiraki. 1986, Clin. Chem. 32, 1464-1467.

8. Imbert, J.L., M. Olle, 1987, Analusis. 15, XL.

9. Welz, B., M.S. Wolynetz, M. Verlinden. 1987, Pure Appl. Chem. 59, (7), 927-936.

10. Welz, B., M. Melcher. 1984. Analyst. 109, 569-572.

11. Huguet, C., P. Chappuis, J. Peynet, F. Thuillier, A. Legrand, F. Rousselet. 1983. Ann. Biol. Clin. 41, 277-283.

12. Sampson., B. 1987. J. Anal. At. Spectrom. 2, (5), 447-450.

13. Saeed, K. 1987. J. Anal. At. Spectrom. 2, (2), 151-155.

14. Hoenig, M. 1987. Analusis. 15, 9.

15. Welz, B., G. Schlemmer, U. Voellkopf. 1984. Spectrochim. Acta. 39B, 501-510.

16. Morisi, G. M. Patriarca, A. Menotti. 1988. Clin. Chem. 34, 1, 127-130.

17. Eckerlin, R.H., D. Hoult, G.R. Carnrick. 1987. Atom. Spectrosc. 8, (1), 64-66.

18. Neve, J., J. Molle. 1986. Acta Pharmacol. Toxicol. (Copenh.), 59, (7), 606-609.

19. Deschuytere, M.A., H. Deelstra, 1987. Trace Elem. Anal. Chem. Med. Biol. Vol. 4, 359-366.

20. Ihnat, M., M.S. Wolynetz, Y. Thomassen, M. Verlinden. 1986. Pure Appl. Chem. 58, 1063-1076.

21. Ihnat, M., Y. Thomassen, M.S. Wolynetz, C. Veillon. 1986. Acta Pharmacol. Toxicol. 59, Suppl. 7, 566-572.

22. Geahchan, A., P. Chambon. 1980. Clin. Chem. 26, 1272-1274.

23. Carnrick, G.R., D.C. Manning, W. Slavin. 1983. Analyt. 108, 1297-1312.

24. Saeed, K. 1986. Acta. Pharmacol. Toxicol. 59, Suppl. 7, 593-597.

25. Norheim, G., A. Haugen. 1986. Acta Pharmacol. Toxicol. 59, Suppl. 7, 610-612.

26. Aadland, E., J. Aaseth, B. Radziuk, K. Saeed, Y. Thomassen. 1987. Fresenius Z. Anal. Chim. 328, 362-366.

27. Suistomaa, U., M. Saaranen, T. Pertulla. 1987. Clin. Chim. Acta. 168, (3), 323-328.

28. Neve, J., S. Charmart, P. Trigaux, F. Vertongen. 1987. At. Spectrosc. 8, 6, 167-169.

29. Kraus, R.J., S.J. Foster, H.E. Ganther. 1985. Anal. Biochem. 147 (2), 432-436.

30. Thompson, J.J., R.S. Houk. 1986. Anal. Chem. 58, (2), 2541-2548.

31. Blotcky, A.J., G.T. Hansen, N. Borkar, A. Ebrahim, E.P. Rack. 1987. Anal. Chem. 59, 2063-2066.

32. Oster, O., W. Prellwitz. 1982. Clin. Chim. Acta. 124, 277-291.

33. Alfthan, G., J. Kumpulainen. 1982. Anal. Chim. Acta. 140, 221-227.

34. Paschal, D.C., N.M. Kimberly. 1986. At. Spectrosc. 7, 3, 75-78.

35. Jowett, P.L.H., M.I. Banton. 1986. Anal. Lett. 19, (11-12), 1243-1258.

36. Neve, J., S. Chamart, L. Molle. 1987. Trace Elem. Anal. Chem. Med. Biol. Proc. Int. Vol. 4, 349-358.

37. Porter, B.L. V.A. Fuavao. J. Sneddon. 1983. At. Spectrosc. 4, 185-187.

DIRECT ANALYSIS OF SELENIUM IN BIOLOGICAL SAMPLES (SERUM, ERYTHROCYTES) BY GRAPHITE FURNACE ATOMIC ABSORPTION SPECTROSCOPY (GFAAS) BY THE METHOD OF STANDARD ADDITIONS.

C. Charlot

Analytical Chemistry Laboratory (Prof. S. Brun), University of Pharmacy, av. Charles Flahault, Montpellier, 34060

D. Rieu

Department of Pediatry, St Charles Hospital, Montpellier, 34060

J. Castel

Pediatric Functional Exploration Laboratory, St Charles Hospital, Montpellier 34060

Introduction

For the last quarter of the century, the quantitative determination of selenium in biological samples by flameles atomic spectroscopy has been used worldwide. Although, electroless discharge lamp (E.D.L.)and zeeman effect background correction system gives good results, the cathode lamp current, matrix modifiers and deuterium arc used with, could afford the same performances. The main matrix modifiers proposed were : NH_4 NO_3 (1), Ni $(NO_3)_2$ (2), Triton X 100 (3) and Palladium (4, 5, 6, 7). In this paper, a method for rapid determination of selenium concentration in human serum and erythrocytes using flameless atomic absorption spectroscopy was described.

Experimental methods

Varian model 1275 AA equipped with a Varian GTA 95 graphite furnace atomizer,deuterium tube and cathode lamp current was used for all measurements. Analyses were performed at 196.0 nm using 1 nm band pass (slit width). A cathode lamp current was set to 5 mA. Pyrolytically coated graphite tubes

Selenium in Medicine and Biology

were used. The purge gas was nitrogen. The graphite furnace settings (drying, charring, atomizing and cleaning cycles) were listed in table 1.

Table 1. Furnace Conditions for the Determination of Serum and Erythrocyte Selenium. Varian GTA 95

Step n°	Temp. °C	Hold s.	Ramp. s.	
1	75	5	3	
2	125	50	3	Dry
3	300	20	3	
4	1200	10	3	
5	1200	15	3	Char
6	1200	1	0	
7*	2700	0.8	0	Atomize
8*	2700	1	0	
9	2700	2	3	Cool
10	40	13	3	

Volume : 20 µl - Temperature of injection : 60°C

* Read command

Replicate : 3

Absorbance peak height - BC on - scale : 0.5

Reagents

- The standard solutions of selenium (Merck) were prepared from an analytical grade (SeO_2) at 1.00 ± 0.002 g in HNO_3 0.5 mol/l for the Se IV.
- Palladium metal (Fluka) powder, purity > 99.9 %.
- Ascorbic acid (Merck).
- Triton X 100 (Fluka) (alkylphenylglycol).
- Dow Corning silicone antifoam emulsion B (Dow Corning Co).
- Human serum control Lyotrol "X.N" (Biomerieux).

Matrix modification technique

a) Triton X 100 reagent : 0.5 % ascorbic acid and 0.25 % antifoam were added to 1 % Triton X 100.

b) Palladium solution 1 g/l in HCl 2 % (m/v).

Sample preparation

Erythrocytes were separated from the serum sample by centrifugation,then washed by 3 ml of NaCl solution (9°/∘∘), centrifugated three times and hemolyzed using an equal volume of water.

1 : An aliquot human serum diluted 1 : 3 with distilled water, was diluted with an equal volume of Triton X 100 R and vortexed vigorously. Then, an equal volume 0.1 % Palladium solution was added and vortexed vigorously. The final dilution of serum was 1 : 9.

2 : For the human erythrocytes diluted 1 : 4 with distilled water,using a volume x 2 of each reagents. The final dilution was 1 : 20.

An equal volume of selenium standard solution (25 - 50 - 100 - 200 - 400 µ/l) was added to the sample before the matrix modifiers addition. 20 µl of the diluted sample was injected into the furnace. The absorbance values of sample and standard reported were obtained from peak height measurements (8).

Results and discussion

The ashing step is very important. It should efficiently convert organic forms of the analyte element to stable inorganic species. Palladium can be used to stabilize elements at several hundred degrees higher than those reached with nickel (1200°C CV 700°C) (6, 9). Palladium is the best modifier when used under reducing conditions : ascorbic acid would be added before Palladium. Charring at 2700°C was necessary when high concentrations of organic species were present.

Accuracy and precision

The linearity was determined from 10 replicates of human serum control and the linear regression equation was calculated.

$$y = 42.49 + 3.3.\ x \qquad x = \text{concentration}$$
$$y = \text{peak height}$$

$\bar{m}$ = mean selenium concentration

$\bar{m} = 1.45\ \mu mole.l^{-1}$ with a standard deviation = 0.09 %.

The limit of detection was 0.113 $\mu mole.l^{-1}$ with an accuracy of 4.9 % and a coefficient of variation of 9 %.

Typical peak height response can be estimated using the following formula:

$$A = \frac{C \times V \times 0.0044\ A.s}{m_o}$$

where A = peak height absorbance

C = concentration of the element in ng/ml

V = volume dispensed in µl

m_o = characteristic mass listed in pg

Characteristic mass is the typical mass of the element which would produce a peak height of 0.0044 absorbance/sec. (10). The characteristic mass (m_o) calculated from the serum control was 20.56 pg/0.0044 A.s. The value listed by the manufacturer in recommended standard conditions, with the graphite furnace when used chemical modifiers, was 20 pg.

Application

With this procedure, some selenium concentrations were determined in serum and erythrocytes of children and listed in table 2. In accordance with several authors (11, 12, 13), selenium concentration in erythrocytes were higher than in serum. Lower levels were found for patients affected by cystic fibrosis, chronic diarrhoea and biliary congenital stenosis.

Table 2. Determination of Selenium in Human Children Serum and Erythrocytes by the Proposed Method.

	n	Serum $\mu mole.l^{-1}$	Erythrocytes $\mu mole.l^{-1}$	Age
Control group	7	1.23 ± 0.20	1.63 ± 0.20	2 to 13 years
Patients cystic fibrosis	9	0.89 ± 0.45	0.97 ± 0.44	5 months to 20 years
Chronic diarrhoea	6	0.48 ± 0.07	1.60 ± 0.19	< 12 months
Biliary congenital stenosis	5	0.57 ± 0.19	0.45 ± 0.17	2 months to 9 years

Conclusion

The proposed method allows a rapid and precise determination of selenium in biological samples (serum, erythrocytes).
The preliminary results reported in this paper will be completed by analysis of other cases.

Acknowledgement

We gratefully acknowledge the technical assistance of the Pediatric Functional Exploration Laboratory staff for his preparation of biological samples.

References

1. Koirtyohann, S.R., M.L. Kaiser. 1982. Anal. Chem. 54, 14, 1515 A.

2. Ediger, R.D.E. 1975. Atomic Absorpt. J. Newl. 14, 127.

3. Welz, B. 1976. Atomic Absorption Spectroscopy. Verlag Chemie, Weinheim, N.Y.

4. Chan, X., Z. Ni. 1979. Acta Chim. Sinica. 37, 261.

5. Schlemmer, G., B. Welz. 1986. Spectro Chim. Acta 41 B, 1175.

6. Voth-Beach, L.M. 1986. Varian Instrument Group Aarc, 205 W.Touhy Ave. Parck Ridge IL 60068.

7. Ekerlin, R.H., D.W. Hoult, G.R. Carning. 1987. Atomic Spectroscopy, 8, 1, 64-66.

8. Hoenig, M., K. Meeus-Verdine, P.O. Scokart. 1984. Analysis 12,6, 279-284.

9. Saeed, K., Y. Thomassen, F.J. Langmyhr. 1979. Analytica Chimica Acta, 110, 285-289.

10. Irwing, H.M.N.H., H. Freiser, T.S. West. 1978. Compendium of Analytical Nomenclature IUPAC, Anal. Chem. Div., Pergamon Press, Oxford.

11. Dickson, R.C., R.H. Tomlinson. 1967. Clin. Chim. Acta, 16, 311-321.

12. Pleban, P.A., A. Munyani, J. Beachum. 1982. Clin. Chem. 28, 311-316.

13. Verlinden, M., M. Van Sprundel, J.C. Van Der Auwera, W.J. Eylenbosch. 1983. Biological Trace Element Res. 5, 92, 102 - 103-111.

DETERMINATION OF SELENIUM IN SERUM BY ELECTROTHERMAL ATOMIC SPECTROPHOTOMETRY

I. MOYNIER, E. BOURRET, L. BARDET

Laboratoire de Physique Industrielle Pharmaceutique
Faculté de Pharmacie, Avenue Charles Flahault, 34060 Montpellier, France

M. FUSSELLIER

Laboratoire Aguettant
1, Avenue Jules Carteret, 69007 Lyon, France

Introduction

The most utilized methods for the determination of Selenium in serum are commonly neutron activation analysis, fluorimetry, gas chromatography, inductively coupled plasma emission spectrocopy and atomic absorption analysis using either electrothermal atomization or hydride formation procedures (1). We opted for electrothermal absorption spectroscopy (ET-AAS) because this technique is rapid, does not need large sample volumes nor laborious sample digestion procedures. Moreover ETAAS is one of the most sensitive techniques (2), overcoming chemical and spectral interferences.

I - Equipment and conditions

A model 2380 atomic absorption spectrophotometer equipped with a HGA 400 graphite furnace, a model 056 stripchart recorder and an AS 40 autosampler, all from Perkin-Elmer Corp., are used. A Perkin-Elmer electrodeless discharge lamp (equipped with a Perkin-Elmer power supply) is chosen. Pyrolytically coated graphite tubes are immersed in a 20 $g.l^{-1}$ Ni $(NO_3)_2$ solution under reduced pressure as described by Oster (3). The instrumental and furnace settings are summarized in Table I.

Selenium in Medicine and Biology

Table I = Instrumental and graphite furnace settings for serum Selenium determination.

. <u>Atomic absorption spectrophotometer</u>

Electrodeless discharge lamp power	6 mA
Wavelength	196,1 nm
Slid Width	0,7 nm
Mode	Absorbance
BG correction	Deuterium
Signal	Peak height

. <u>Graphite furnace</u>

Step	DRY		CHAR			ATOMIZE	CLEAN	COOL
Temperature (°C)	90	120	300	500	1200	2600	2700	20
Ramp (S)	20	20	15	15	15	1	1	1
Hold (S)	10	5	10	10	10	4	2	2
Read (S)	-	-	-	-	-	5	-	-
Record	-	-	-	-	-	x	x	-
Flow Interrupt (argon purge gas)	-	-	-	-	-	x	-	-
Sample volume						20 µl.		

II - Reagents and sample preparation

. Reagents : Selenium standard (Merck Titrisol, 9915), HNO_3(Merck suprapure reagents, 456), nickel nitrate (Merck, 6721) and triton X 100 (Merck, 12298).
. Serum collection : 30 presumed healthy subjects aged 27 to 46 years take part in this study. Fasting morning blood is drawn by venepuncture. After clotting and centrifugation, serum samples are stored at -20°C. Hemolysed samples are rejeted.
. Method : serum samples are dilued ten fold with a Selenium working solution and a modifier solution by a standard addition method to yield 20, 50 and 100 PPb Se in 0,1% Ni (II), 0,1% triton and 0,072 M HNO_3 solution.

III - Method and results

The major difficulty lies in part in the small amounts of Selenium present in serum sample and in the element's semimetallic character (4). So we opt on one hand for the method of standard additions and on the other hand, to prevent Selenium losses by premature volatilization during the ashing step, nickel ions are added to the sample solutions (5) . Enhancement of Selenium sensitivity is noted with nickel addition (0,1% Ni) . The concentration of nitric acid sufficient to keep dilued samples clear is found to be 0,072 M (6). Signal intensity and reproductibility are improved by addition of triton X 100 (0,1%). More reproducible results are obtained with graphite pyrocoated tubes treated by a nickel nitrate solution. Figure 1 shows a recorder tracing of a typical standard-additions curve for the determination of Selenium in a serum.

Fig.1 - Recorder tracing of a standard-additions curve.

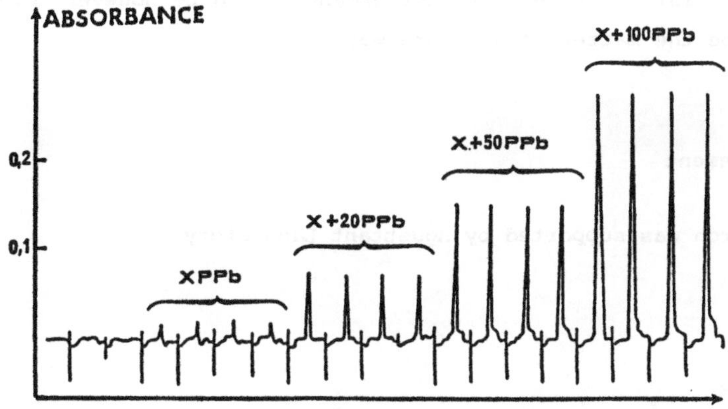

The validity of the method is checked by statistical tests for linear response and by research of a systematic error with the help of recovery test. Under these conditions the linearity is obtained in the concentration range up to 150 PPb for a ten fold dilued serum (Figure 2).

Fig.2 - Linearity of the calibration curve of Selenium in serum.

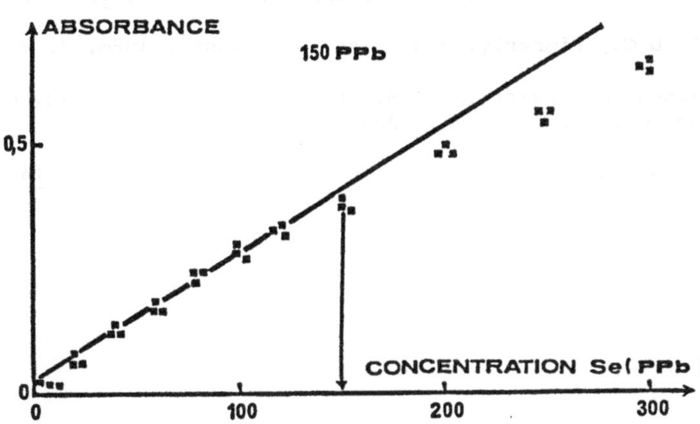

The detection limit of 3,5 PPb Se serum is calculated. Selenium added to serum is recovered completely in this procedure since the recovery is always found to be in the range of 90-110% (Table II).

Table II = Accuracy

Se added (PPb)	Recovery (%)
10	94
20	109
30	97

The present method is therefore applied to the determination of Selenium in 30 healthy adults. The mean concentration in serum is found to be 96 PPb (S.D. 13). Different patient groups should be however investigated which will be the matter of a future work.

Acknowledgement

This research was supported by Aguettant Laboratory.

References

1 . Veillon, C., Lewis, S.A., Patterson, K.Y. - Acta Pharmacol. Toxicol. 1986, 59, suppl.7, p573.

2 . Welz, B., Schlemmer, G., Voellkopf, U. - Acta Pharmacol. Toxicol., 1986, 59, suppl.7, p.589.

3 . Oster, O., Prellwitz, W. - Clin. Chim. Acta, 1982, 124, p.277.

4 . Paschal, D.C., Kimberly, M.M. - At. Spectrosc., 1986, 7, 3, p.75.

5 . Kumpulainen, J., Raittila, A.M., Lehto, J., Koivistoinen, P. - J. assoc. off. anal. chem., 1983, 66, 5, p.1129.

6 . Alfthan, G., Kumpulainen, J. - Anal. chim. acta., 1982, 140, p.221.

SELENIUM DETERMINATION IN DIFFERENT BIOLOGICAL FLUIDS BY STABLE ISOTOPE DILUTION GAS CHROMATOGRAPHY MASS SPECTROMETRY

V. Ducros, D. Ruffieux, N. Belin, A. Favier
Laboratoire de Biochimie C
C.H.R.U.G. - B.P 217 X - 38043 Grenoble Cédex - France

Introduction

Many analytical methods for selenium determination are now available but stable isotope dilution gas chromatography mass spectrometry allows an accurate, sensitive and precise measurement of selenium levels in various biological fluids (1).

A method has been developed for analyzing selenium in serum (2), urine and red blood cells by GC/MS.

Blood is collected by peripheral venipuncture with disposable seringues and transfered in a selenium free polystyrene tube containing heparine trace elements free (Prolabo). After centrifugation, R.B.C and plasma are separated and frozen at -20°C until used for analysis. Urine are collected in a normal plastic bottle and an portion (about 50 ml) is frozen and stored for further analysis.

Materials

Reagents :

- ^{76}Se (atomic abundance = 98.5%) is obtained in elemental form from "Commissariat à l'Energie Atomique" (CEA Saclay - France) dissolved in concentrated nitric acid, completed to 100 ml by HCl 0.1 N and further diluted in dezionized water.
- Standards prepared from a solution of SeO_2 (Titrisol-Merck) 1 g and diluted to contain 0 - 0.2 - 0.5 - 1 - 1.5 - 2 mg/l of selenium.
- Derivatizing reagent used is 4-nitro-O-phenylene diamine (NPD) 4 g/l, purified by cyclohexane extraction.

Sample preparation :

- Serum (500 μl) is spiked with 50 μl of ^{76}Se (1 mg/l) and digested by nitric and perchloric acids (1:1). After mineralisation (48 h at 180°C) and elimination of residual nitric acid, selenate is reduced to selenite by HCl 0.1 N. Selenite is chelated with 4-nitro-O-phenylene diamine to form stable nitropiazselenol Se-NPD (scheme 1) which is extracted into chloroform. A few microliters of this extract are introduced into the gas chromatograph.
- Urine (1 ml)
- R.B.C (500 μl)) same preparation like serum

Scheme 1 : Formation of nitropiazselenol from selenite and 4-nitro-0-phenylenediamine.

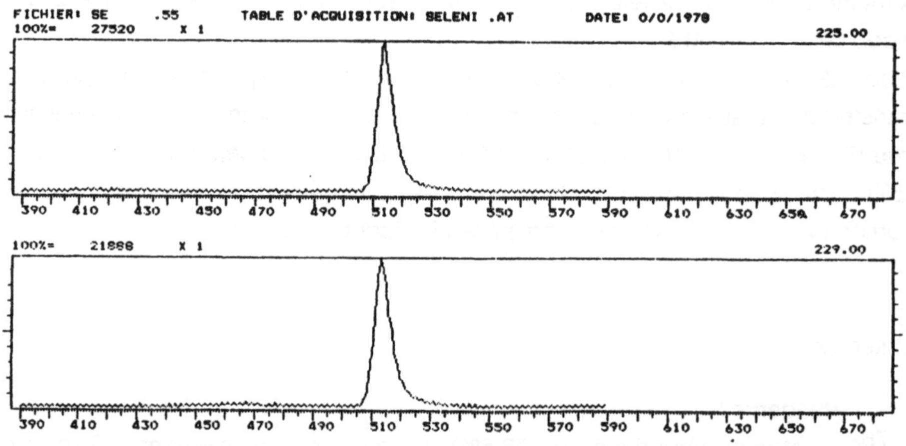

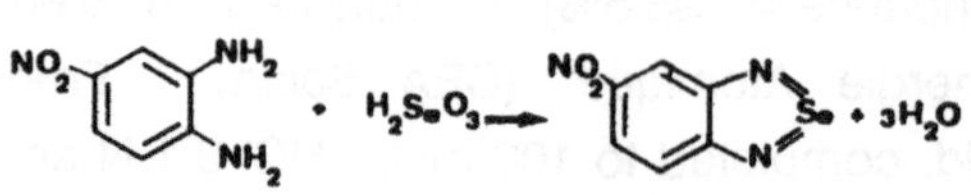

Methods

Operating parameters :

By means of electron impact ionization, we obtain the Se-NPD$^+$ parent ion cluster (scheme 2) and we measure the $^{80}Se/^{76}Se$ isotope ratio.

Scheme 2 : Relative abundance of Se-NPD Isotopes

Se-NPD Isotopes	Se-NPD % (Natural Se)	Se^{76}-NPD % (Enriched Se)
223	1.73	1.19
225	9.01	90.31
226	8.44	7.10
227	23.88	0.88
229	48.41	0.47
231	8.64	0.05

Isotope ratio measurement is made on a quadrupole mass spectrometer (Nermag-R10-10C) coupled to a gas chromatograph (Girdel 32).

GC conditions

- solid injector 250°C
- capillary column CP Sil 5 (Chrompack) lenght 9 m, I.D 0.32 mm, film 1 μm
- helium pressure 0.5 bar, fuite 20 ml/mn.

MS conditions

- electron energy 70 eV
- filament intensity 200 μA
- electronic gain 10^7 V/A
- time of integration :
  - 400 ms for ion 225 (^{76}Se)
  - 400 ms for ion 229 (^{80}Se)

Results

	Serum	R.B.C	Urine
Normal values	66 ± 13 /ug/l	105 ± 15 /ug/l	23 ± 10 /ug/day
	0.83 ± 0.16 /umol/l	1.33 ± 0.19 /umol/l	0.29 ± 0.13 /umol/day
Precision	1.6 % (n = 30)	8.4 % (n = 10)	8.3 % (n = 10)

Discussion

Normal values are in agreement with those of literature.
In serum, the precision is good, better than that obtained by Electrothermal Atomic Absorption Spectrometry (E.A.A.S.) in our laboratory (16%).
In urine and R.B.C, precision is satisfactory but not so good as in serum. That can be explained by the chemical form of Se in urine (Trimethylselenonium TMSe) or in R.B.C (Se incorporated in protein) that is more difficult to ash than serum Se. Another factor decreasing precision is the small number of samples (n = 10) used to determinate the precision in urine and R.B.C.

References

1. Reamer D.E. and C. Veillon. 1981. Anal. Chem. 53, 2166-2169.

2. Ducros V., J. Arnaud, A. Favier. 1987. In : Trace Elements in Human Health and Disease, Second Nordic Symposium, Odense, Denmark - K 3.

DIRECT DETERMINATION OF SELENIUM IN HUMAN AND ANIMAL PLASMA BY ENERGY DISPERSIVE X-RAY FLUORESCENCE

Maier E.A.[1,2], Rastegar F.[1], Ruch C.[3], Heimburger R.[1], Christophe C.[4] and Leroy M.J.F.[1,5].

(1) Ecole Européenne des Hautes Etudes des Industries Chimiques de Strasbourg (EHICS). Laboratoire de Chimie Minérale, Unité 405 du C.N.R.S., 1 rue B. Pascal, 67008 Strabourg Cedex, France.

(2) Presently at the Commission of the European Communities - Joint Research Center - Central Bureau for Nuclear Measurements, 2440 Geel, Belgium.

(3) Siemens A.G. Gerätewerk, Karlsruhe, F.R.G..

(4) Laboratoire de Biologie Médicale du Dr. V. Schuh, 1 Quai des Bateliers, 67000 Strasbourg, France.

(5) To whom correspondence should be addressed.

Selenium has been recognized for a long time to be an essential trace element. It seemed to be interesting to characterize the exact amount of selenium in different media in order to stress the differences between humans and various animal models used for research. More generaly , it seemed to be important to determine the normal "trace element profile" for these animals . Unfortunately, it is still difficult to quantitate selenium in biological media because of its low concentrations and the small sample sizes usually available. The latter is also a limiting factor for small animals like rats or mice especially when the operator cannot sacrifice the animals. We briefely present the results which we obtained using energy dispersive x-ray fluorescence on samples from healthy humans, from subjects showing various disease, and from rats, cats, dogs, and horses.

MATERIAL AND METHOD

The energy dispersive x-ray spectrometer used in this work has been previously described (1-6). We will just review its most important characteristics.

The low detection limits attained with this prototype are mainly attributed to the fact that the irradiation is performed with a high power x-ray tube (1 kW for plasma), and to the very fine geometry of the irradiation chamber which allows direct irradiation of the sample without saturating the detector with the

Selenium in Medicine and Biology

x-rays emerging from the tube . The Si(Li) detector collects simultaneously the x-rays of the different elements present in the sample. An automatic sampler, controlled by computer, allows the succesive analysis of a batch of 40 samples (6). All the serum samples were prepared following the same procedure. Blood was collected in tubes tested for their trace element content in order to avoid any contamination. Sera from rats were collected in a physiology laboratory. The blood was taken from the corner of the eye of the animal using a capillary tube. The sera from the other animals were collected in the biological laboratory which also participated in this study. Human sera were analysed over several years and 3,500 samples from healthy subjects were used to establish the normal reference values. The various diseases were characterised by using at least 100 samples for each clinically certified disease. The sample preparation and the description of the entire analytical procedure were previously described and had not changed significantly (1,6).

RESULTS

Calibration and precision: The detection limit for selenium in water was found to be 100 pg (or 2 ng/ml for a deposit of 50 µl) for 1,000 seconds irradiation time (2). In plasma the detection limit for 500 seconds (routine conditions) is increased to 1 ng or 20 ng/ml for a deposit of 50 µl (1).

For powder pellets of coal or animal tissues, an accumulation of data for 10,000 seconds leads to a detection limit of 0.2 µg/g (Table 2). Table 1 shows the characteristics of the calibration curve of selenium prepared by the method of known additions to a standard serum (Seronorm, batch 161, Nyegaard & Co, Oslo, Norway). The quantitation is performed after normalizing the peak intensity of selenium with respect to the signal of yttrium which is added as an internal standard to each sample. To evaluate the precision of the method we measured the same sample five times. To assess reproducibility, we analysed five different aliquots of the same serum (6). All these measurements were carried out under routine conditions: 500 seconds irradiation time, 40 kV, 25 mA, the x-ray

TABLE 1: PRECISION AND REPRODUCIBILITY FOR SELENIUM IN HUMAN PLASMA

µg/ml	precision C.V.%	reproducibility C.V.%
0.112	11.2	11.8
0.212	6.05	3.87
0.312	8.20	10.9
0.412	4.10	2.64
0.512	3.20	2.86
0.612	4.50	5.07

continuum of the x-ray tube being eliminated with a 150 µm thick molybdenum filter. To improve the sensitivty it is only necessary to increase the irradiation time. Table 2 shows the results for coal and animal powder pellets after 10,000 seconds accumulation of data. The samples have been certified or are under certification for selenium by the NBS (National Bureau of Standards) or the BCR (Bureau Communautaire de Références). Another advantage of energy dispersive x-ray fluorescence lays is its ability to perform multielemental analysis. It is possible to determine simultaneously up to 70 elements with detection limits under 1 µg/ml.

TABLE 2: DETERMINATION OF SELENIUM IN CERTIFIED REFERENCE MATERIALS

MATERIAL	VALUE DETERMINED BY EDXRF µg/g (2s)	CERTIFIED VALUE µg/g (2s)
COALS		
BCR 180	1.2 +/- 0.2	1.32 +/- 0.06
BCR 181	1.0 +/- 0.2	1.15 +/- 0.06
NBS 1632a	2.6 +/- 0.4	2.6 +/- 0.7
MUSSEL TISSUE		BCR provisory result
	1.6 +/- 0.2	1.71 +/- 0.12
ESTUARY SEDIMENTS		BCR provisory result
	1.92 +/- 0.82	1.91 +/- 0.92
DETECTION LIMIT IN 10,000 SECONDS : 0.2 µg/g		

Distribution of trace elements: Table 3 lists the mean values of the studied trace elements for the different animals and for humans without sexual segregation. The mean value for selenium differs significantly from one animal to another. The mean values for the other elements, except copper, are also very different. It seems to be difficult to select, with these criteria and between

TABLE 3: CONCENTRATIONS OF ELEMENTS IN VARIOUS PLASMA SAMPLES (in mg/L)

Element	HORSE	DOG	CAT	RAT	MAN
K	180+/-23	175+/-35	180+/-34	218+/-42	192+/-23
Ca	132+/- 7	111+/- 8	104+/- 5	109+/- 9	91+/- 4
Fe	2.8+/-0.8	2.7+/-1.1	3.6+/-1.2	4.4+/-1.0	1.5+/-0.3
Cu	1.3+/-0.3	0.8+/-0.2	1.3+/-0.2	1.2+/-0.2	1.1+/-0.2
Zn	0.7+/-0.1	0.8+/-0.2	0.9+/-0.1	2.1+/-0.3	0.9+/-0.1
Br	6.6+/-1.9	7.6+/-2.7	8.1+/-1.4	13.2+/-1.6	4.2+/-1.2
Rb	0.10+/-0.02	0.20+/-0.06	0.30+/-0.12	0.18+/-0.05	0.15+/-0.03
Se	0.13+/-0.04	0.26+/-0.08	0.52+/-0.09	0.56+/-0.09	0.083+/-0.015
number of individuals	62	40	15	60	3,500

the studied animals, a model for extrapolation to humans. This fact is confirmed by the variation of the trace element distribution for various diseases. Table 4 shows that for two pathologies like hypertension (162 subjects studied) and infections (213 subjects) the trace element amounts differ greatly. These variations are characteristic and signify a stable "trace element profile" for the above diseases. It must be stressed that for hypertension the selenium level is only slightly affected but in the case of infections the amount decreases markedly. That indicates that the problem should be considered under a multi-elemental point of view taking into account the differences between the various animals and the variations due to the diseases. In order to investigate this field a multi-element technique, which only requires small samples, is a necessity. Today only energy dispersive x-ray fluorescence spectrometry, neutron activation and particle induced x-ray emission are suitable for these kind of investigations. The x-ray fluorescence prototype used in our laboratory to perform the determinations, does not require any more space or supplies than the classical atomic absorption or emission spectrometers.

TABLE 4: MULTI-ELEMENTAL STUDIES FOR HUMAN DISEASES

HYPERTENSION					STUDIED POPULATION: 162 individuals								
Whole blood					Plasma								
Fe	Cu	Zn	Rb	Mg	Ca	Fe	Cu	Zn	Se	Br	Rb	Mg	I
2+	0	3+	1+	3+	1-	2-	2+	2-	1+	0	1-	3+	0
INFECTIONS					STUDIED POPULATION: 213 individuals								
Whole blood					Plasma								
Fe	Cu	Zn	Rb	Mg	Ca	Fe	Cu	Zn	Se	Br	Rb	Mg	I
1-	2+	2-	2+	1+	1-/0	1-	3+	3-	2-	2+	1+	1+	0

+/-: standard deviations over or under the mean value of healthy subjects
I: characteryzed as tetraiodothyroxine and triiodotyroxine

(1) Rastegar, F., Maier, E.A., Heimburger, R., Christophe C. et al.. Clin Chem 1984;30:1300-3.

(2) Ruch, C., Rastegar, F., Heimburger, R., Maier, E.A., et al..Anal Chem 1985;57:1691-4.

(3) Maier, E.A., Rastegar F., Heimburger, R., Ruch, C., et al.. Clin Chem 1985;31:551-5.

(4) Maier, E.A., Dietemann-Molard, A., Rastegar, F., Heimburger, R., et al.. Clin Chem 1986;32:664-8.

(5) Ruch, C., Heimburger, R., Maier-Sargentini, L., Maier, E.A., et al.. Analusis 1987;15(4):159-67.

(6) Rastegar, F., Maier, E.A., Heimburger, R., Leroy, M.J.F., et al..In: Trace Element Analytical Chemistry in Medicine and Biology, Vol.4, (P. Brätter and P. Schramel, eds.), Walter de Gruyter & Co, Berlin New-York, 1987.

EXPRESSION OF URINARY SELENIUM LEVELS IN HUMANS

J. Nève[+] and A. Peretz[++]

Free University of Brussels, [+]Toxicology and Bioanalytical Chemistry Unit, Campus Plaine, 205-1, B-1050 Brussels; [++]Saint-Pierre Hospital, Department of Rheumatology, rue Haute, 322, B-1000 Brussels, Belgium.

Introduction

Urinary selenium (Se) output represents, under a wide range of Se intake, the majority (50-60 %) of the total Se losses. High or toxic intake results in a greater excretion by other routes such as faeces, sweat and breath(1-4). Efforts have been made to characterize the different forms of Se in urine, and to precise their differential renal handling (1-6).

The use of urine Se levels in the assessment of human Se status has been largely confined to diagnosis and monitoring of increased exposure to the element, where significant correlations with high intake levels have been disclosed (1-4). In other states such as normal exposure or deficiency, results are more controversial, and relations with dietary intake or blood levels are poor (1-4).

Determination of Se excretion in urine involves several difficulties, for example in the collection of samples, and in the way of expressing or interpreting the results.Twenty four hours urine samples have been found more correct as an estimation of Se excretion, but are difficult to collect. Results from random samples fluctuates widely due to variations in Se intake, and also are subject to dilution effects (5). HOJO recently demonstrated that these problems can be overcome by expressing results in respect to creatinine content (7).

The purpose of this study is to further evaluate the different ways of expressing the results, and to identify among the different urine fractions collected during a day which is the most suitable.

Subjects and Methods

Fractional urine samples were collected during a whole day from 6 healthy adults, 5 males and 1 female, 27-35 years old. Se concentration was determined in each fraction and in the assembled total amount by graphite furnace atomic absorption spectroscopy with nickel as a modifier, after wet ashing

Selenium in Medicine and Biology

and extraction with an aromatic o-diamine into toluene (8). Quality of the method was established by analysis of urine reference material(Nycomed, Oslo). Creatinine content (Creat) was measured by the method of JAFFE. Other urine samples corresponding to the first or second urine mictions of the day were also collected 7 times during a period of 60 days in 20 other healthy adults, 11 males and 9 females, 26-40 years old.

Results

Table 1 shows in the six investigated subjects the results for urine Se levels expressed in three different units . Data for successive single void urines showed the highest variability when expressed in µg (mean CV= 51 %, range= 37-66) and in µg Se/L (37 %, 17-64). Data expressed in µg Se/ g creat. remained rather constant during the whole day in the six individuals (20 %, 5-35). A strong correlation could be calculated in these 6 subjects between Se and creatinin levels in all urine fractions ($r = 0.95$, $n = 35$, $p < 0.001$). A similar strong relation was observed in urine samples from the 20 other adults ($r = 0.84$, $p < 0.001$).

Urine Se levels expressed per creatinine content in single void urines were satisfactory estimators of daily urinary Se excretion, particularly the values from the first morning sample. Despite of the small number of subjects, correlations could indeed be established between this value and 24-hours µg Se ($r = 0.81$, $p < 0.05$) or 24-hours µg Se/g creat ($r = 0.73$, $p < 0.1$). Time to time variability of this parameter examined 7 times in 20 individuals during a period of 60 days was low (mean CV= 23 %, 11-39).

Discussion

This investigation confirms the results of HOJO (7) supporting the interest of expressing the concentration per creatinine levels for Se in single void urines obtained from various individuals under normal exposure conditions. Effects of dilution are indeed minimized in these conditions. The values determined remained quite constant, not only during one day, but the time-to-time variability was also very low . Expressing the data in µg Se/L resulted in a rather poor fraction to fraction reproductibility, implicating that this way leads to erroneous estimation of urinary Se output. However, this may be acceptable to a certain extent, unless in the assessment of Se exposure. Normal Se values in urine are indeed around or below 30 µg Se/L,

Table 1: Urine Se levels in 6 subjects expressed in different units in fractional urine mictions or per day.

Subject	Expression of results	Selenium excreted per miction (1= first morning void) 1	2	3	4	5	6	7	CV %	Daily excreted selenium in µg	µg/L	µg/g creat.
I	µg	9.92	5.25	3.25	5.04	6.65			41	30.1		
	µg/L	32	42	26	28	19			29		27.6	
	µg/g creat	17.5	18.3	17.5	17.3	19.4			5			18.0
II	µg	6.80	3.19	4.79	3.52	3.15	7.10		37	28.6		
	µg/L	20	22	33	32	21	20		25		23.0	
	µg/g creat	21.1	26.8	20.5	19.2	19.6	18.4		14			20.3
III	µg	6.45	1.49	1.80	1.32	3.75	1.62	5.78	71	22.2		
	µg/L	43	11	18	11	25	6	35	64		20.4	
	µg/g creat	15.9	12.6	13.8	12.1	14.9	9.5	18.9	22			14.9
IV	µg	9.00	1.89	2.87	3.06	4.03			66	20.9		
	µg/L	45	13	41	34	35			36		33.7	
	µg/g creat	14.3	9.2	17.5	16.1	15.5			22			14.4
V	µg	8.60	3.80	1.80	3.36	2.88			52	20.5		
	µg/L	43	10	9	28	32			53		22.1	
	µg/g creat	15.0	15.2	27.9	20.4	19.3			24			17.0
VI	µg	2.93	2.75	1.49	2.59	1.37	1.00	2.10	37	14.2		
	µg/L	9	11	9	11	7	10	12	17		16.5	
	µg/g creat	14.1	28.8	17.3	16.4	22.6	9.6	25.4	35			17.9

while the maximum allowable concentration is fixed at 100 µg Se/L (1-4).

Single void urine Se level in respect to creatinine content, particularly the value determined in the first morning void, is a satisfactory parameter for estimating daily urinary excretion. If daily collections are not possible, this fraction is therefore to be selected and the result expressed in the recommended way.

The strong correlations observed between Se and creatinine content in all collected urine fractions indicate that urinary Se level depends on glomerular filtration, unless under normal exposure conditions (1,2). Such a relation suggests that Se urinary excretion is a function of the muscular mass of individuals, which is directly related to the urinary creatinine content. This was confirmed in the 6 subjects examined by the correlations obtained between 24-hours urine Se levels and either the calculated Body Mass Index ($r = 0.91$, $p < 0.01$) or the calculated Lean Body Mass ($r = 0.80$, $p < 0.05$). A reexamination of these results in a larger population group is necessary in order to confirm the influence of muscular mass on Se urinary excretion.

References

1. Nève,J., Vertongen, F., Molle,L.1985. Clin. Endocrinol. Metab. 14, 629.
2. Nève,J., Henry, M., Peretz,A., Mareschi,J.P.1987. Cah.Diet.Nutr.12, 145.
3. Robberecht, H.J., Deelstra,H.1984. Clin.Chim.Acta 136, 107.
4. Robberecht, H.J., Deelstra,H.1984, Talanta, 31, 497.
5. Levander,O.A.1986. J.Am.Coll.Nutr.,5,37.
6. Robinson,J.R., Robinson,M.F.,Levander,O.A.,Thomson,C.D.1985.Am.J.Clin.Nutr. 41,1023.
7. Hojo,H. 1982.Bull.Environm.Cont.Toxicol.29,37.
8. Nève,J., Hanocq,M., Molle,L., Lefebvre,G. 1982.The Analyst 107,934.

AN AUTOMATED METHOD (RANSEL KIT) FOR THE DETERMINATION OF GLUTATHIONE PEROXIDASE IN ERYTHROCYTE FROM HEALTHY ADULT PATIENTS.

P. Thérond, M. Marchand, D. Biou, J.F. Demelier

Laboratory of Biochemistry, Bretonneau Hospital, 2, rue Carpeaux, 75018 Paris, France

Introduction

Glutathione peroxidase (GPx) is a selenoenzyme found in most human tissues (1). Decreased activity of this enzyme and selenium in plasma and/or erythrocyte, have been linked to hemolytic anemia in premature infants, child cardiomyopathy, cystic fibrosis (2, 3). The enzyme assay is less time consuming and easier to perform than the chemical analysis of selenium. At the present time it is not possible to give typical values for GPx activity because little or no standardization of this measurement has been established. Thus, it is very important to develop a standard method to assess plasma and erythrocyte glutathione peroxidase activities. This determination is usually measured by the method of Paglia and Valentine (4) with or without modifications. Most clinical laboratories have established their method to determine GPx, however because of differences in the concentrations of the various constituents used there has been a problem comparing GPx results between laboratories. For example, increasing the concentration of glutathione or peroxide in the test will increase the apparent GPx level. Poor stability of working reagent means that the method can be laborious due to having to frequently make up a complex reagent preparation. We propose to study the interest of a new diagnostic kit to assess GPx and to compare it with the method of Paglia and Valentine.

Material and method

We used Ransel glutathione peroxidase kit from Randox laboratories LTD (Ardmore, Diamond Road, Crumlin, BT 294QY Co., Antrim, N. Ireland). This method is based on the method of Paglia and Valentine. Glutathione (GSH) is oxidised by cumene hydroperoxide catalysed by GPx present in erythrocyte. The decrease in absorbance at 340 nm is measured. To determine enzyme activity, we used a "Flexigem" microcentrifugal analyzer (ENI, Inc., Fairfield, NJ 07006, USA). The kit comprises : reagent 1 : NADPH (0.28mM), glutathione reductase (500 UI), EDTA (4.3 mM), reduced glutathione (4 mM); reagent 2 : phosphate buffer 50 mM, pH = 7.2; reagent 3 : cumene hydroperoxide (0.18 mM); reagent 4 : reduced glutathione 2 mM in phosphate buffer (R2). All reagents are lyophilised except R3. R1 is reconstitued with R2.

Erythrocytes, separated from plasma by centrifugation and then 3-fold washed in NaCl 0.15 M, are diluted in R4 as follows : 50 µl of erythrocyte in 2 ml of R4. This dilution in reduced glutathione restores GPx levels to their original values when blood samples were stored at -20°C. Erythrocytes were thus diluted 41 times in R4 diluent and allowed for at least 10 mn before analysis. For automated analyses centrifugal analyzer conditions are : filter : 340 nm; temperature : 37°C; incubation time : 60 s; first read : 10 s; second read : 20 s; measuring time : 210 s; multiplication factor : 8681. The sequences performed by the analyzer were as follows : in the inner disc : 500 µl of R1, 10 µl of water, 10 µl of diluted erythrocytes; in the outer disc : 20 µl of R3. Hemoglobin levels are measured on the washed erythrocytes using the cyanmethaemoglobin method. The non enzymatic rate is obtained by substituting water for haemolysate. The blank reading is then sustracted from the assay and the following formula is applied to calculate the GPx activity : GPx, UI/l = 8681 x OD/mn, then multiplied by the dilution factor (41) and refered to haemoglobin. Finally the erythrocyte GPx activities were expressed as UI/g Hb.

Results

The precision of the assay was examined by running low, medium and high GPx erythrocyte samples for within run precision and medium level for day to day precision (Table 1). The lower limit of detection is determined from the within day assay precision (blank + 3 SD); the value in erythrocyte is 4 UI/g Hb. The stability of the working reagents after reconstitution is for up 2 days for R1 and 2 weeks for R4. R3 is prepared each day. The comparison with the reference method of Paglia and Valentine was obtained by testing independently by both methods 41 fresh samples. We have found an excellent correlation between the two : $r = 0.975$. The regression line was : $y = 1.16\ x + 0.99$. The mean value obtained on our 41 samples from healthy adult patients was 45 ± 8 UI/g Hb.

Table 1 : Precision of the Ransel kit for the determination of GPx in erythrocytes (UI/g Hb).

		Mean	SD	CV, %
Within-run : n = 10				
	Low level	11	1	8.8
	Medium	39	2	3.3
	High	64	4	2
Day to day : n = 12				
	Medium	41	2.5	5.8

Discussion

The proposed kit gives total satisfaction : within day and day to day precision assays are satisfactory. If we compare this kit with the reference method of Paglia and Valentine this method is more convenient because the working reagent is stable up 2 days at fridge temperature

and the concentrations of the various components are constant, thus enabling very accurate inter-laboratory comparisons. In the original method of Paglia and Valentine haemolysates were prepared using Drabkin's reagent. We have observed atdouble strength Drabkin's reagent a loss of GPx activity particularly in stored samples. This observation was also reported in a recent publication (5). On the other hand since hydrogen peroxide has been replaced by cumene hydroperoxide this is no longer applicable in the present method.
In conclusion we have found this first enzymatic kit for GPx determination very useful, precise and sensitive. From a practical point of view we strongly advise the different laboratories to use it to suppress the disparity of results observed in the different publications.

References

1. Ganther, H.E., D.G. Hafeman, R.A. Laawrence. 1976. In : Trace Elements in Human Health and Disease (A.S. Prasad Eds.). Academic Press, New York, NY, p.165.

2. Thomson, C.D., H.M. Rea, V.M. Doesburg, M.F. Robinson. 1977. Br. J. Nutr. 37, 457.

3. Gross, S. 1976. Sem. Hematol. 13, 187.

4. Castillo, R., C. Landon, K. Eckhardt, V. Morris. 1981. J. Pediatr. 99, 583.

5. Blanchflower, W.J., D.A. Rice, W.B. Davidson. 1986. Biol. Tr. El. Res. 11, 89.

Section IV

Selenium in human diseases

SELENIUM IN NEONATES AND CHILDREN

I. Lombeck, H. Menzel
University Children's Hospital and Institute of Toxicology,
University of Düsseldorf, FRG

Selenium status of infants and young children

During childhood selenium status shows a marked age relationship. At birth the selenium content of the liver is higher than in later infancy suggesting the accretion of selenium stores during gestation (2, 29). Other tissues show similar phenomena, f.e. hair selenium is very high at birth (21). There is no difference between values from premature and mature infants. In all newborns the hair values are about double as high as in the mothers (19). Within the first year of life they decrease to adult values. In contrast the selenium concentration of plasma is low in newborns (14, 27), reflecting probably a reduced selenium placental transport during the last days of pregnancy.

During the first months of life the selenium concentration of plasma or serum decreases rapidly (14, 27). This decrease occurs in breast-fed and in bottle-fed infants. Usually it is more pronounced in preterm than in term infants (1), probably depending on the lower body stores. Bayliss et al. (2) reported that although the liver selenium content is higher in premature infants the selenium store is much smaller because of the smaller liver size. Like other investigators (1, 8) we (19) recently found that preterm infants (birth weight: 1700 $\pm$ 139 g) fed with human milk showed a significant decline of their serum selenium concentration during the first four weeks. In formula-fed infants this decrease was even more pronounced (fig. 1). Plasma glutathione peroxidase activity showed a parallel decline. A significant difference between the enzyme activities of the two groups after 4 weeks could not be evaluated because

Selenium in Medicine and Biology

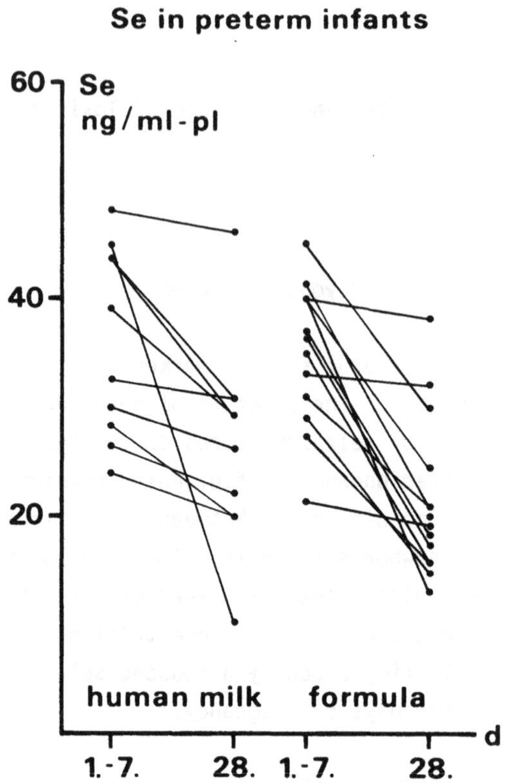

Fig. 1 Plasma selenium concentration in preterm infants fed human milk or formula in the first 4 weeks of life

of the broad variance and the smaller sample size. Rudolpf et al. (25) found in low-birth weight infants fed different milk-based formula a decline of plasma glutathione peroxidase activity but no significant change in plasma selenium.

Selenium intake

It is not clear if the decrease of the selenium concentration in glutathione peroxidase activity in plasma of term and preterm infants within the first months of life is a physiological phenomenon or caused by a too low intake due to the low selenium content of milk. Milk formula, casein-based formula or soy-based formula (16, 24, 30) have about 40 - 70 % less selenium than human milk (13, 16, 24, 30). 50 % of the breast-fed infants and 90 % of the formula-fed infants in many European and overseas countries get less than 10 µg Se/day, that means less than which was recommended for American infants in the recommended dietary allowances from 1980 (22). Because of the discrepancy between the recommended intake and true intake of most infants it is questionable if this recommendation is right and feasible.

In the 2nd half year of life the selenium intake increases, consequently the selenium status rises. The selenium intake depends on the Beikost chosen (17). The selenium content of different constituents of the Beikost varies extensively. Fruit and vegetable in general contain less than 5 ng/g in FRG, also canned infant food: vegetable + meat are not rich in selenium (6). In FRG milk paps are the meals with the highest selenium content for infants. Besides this other whole grain products like bread and biscuits belong to the nutrients which are rich in selenium. About 40 % of the daily selenium intake of small German infants derive from cereals, 15 % from milk or from canned food, eggs and other products. The selenium content of cereals exhibit great regional variations. These variations are greater than that of dairy or meat products. Thus the selenium intake of young children is mainly influenced by the origin and amount of cereals given.

Low selenium status

A very low selenium intake exhibit infants on parenteral nutrition (3) or dietetically treated patients with inborn errors of the amino acid metabolism, e.g. phenylketonuria or maple syrup urine disease. Their diet consists of amino acid mixtures or protein hydrolysates which are supplemented with minerals and trace elements. Because of the narrow range between essentiality and toxicity and scarce knowledge about the bioavailability of selenium compounds no selenium is added to the mixtures. Furtheron their nutrition comprises vegetables and fruits with low protein content. Protein-poor cereals but no whole grain products are included. Therefore their selenium intake amounts to 20 % of that of healthy children of the same age (17). During the first 3 months of dietotherapy the low selenium intake leads to a drastic decline of the serum selenium concentrations (15) and also to that of erythrocytes, platelets and urine. Not only the selenium values are reduced in the blood constituents but also the glutathione peroxidase activities in plasma, erythrocytes and platelets. The hair as long-term indicator of the selenium status shows also a reduced selenium content. The values in the dietetically treated patients are as low as in Chinese children from areas with endemic cardiomyopathy (fig. 2).

In a long mountainous belt from Northeast to Southwest of China a cardiomyopathy, called Keshan's disease, was known since 1935. In some years 11 % of the children and women of child-bearing age suffered from this disease. 50 % of the affected children died. Symptoms of the acute, subacute, chronic or latent type of cardiomyopathy comprise disturbance of the rhythm, enlargement of the heart, heart failure or cardiogenic shock (4). During 4 decades the cause of this fatal disease was unknown. In the 1970th first low hair selenium values were estimated. In the low selenium status, besides hair, blood selenium values and erythrocyte glutathione peroxidase activities were reduced, resulting from a very low selenium intake in peasants who live on locally grown food. Supplementation with sodium selenite lead to a drastic decline of the morbidity.

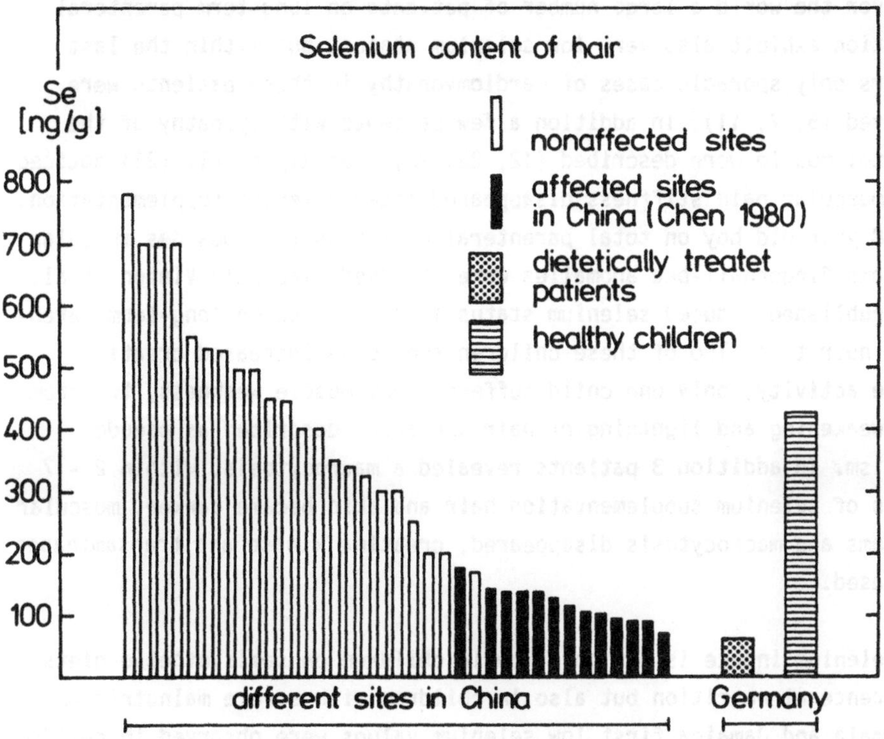

Fig. 2 Selenium content of hair in Chinese children from Keshan-affected and non-affected areas and from healthy German children and dietetically treated patients with phenylketonuria

Not only in Keshan's disease but also in Kaschin-Beck disease (20) selenium deficiency seems to play a role in the etiology or pathogenesis. Kaschin-Beck disease is a chondroosteoarthropathy, where the enchondral ossification is disturbed, which leads to a deformity of affected joints, cripling and dwarfism. It is unclear whether Keshan disease and Kaschin-Beck disease are poorly selenium deficiency diseases or whether other nutritional, environmental or infectious factors contribute to the pathogenesis. The dietetically treated patients with inborn errors of the amino acid metabolism have similar low selenium values, but they do not show signs of cardiomyopathy or changes of the electrocardiogram.

All over the world a large number of patients on long-term parenteral nutrition exhibit also very low selenium status, but within the last 8 years only sporadic cases of cardiomyopathy in these patients were observed (5, 7, 11). In addition a few patients with myopathy of the sceletal muscle were described (12, 23, 28). Van Rij et al. (23) noticed that muscular pain stiffness disappeared after selenium supplementation. In a 8 year old boy on total parenteral nutrition (12) besides muscular symptoms fingernail-bed anomalies were observed. Recently Vinton et al. (28) published reduced selenium status in 4 children on long-term parenteral nutrition. Two of these children showed an increased creatine kinase activity, only one child suffered from muscle weakness, two from nail weakening and lightning of hair and skin, described as pseudo-albinism. In addition 3 patients revealed a macrocytosis. Within 2 - 7 months of selenium supplementation hair and skin became darker, muscular symptoms and macrocytosis disappeared, creatine kinase and transaminases decreased.

The selenium intake is not only low in children on semisynthetic diets or on parenteral nutrition but also in children with severe malnutrition. In Guatemala and Jamaica first low selenium values were observed in children with Kwashiorkor and an amelioration of the clinical course could be shown after selenium supplementation. Our recent investigation in Sudan (10) revealed a reduced selenium status also in patients with marasmic Kwashiorkor from Sudan. There was no correlation to sex, age, region or breast-feeding time. In marasmic children with normal selenium values there was a drastic decline of plasma selenium during rehabilitation, whereas other marasmic patients who had already low values on admission kept the low selenium status during rehabilitation (fig. 3). Their diet consisting of skimmed milk powder, sugar and sesame oil provided only 10 - 20 µg Se/d. Myopathies of the skeletal or cardiac muscle were not observed. Recently low selenium values were measured in Kwashiorkor patients also from Jamaica. There these patients die more often than in Sudan from cardiac insuffieciency. Golden et al. (9) discuss that the low selenium status in combination with high ferritine values may lead to a cardiomyopathy because of inadequate antioxidative defence.

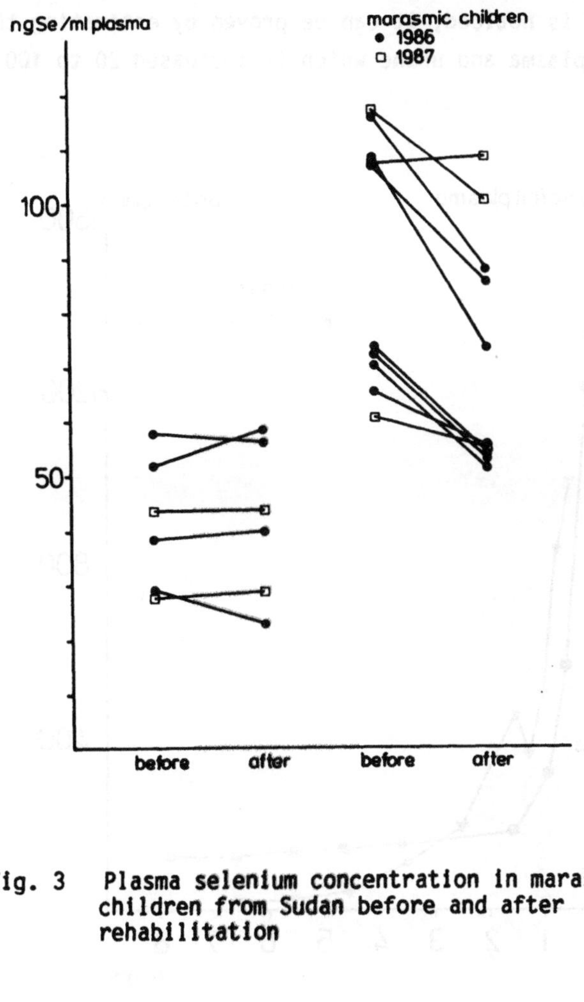

Fig. 3 Plasma selenium concentration in marasmic children from Sudan before and after rehabilitation

Acute selenium poisoning

Before the essential role of selenium was evaluated it was known for its toxic properties. In children until now 5 cases of acute selenium poisoning were observed. The clinical symptoms were not characteristic: restlessness, hypersalivation, vomiting, diarrhea, muscular spasm and arrhythmia. 3 from the 5 children died. The diagnosis can be suspected, if a garlic odour is noticed, it can be proven by estimating the selenium contents in plasma and urine which is increased 20 to 100 fold (fig. 4) (18).

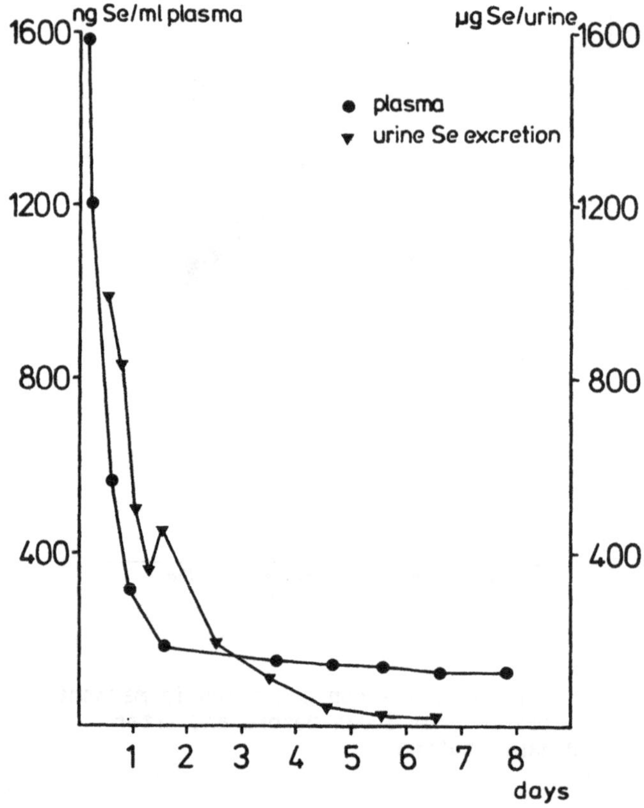

Fig. 4 Plasma selenium concentration and urinary selenium excretion in a girl after acute selenium poisoning

Weeks later the selenium levels in erythrocytes and hair are still elevated. No specific therapy exists and we do not know at which level the acute intoxication will be life-threatening.

Conclusions

If we talk about selenium status in children we must be aware that besides low selenium status also selenium poisoning may occur. Groups at risk for a low selenium status are formula-fed infants, patients on total parenteral or tube feeding, patients on semisynthetic diets and children from selenium-poor areas. Low selenium intake leads to low selenium glutathione peroxidase activity in many tissues. As far as we know today this may cause in some children under distinct conditions cardiac or skeletal myonecrosis probably bone formation may be altered too.

References

1. Amin, S., S.Y. Chen, P.J. Collipp, M. Castro-Magana, V.T. Maddaiah, S.W. Klein 1980. Nutr. Metab. 24, 331.

2. Bayliss, P.A., B.E. Buchanan, R.G.V. Hancock, S.H. Zlotkin 1985. Biol. Trace Elem. Res. 7, 55.

3. van Caillie-Bertrand, M., H.J. Degenhart, J. Fernandes 1984. Acta Paediatr. Scand. 73, 816.

4. Chen, C., Y. Guanggu, J. Chen, Y. Chen, Z. Wen, K. Ge 1980. Biol. Trace Elem. Res. 2, 91.

5. Collipp, P.J., S.Y. Chen 1981. New Engl. J. Med. 304, 1304.

6. Ebert, K.H., I. Lombeck, K. Kasperek, L.E. Feinendegen, H.J. Bremer 1984. Z. Ernährungswiss. 23, 230.

7. Fleming, C.R., J.T. Lie, H.T. McCall, J.F. O'Brien, E.E. Baillie, J.L. Thistle 1982. Gastroenterology 83, 689.

8. Friel, J.K., R.S. Gibson, A. Peliowski, J. Watts 1984. J. Ped. 104, 763

9. Golden, B., M.D. Ramdath personal communication

10. Hassan, M. et al. in publication

11. Johnson, R.A., S.S. Baker, J.T. Fallon, E.P. Maynard, J.N. Ruskin, Z. Wen, K. Ge, H.J. Cohen 1981. New Engl. J. Med. 304, 1210

12. Kien, C.L., H.E. Ganther 1983. Am. J. Clin. Nutr. 37, 319.

13. Kumpulainen, J., E. Vuori, P. Kuitunen, S. Mäkinen, R. Kara 1983. Internat. J. Vit. Nutr. Res. 53, 120.

14. Lombeck, I., K. Kasperek, H.D. Harbisch, L.E. Feinendegen, H.J. Bremer 1977. Eur. J. Pediatr. 125, 81.

15. Lombeck, I., K. Kasperek, H.D. Harbisch, K. Becker, E. Schumann, W. Schröter, L.E. Feinendegen, H.J. Bremer 1978. Eur. J. Pediatr. 128, 213.

16. Lombeck, I., K. Kasperek, B. Bonnermann, L.E. Feinendegen, H.J. Bremer 1978. Eur. J. Pediatr. 129, 139.

17. Lombeck, I., K.H. Ebert, K. Kasperek, L.E. Feinendegen, H.J. Bremer 1984. Eur. J. Pediatr. 143, 90.

18. Lombeck, I., H. Menzel, D. Frosch 1987. Eur. J. Pediatr. 146, 308

19. Lombeck, I. unpublished data

20. Mo, D. 1984. 3rd Intern. Symp. on Selenium in Biology and Med., May 28. Beijing, PR of China.

21. Musa Al-Zubaidy, I., I. Lombeck, K. Kasperek, L.E. Feinendegen, H.J. Bremer 1983. Z. Ernährungswiss. 22, 1.

22. Recommended dietary allowances, food and nutrition board / national research council, 9th revised ed., National Academy of Sciences, Washington, D.C., 1980.

23. van Rij , A.M., C.D. Thomson, J.M. McKenzie, M.F. Robinson 1979. Am. J. Clin. Nutr. 32, 2076.

24. Roekens, E., H. Robberecht, M. van Caillie-Bertrand, H. Deelstra, R. Clara 1985. Eur. J. Pediatr. 144, 45.

25. Rudolph, N., O. Preis, E.J. Bizzos, M.M. Reale, S.L. Wong 1981. J. Pediatr. 99, 57.

26. Schwarz, E. 1965. Lancet I, 1335

27. Verlinden M., M. van Sprundel, J.C. van der Auwera, W.J. Eylenbosch 1983. Biol. Trace Elem. Res. 5, 103.

28. Vinton, N.E., K.A. Dahlstrom, C.T. Strobel, M.E. Ament 1987. J. Pediatr. 111, 711.

29. Westermarck, T. 1977. Acta pharmacol. et toxicol. 41, 121.

30. Zabel, N.L., J. Harland, A.T. Gormican, H.E. Ganther 1978. Am. J. Clin. Nutr. 31, 850.

SELENIUM AT DELIVERY

D. Bouglé, J. Voirin, R. Venezia, J.F. Duhamel
Service de Néonatologie, CHRU, 14000 Caen, France

G. Muller
Service d'Obstétrique, CHRU, 14000 Caen, France

F. Bureau, M. Drosdowsky
Laboratoire de Biochimie A, CHU, 14000 Caen, France

Introduction

The main role of Selenium (Se) is as a component of glutathion Selenium Peroxidase which protects the cells against free radical induced peroxidation (1).

Several neonatal events are related to an increased peroxidation, i.e. ischemia-reperfusion in acute fetal distress or hyperoxia during ressuscitation or assisted ventilation.

In the animal, Se deficiency decreases the tolerance to oxygen (2). Erythrocyte fragility and hemolytic anemia of the neonate are increased when plasma levels are low (3).

Blood concentrations are strongly correlated to dietary intakes (1). Therefore the fetus is faced with wide variations in maternal levels. His ability to regulate the constitution of his own reserves is not well known. We planned to study the relationship between maternal and cord blood levels and to assess the influence of obstretical events upon these values.

Subjects and methods

Blood samples were obtained from 17 unselected mothers in the West of France during delivery and from the cord blood of their term newborns.

Selenium in Medicine and Biology

Parity ranged from 1 to 4, Apgar score from 5 to 10 and birth weight from 2270 to 4180 g.
Blood was drawn on polypropylene tubes to avoid contamination, then centrifugated and frozen until analysis.
Selenium was measured by atomic absorption spectrophotometry (Perkin Elmer 3030, Perkin Elmer Division Instruments, B.P. 76, 79391 Bois d'Arcy, France) at 196 nm, with a programmed temperature of 2300°C.

Data were analyzed by Student's t test and multiple regression analysis for statistical significance.

Results

Mean plasma levels of mothers (M) and cord bloods (CB) were ($\bar{x} \pm 1$ SD) 0.54 ± 0.12 µmol/l and 0.47 ± 0.12 µmol/l. Mean CB/M ratio was 0.87. There was no correlation between M and CB values nor between M or CB and gestational age, parity, or occurence of acute fetal distress as assessed by the Apgar Score.
Birth weight and CB levels displayed a significant correlation ($r = 0.64$; $p = 1.9$ %) (fig. 8). Birth weight and M levels displayed no relationship.

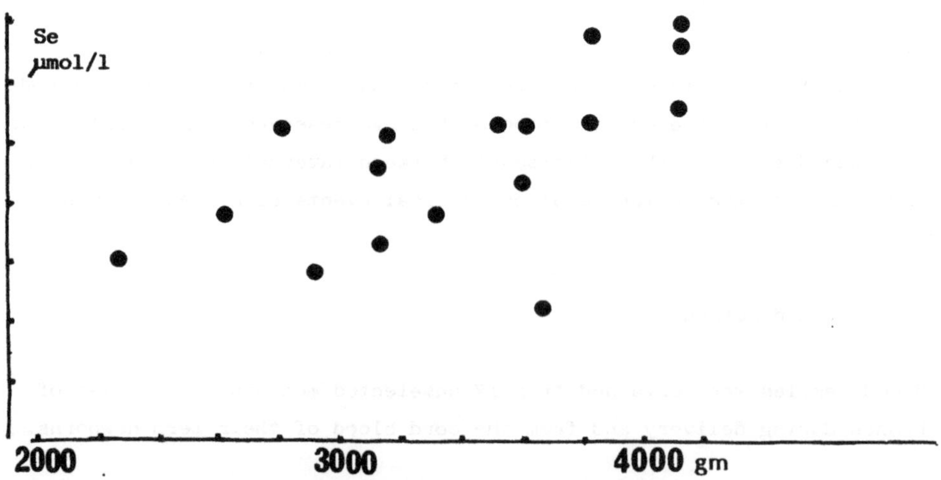

CORRELATION BETWEEN BIRTHWEIGHT AND CORD BLOOD Se LEVEL
$r = 0,64$ $\alpha = 1,9$ %

Discussion

Plasma levels depend on dietary intakes (1) and therefore normal limits vary widely according to geographic areas.

The lower normal limit seems to be 40-45 µg/ml (0.51 - 0.57 µmol/l) below which clinical signs of deficiency (1) and increased erythrocyte fragility (3) may develop.

Se concentrations of fetal organs are similar to those of adults (4). The liver concentration of fetuses from New-Zealand, which is a Se deficient area, has been reported to be lower than in material from Western Germany. During pregnancy plasma levels may display a slight decrease (about 16 %) (5) or remain unchanged (6).

Our results suggest that Se deficiency is common in pregnant women of our region . Wether this reflects a specific low avaibility of Se or non specific malnutrition remains debatable.

Plasma levels at birth have been reported to vary widely with respect to the overall Se supply of the geographic area (1, 6-9). They are lower than those of adults. The Se cord blood level varies from 50 % (1,7) to 64 % (6) of the mother concentration. In these studies the mean mother levels are 1.23, 1.29 and 1.39 µmol/l, far higher than ours (0.54 ± 0.12 µmol/l). On the other hand the cord blood levels we found are in the range of previous studies ; CB/M ratio is therefore higher than previously reported. This suggests that the fetus is able to regulate his own levels in spite of impaired mother status.

Among the clinical data analysed, only birth weight was correlated to Se levels. Yet AMIN (8) did not find such a relation. The relation between Se and growth has been previously suggested : Se deficiency has been related to decreased growth hormone in the anterior pituitary in the rat (10)

Conclusions

The plasma levels of pregnant women from Western France showed a high incidence of deficiency values. However cord blood levels were close to normal suggesting than the fetus is able to regulate the constitution of his own reserves. The significance of the relationship found between cord blood level and birth weight remains to be explained.

References

1. Nève, J., M. Henry, A. Peretz, J.P. Mareschi, 1987. Cah. Nutr. Diet. 22 145-162.

2. Cross, C.E., G. Hasegawa, K. Reddy. 1977. Res. Commun. Chem. Pathol. Pharmacol. 16, 695-706.

3. Gross, S. 1976. Semin. Hematol. 13, 187-199.

4. Casey, C.E., K.M. Hambidge. 1985. In : Vitamin and mineral requirements in preterm infants (R.C. Tsang, ed.). Marcel Dekker Inc, New York. p. 153-184.

5. Behn, E., W. Wolters. 1979. J. Clin. Chem. Clin. Biochem. 17, 133.

6. Haga, P., G. Lunde. 1978. Acta Paediatr. Scand. 67, 735-739.

7. Lombeck, I., K. Kasperek, H.D. Harbisch, L.E. Feinendegen, H.J. Bremer. 1977. Europ. J. Pediat. 125, 81-85.

8. Amin, S., S.Y. Chen, P.J. Collipp, M. Castro-Magnana, V.T. Maddaiah, S.W. Klein. 1980. Nutr. Metab. 24, 331-340.

9. Rudolph, N., S. Wong. 1980. Pediat. Res. 14, 461.

10. Ewan, R.C. 1976. J. Nutr. 106, 702-709.

SERUM SELENIUM LEVELS AND SELENIUM INTAKES IN NEWBORN

J.M. Fraga, J.R. Cervilla, J. Varela-Iglesias, J. Peña
Unidad Neonatal, Departamento de Pediatria, Hospital General de Galicia, Santiago de Compostela, Spain

J.A. Cocho
Laboratorio de Alteraciones Metabolico-Geneticas y Nutricionales, Departamento de Pediatria, Facultad de Medicina, Hospital General de Galicia, Santiago de Compostela, Spain

Introduction

Selenium is a normal constituent of mammalian milk(1,2). Its highest levels are found in colostrum (1,3) and is the only source of this trace element for the suckling mammals (1,2,3). The evaluation of Selenium nutrition in newborns and suckling infants, is hindered by the absence of reliable indicators(4). Serum Selenium levels are very variable and furthermore reflect daily Selenium intake rather than nutritional state (4,5,6). The purpose of this report is to compare the serum levels and intakes of Selenium in newborn babes with diferents weight and gestational ages feeding maternal or formula milk and with those who suffer clinical disorders and were fed with total parenteral nutrition (TPN).

Material and Methods

We studied four groups of newborns in Galicia (N.W. Spain) during the first 45 days of life : Group I (the control group) comprised 71 babes born at term in a normal nutritional state; Group II 75 apparently non-pathological babes born before the 38th week; Group III 66 babes born with intrauterine growth

Selenium in Medicine and Biology

retardation and weighing less than percentile 10 for gestational age; and Group IV 89 babes with a variety of cardiac, respiratory or digestive problems.

Groups I - III were being fed human or formula milk enterally <u>ad libitum</u> .

Serum and milk Selenium content were determined fluorometrically with slight modification (6), and the Selenium intake by measuring daily ingested milk or in the case of parenteral nutrition the Selenium provided by blood or plasma transfusions weekly.

Statistical analyses of the data were carried out by Student or Wilcoxon test and the values expresed as mean ± standard deviation

Results

Serum Selenium levels (µg/L) shown in table 1 were less in Group IV (29.8 ± 16.0) than in Group I (40.3 ± 18.9) $p<0.001$, Group II (34.5 ± 21.6) $p>0.05$, or Group III (38.2 ± 19.1) $p=0.01$, and less in Group IIthan in Group I $p=0.02$.

Selenium intake (µg/day) was 66.6% greater $p<0.001$, among the 31 Group I newborns fed human milk than among the 40 fed formula milk (4.0 ± 1.0 as against 2.4 ± 0.8).

Table 1: Serum Selenium levels (µg/L) in Galician Newborns (N.W. Spain)

Group	N	$\bar{X}$ ± S.D.	Range
I	71	40.3 ± 18.9	18 - 106
II	75	34.5 ± 21.6	6 - 99
III	66	38.2 ± 19.1	7 - 93
IV	89	29.8 ± 16.0	5 - 102

IV<I ($p<0.001$), III ($p=0.001$) and II ($p>0.05$)

The members of Group IV had been under TPN for 6-42 days (mean 20 days) with 1 or 2 blood or plasma transfusions a week, received a total Selenium intake of 0.5 - 0.9 µg/day.
The serum Selenium levels of Group I (see Table 2) were significantly higher in those fed with maternal milk (42.5 ± 20.7) as against (28.9 ± 8.9) µg/L of those who received formula milk.

Table 2: Serum Selenium levels (µg/L) in Galician Newborns (Control Group) (N.W. Spain)

	N	$\overline{X}$ ± S.D.	Range
Human milk	31	42.5 ± 20.7	20 - 106
Formula milk	40	28.9 ± 8.9	18 - 49

Formula < Human (p<0.01)

Discusion and Conclusions

The highest serum Selenium levels were shown by babes born at term after normal intrauterine growth. Of these, those fed human milk had greater intake and higher serum levels than those fed formula milk, though intake was always much lower than the 10 µg/day recommended by the U.S. National Research Council (7).
In keeping with their low intake, newborns under TPN had the lowest serum Selenium levels and can be considered to be suffering <u>marginal deficiency</u> (3,8,9), producing a risk of biochemical and/or subtle clinical abnormalities due to excesive intracellular oxidation in a period of rapid growth and nutritional demand (10,11).

References

1. Fraga, J.M., J.A. Cocho, C. Parrado, C. Colon, J.R. Cervilla, C. Vidal, J.R. Alonso. 1985. Society for Study of Inborn Errors of Metabolism. 23rd Annual Symposium, Liverpool 4th-6th Sept, p. 54.

2. Bratter, P., V.E. Bratter de Negretti, U. Rosick, H.B. Stockhausen. 1987. In: Trace Element Analytical Chemistry in Medicine and Biology, Vol 4 (P. Bratter and P. Schramel, eds.). Walter de Gruyter, Berlin-New York. pp. 67-77.

3. Campfield, T., G. Hepner, L. Werhan, B. Shah. 1987. Pediatr. Res. (Abstract) 21, 424.

4. Commitee on Nutrition, American Academy of Pediatrics.1985. In: Pediatric Nutrition Hanbbook (G.B. Forbes and C.W. Woodruff, eds.). American Academy of Pediatrics, Illinois. pp. 129-130.

5. Pleban, P.A., B. Numerof, B.A. Chrenka, F. Wirth. 1983. In: Chemical Toxicology and Clinical Chemistry of Metals (S.S. Brown and J. Savory, eds.). Academic Press Inc, London. pp. 343-346.

6. Cocho, J.A., C. Parrado, J.R. Cervilla, J.R. Alonso-Fdez, J.M. Fraga. 1984. Quimica Clinica 3, 19-22.

7. Recommended Dietary Allowances. 1980. Nutr. Rev. 38, 290-294.

8. Marchand, M., P. Therond, J. Navarro, D. Biou, J.F. Demelier. 1987. Clin. Chem. 33, 934.

9. Kelly, D., A. Shenkin, A. Coe, J.A. Walter-Smith. 1987. Pediatr. Res. 22, 85.

10. Stockhausen, H.B.. 1987. In: Trace Element Analytical Chemistry in Medicine and Biology, Vol 4 (P. Bratter and P. Schramel, eds.). Walter de Gruyter, Berlin-New York. pp. 61-67.

11. Nair Madhavan, P.N., S.A. Schwartz. 1987. Padiatr. Res. (Abstract) 21, 315.

GLUTATHIONE PEROXIDASE ACTIVITY IN ERYTHROCYTES OF NEWBORNS FED MATERNAL OR FORMULA MILK

J.R. Cervilla, J.R. Fernandez-Lorenzo, J.M. Fraga

Unidad Neonatal, Departamento de Pediatria, Hospital General de Galicia, Santiago de Compostela, Spain

J.A. Cocho

Laboratorio de Alteraciones Metabolico-Geneticas y Nutricionales, Departamento de Pediatria, Facultad de Medicina, Hospital General de Galicia, Santiago de Compostela, Spain

J.I. Ramos-Martinez

Departamento de Bioquimica, Facultad de Farmacia, Santiago de Compostela, Spain

Introduction

Glutathione Peroxidase (GPx) activity is considered by some investigators to be one the most sensitive indicator of Selenium nutrition (1,2,3,4).

The activity of enzyme in mammalian cells, particularly in erythrocyte, has been found to reflect Selenium function (5, 6,7) and is one of the best indices of the availability of Selenium in the diet (2,3,5).

The aim of the present investigation was to compare antioxidant erythrocyte GPx activity in normal full-term, premature and growth retarded neonates of both sexes. By other side, our purpose was to look for if there is any difference in GPx activities between those newborns who were fed with maternal or formula milk.

Selenium in Medicine and Biology

Material and Methods

Erythrocyte GPx activity was measured in three groups of newborns during the first month of life in Galicia (N.W. Spain). Group I (the control group) comprised 46 babes born at term after normal intrauterine growth, Group II 44 babes born before the 38th week of gestation and Group III 45 babes born with intrauterine malnutrition and weighing less than the percentile 10 for gestational age. All were fed maternal or formula milk ad libitum.
The activity of erythrocyte GPx was determined by the method of Paglia and Valentine slightly modified (8), and the results are given in Units per gram hemoglobin.
The Selenium intakes were determined by measuring the Selenium content of human and formula milk fluorometrically and the results expresed in µg per day.
All the data are subjected to statistical evaluation by Student or Wilcoxon test and expressed as mean ± S.D. .

Results

The mean erythrocyte GPx activities (U/gHb) of the three groups in descending order were, Table 1, 8.4 ± 2.8 (Group II), 8.1± 2.9 (Group I) and 7.8 ± 2.5 (Group III). None of the intergroup differences were statistically significant ($p>0.05$).
The mean Selenium intake (µg/day) of 26 term newborns fed maternal milk was significantly higher ($p<0.001$) than of the 20 members of same group fed formula milk (4.0 ± 1.0 versus 2.4 ± 0.8).
The mean erythrocyte GPx activities in the members of Group I fed maternal milk were significantly greater ($p<0.001$) than those of the same group who were fed with formula milk. The activities are compared in Table 2.

Table 1: Erythrocyte Glutathione peroxidase activity (U/gHb) in Newborn infants.Galicia (NW.Spain)

Group	N	$\bar{X} \pm$ S.D.	Range
I	46	8.1 ± 2.9	3.1 - 15.2
II	44	8.4 ± 2.8	2.5 - 14.6
III	45	7.8 ± 2.5	3.3 - 13.7

I, II, III differences $p > 0.05$

Table 2: Erythrocyte Glutathione peroxidase activity (U/gHb) in Newborns fed Maternal or Formula Milk (Galicia, NW. Spain)

	N	$\bar{X} \pm$ S.D.	Range
Maternal milk (MM)	26	8.5 ± 2.8	4.6-15.2
Formula milk (FM)	20	6.6 ± 2.3	3.0-10.8

MM > FM, $p < 0.001$

Discussion and Conclusions

Since no significant differences were found among the erythrocyte GPx activities of milk-fed babes born at term, prematurely or with malnutrition, this parameter discriminates poorly among different states of maturity or gestational development in the newborn (5,9,10,11). Other authors however have found a decrease erythrocyte GPx activities in very premature newborns with gestational age less than 32th week (3,4) .

Though erytrocyte GPx activity appears not to be a good measure of immediate intake of Selenium in the diet (6,7), the fact that the greatest values were found in newborns fed maternal milk suggests that the Selenium in maternal milk is absorbed better, which in conjunction with more efficient incorporation in GPx molecule results in better overall bioavailability (1,2,3,6).

Feeding with maternal milk is therefore recommended for optimal

Selenium nutrition of newborns and suckling infants, or if that were not possible, feeding a formula milk supplemented with Selenium to amount the human milk content (12).

References

1. Lombeck, I., K. Kasperek, H.D. Harbisch, L.E. Feinendengen, H.J. Bremer. 1977. Eur. J. Pediatr. 125, 81-88.

2. Lombeck, I., K. Kasperek, H.D. Harbisch, K. Becker, E. Schumann, W. Schruter, L.E. Feinendengen, H.J. Bremer. 1978. Eur. J. Pediatr. 128, 213-223.

3. Gross, S.. 1976. Sem. Hematol. 13, 187-199.

4. Rudolph, N., S. Wong. 1980. Pediatr. Res.(Abstract) 14, 461.

5. Rudolph, N., O. Preis, E.I. Bitzos, M.M. Reale, S.L. Wong. 1981. J. Pediatr. 99, 57-62.

6 Lombeck, I., H. Menzel, G. Steiner, K. Kasperek. 1982. Klin. Padiat. 194, 303-305.

7. Kitsunezuka, Y.. 1982. Kobe Daigaku Igakubu Kiyo 43, 1013-1021.

8. Cocho, J.A., J.R. Cervilla, J.I. Ramos-Martinez, J.M. Fraga. Quimica Clinica 4, 9.

9. Emerson, P.M., D.Y. Mason, J.E. Cuthbert. 1972. Br. J. Hematol. 22, 667-680.

10. Varga, Sz. J., B. Matkovics, L. Pataki, A. Molnar, Z. Novak. 1985. Clin. Chim. Acta 147, 191-195.

11. Necheles, T.F., M.H. Steinberg, D. Cameron. 1970. Br. J. Hematol. 19, 605-612.

12. Cocho, J.A., C. Parrado, J.R. Cervilla, J.M. Fraga. 1984. Ann. Biol. Clin. 42, 157.

SELENIUM AND OXIDANT INJURY OF CYSTIC FIBROSIS CHILDREN

M.J Richard, J. Arnaud, A. Favier

Laboratoire de Biochimie C, C.H.R.U.G., BP 217 X, 38043 GRENOBLE Cédex, FRANCE

B. Aguilaniu

Pavillon D, C.H.R.U.G, BP 217 X, 38043 GRENOBLE Cédex, FRANCE

J.P Gout, G. Decoux

Centre Hospitalier, 38500 VOIRON, FRANCE

Introduction

Cystic fibrosis occurs in about 1/2000 of live births in Caucasian. Symptoms are mainly a consequence of pulmonary infections and pancreatic insufficiency. Malnutrition can be a major complication of this disease (1)(2). WALLACH (3) explained cystic fibrosis of the pancreas in humans by a selenium deficiency. There is no confirming evidence for this relationship (4)(5). Netherless in the chick, the Se deficiency is characterized by an atrophy and fibrotic degeneration of the pancreas (6).
In the present work, we investigated plasma selenium (Se) and examined its relationship with erythrocyte Se glutathione peroxidase (RBC GPX) and lipid peroxidation products in cystic fibrosis (CF) patients.

Patients and methods

The study population consists of 14 CF subjects aged 2 to

Selenium in Medicine and Biology

19 years in all degree of illness severity. Age matched controls were investigated. Blood was collected into heparinized trace element free tubes. Plasma and RBC GPX activity were immediatly measured (method of GUNZLER and al.(7)). Trace elements : selenium, zinc, copper, and manganese were measured by direct electrothermal atomic absorption spectrometry. Iron level was determined according to standard laboratory method. Vitamins E and A were assayed after specific extraction by a fluorometric method. Plasma caroten was measured spectrophotometrically (after cyclohexane extraction). For vitamins B9 and B12 microbiologic methods were used. Among plasma lipid peroxidation products, we have estimated malonaldehyde (MDA) by a fluorometric micromethod using thiobarbituric acid test (unpublished). Lipid hydroperoxides (HPO) were performed by an enzymatic method (8).
A "t" student test was used to compare the results. Correlations were calculated using Pearson's correlation coefficient.

Results

Trace elements : Table 1 shows that the mean Se, Zn and Fe plasma levels are all significantly decreased in CF patients compared with the controls.
Both plasma GPX and RBC GPX are decreased. Plasma Se in CF children is correlated with RBC GPX ($r = 0{,}459$, $p < 0.02$).
Vitamin status : Despite oral supplementation, patients have low serum retinol, caroten and vitamin E concentrations. Vitamin B9 and B12 were unchanged.
Plasma lipoperoxide values : MDA and HPO levels are significantly higher than those in the controls. Attention has been drawn to the relationship between lipoperoxides and GPX activity, but plasma GPX and RBC GPX did not correlate with the plasma lipoperoxide values.

Table 1 : The level of lipoperoxide and the antioxidant system in cystic fibrosis patients. Results are expressed as mean ± 1 standard deviation.

			CF Patients (n = 14)	Controls (n = 30)	
TRACE ELEMENTS	Selenium	$\mu mol.l^{-1}$	0.44 ± 0.22	0.97 ± 0.17	*
	Zinc	$\mu mol.l^{-1}$	14.7 ± 1.8	16.7 ± 2.0	*
	Copper	$\mu mol.l^{-1}$	22.3 ± 5.2	21.2 ± 3.9	
	Manganese	$nmol.l^{-1}$	24.9 ± 10.0	20.5 ± 4.0	
	Iron	$\mu mol.l^{-1}$	12.0 ± 5.7	22.0 ± 10.0	*
VITAMINS	Vitamin A	$\mu mol.l^{-1}$	2.2 ± 0.5	3.2 ± 0.9	*
	Beta Carotene	$\mu mol.l^{-1}$	0.27 (0-0.9)	3.3 ± 1.2	*
	Vitamin E	$\mu mol.l^{-1}$	22.1 ± 9.8	30.0 ± 9.0	**
	Vitamin B 9	$nmol.l^{-1}$	18.3 ± 8.2	22.5 ± 7.0	
	Vitamin B 12	$pmol.l^{-1}$	469 ± 240	369 ± 110	
Glutathione peroxidase					
	Plasma	UI/l	239 ± 60	335 ± 19	*
	Erythrocytes	UI/g Hb	10.9 ± 2.3	20.0 ± 1.7	*
Lipoperoxidation products					
	MDA	$\mu mol.l^{-1}$	3.1 ± 0.6	2.5 ± 0.2	*
	HPO	$\mu mol.l^{-1}$	181 ± 34	120 ± 19	*

T-test Student : * ($p<0.001$) ** ($p<0.02$)

Discuss

This study confirms very low serum selenium level in CF patients. These results are in agreement with other studies (5). Concomittantly with the low Se supply, there is a decrease in the RBC and plasma GPX. In addition, all the antioxidant system is decreased especially vitamins A and E as described by other authors (9). We find correlations between plasma Se and vitamin E ($r=0.572$, $p<0.01$), and between RBC GPX and vitamins (vit. E : $r=0.593$, $p<0.001$; vit. A : $r=0.490$, $p<0.01$; caroten : $r=0.638$, $p<0.001$; folate : $r=0.785$; $p<0.01$). Is it due at a global malabsorption ? Some authors reported an interaction between Se and vitamin metabolism.

Our findings indicate a preclinical Se deficiency state accompagnied by a raised lipidoperoxidation. This net increase of plasma lipid peroxides may be explain by pulmonary insufficiency and chronic infection injury. The products e.g. MDA, HPO destroy cellular membranes, leading to change in fluidity and permeability of pulmonary parenchym (10). The oxygen species inactivate a lot of proteins such as α1 Anti trypsine inhibitor (11). By their deleterious effects they would be a cause of connective tissue elastase degradation.

More, Se regulates oxygenated metabolites production of arachidonic acid. A GPX dysfunction leads a modification in the amount and type of metabolites that are produced. Increased production of thromboxane A2 (proaggregatory, vasoconsctrictory and proinflammatory product) and decreased prostaglandin PG I2 synthesis were disclosed in Se deficiency states (12).

Evidence of nutritional depletion is common in patients with chronic obstructive lung disease (13). It can affect pulmonary function by impairment of respiratory muscle function, ventilatory drive and pulmonary defense mechanisms. Our findings speculate that antioxidative micronutriments play an important role in the course of CF disease. By its preventive effect against oxygen species, Se limits peroxide induced damages on lipid structure and metabolite integrity in arachidonic acid metabolism. Se is also a regulator of immunity by the maintenance of immunocompetent cells membrane integrity. On the basis of this data, a selenium supplementation in CF malnourished patients is actually performed by cyclic parenteral nutrition.

References

1. Kraemer, R., A. Rudebery, B. Kadain, and al. 1978. Acta Paedriatr. Scand. 67, 33.

2. Aguilaniu, B. : Rev. Fr. Mal. Resp. (in press)

3. Wallach, J.D. 1979. In : Trace Substances in Environmental Health, University of Missouri, 8.

4. Castillo, R., C. Landon, N. Lewiston, V. Morris, O. Levander. 1981. In : Selenium in biology and Medicine (J.E. Spallhotz, J.L. Martin and H.E Ganther, eds). AVI, Wesport, p. 464.

5. Nève, J., R. Van Geffel, M. Hanocq, L. Molle. 1983. Acta Paediatr. Scand. 72, 437.

6. Combs, G.F. Jr., M.J. Bunk. 1981. In : Selenium in Biology and Medicine (J.E. Spallhotz, J.L. Martin and H.E Ganther, eds). AVI Wesport, p. 70.

7. Günzler, W.A., H. Kramers, L. Flohe. 1974. Z. Klin. Chem. Klin. Biochem. 12, 444.

8. Heath, R.L., A.L. Tapel. 1976. Anal. Biochem. 76, 184.

9. Congden, P.J., G. Bruce, M.M. Rothburn, P.C.N. Clarke, J.M. Littlewood, J. Kelleher, M.S. Losowsky. 1981. Arch. Dis. Child. 56, 708.

10. Halliwell, B., J.M.C. Gutteridge. 1984. Biochem. J. 219, 1.

11. Wolf, S.P., A. Garner, R.T. Dean. 1986. TIBS 11, 27.

12. Guidi, G.S., R. Schiavon, A. Biasioli and al. 1984. J. Lab. Clin. Med. 104, 574.

13. Askanazi, J., 1986. Clinics in Chest Medicine.

OXYGEN FREE RADICALS AND ANTIOXIDANTS

John M.C. Gutteridge
National Institute for Biological Standards and Control,
Blanche Lane, South Mimms, Potters Bar, Hertfordshire, EN6 3QG,
UK.

A. The Toxicity of Oxygen

During the chemical and biological evolution of the Earth its atmosphere changed from a highly reducing state to the oxygen-rich state that we know today. Free molecular oxygen (dioxygen) probably appeared on the Earth's surface some 2×10^9 years ago as a result of photosynthetic micro-organisms acquiring the ability to split water. Oxygen is now the most abundant element in the Earth's crust and the second most abundant element in the biosphere. The concentration of molecular oxygen (hereafter referred to as oxygen) in dry air has risen to 21%. Iron, the element of equal importance in this story, is the fourth most abundant element in the Earth's crust.
The author defines a free radical as an atom or molecule capable of independent existence that has one or more unpaired electrons. Within this broad definition we can include the oxygen molecule, a hydrogen atom and most of the transition metal ions (1). The importance of metal ions as free radicals is discussed in detail below. Oxygen is a bi-radical with two unpaired electrons of parallel spin in antibonding orbitals. Any incoming pair of electrons, from an atomic or molecular orbital, attempting to react with oxygen would have anti-parallel spins and require one of the electrons to undergo a spin inversion. To avoid this 'spin restriction', oxygen prefers to accept its electrons one at a time. This has a

Selenium in Medicine and Biology

major advantage for life in an aerobic environment because it considerably slows down the reaction of oxygen with non-radicals. A major disadvantage is, however, that electrons when added singly to oxygen lead to the formation of reactive intermediates, two of which are free radicals. The 4-electron reduction of oxygen to water is shown in Equations 1-4 giving rise sequentially to the superoxide radical (O_2^-), hydrogen peroxide (H_2O_2) and the hydroxyl radical ($\cdot OH$).

$$O_2 + e + H^+ \longrightarrow HO_2\cdot \xrightarrow{pH\ 7.4} H^+ + O_2^- \quad [1]$$
$$O_2^- + e + 2H^+ \longrightarrow H_2O_2 \quad [2]$$
$$H_2O_2 + e \longrightarrow OH^- + \cdot OH \quad [3]$$
$$\cdot OH + e + H^+ \longrightarrow H_2O \quad [4]$$

The superoxide radical is produced in numerous biological processes, particularly the electron transport chains of mitochondria and the endoplasmic reticulum, where electrons can 'leak' onto oxygen. The production of superoxide by activated phagocytic cells is one of the most studied radical-producing systems (2,3). When opsonised particles are contacted by neutrophils a 'respiratory burst' takes place with the uptake of oxygen and the release of superoxide radicals into the phagocytic vacuole and outside the phagocyte. These radicals are usually beneficial to the host because they participate in the killing of ingested micro-organisms.

Many studies have shown that the superoxide radical is not a particularly reactive species in aqueous solution and cannot account for most of the damage observed in systems in which it is generated (4-6). However, there are a few specific sites within cells at which O_2^- can do some direct damage (7). Superoxide is recognised to be a fairly good reductant in aqueous solution. Its protonated form $HO_2\cdot$ is considerably more reactive both as a deductant and as an oxidant than is O_2^- in aqueous solution. It is now well established that the superoxide radical can assist the formation of a more damaging radical species by 'Fenton Chemistry'. Superoxide-dependent

Fenton Chemistry (also called the 'iron-catalysed Haber-Weiss reaction') is shown in Equations 5-7.

$$2O_2^- + 2H^+ \longrightarrow H_2O_2 + O_2 \quad [5]$$
$$O_2^- + Fe^{3+} \rightleftharpoons (Fe^{3+}\text{-}O_2^- \leftrightarrow Fe^{2+}\text{-}O_2) \rightleftharpoons Fe^{2+} + O_2 \quad [6]$$
$$Fe^{2+} + H_2O_2 \xrightarrow[\text{Fenton Reaction}]{} Fe^{3+} + OH^- + \cdot OH \quad [7]$$

The superoxide radical dismutes to form hydrogen peroxide (Equation 5) and also reduces, to the ferrous state, any iron complexes present (Equation 6). This reduction proceeds with the intermediate formation of perferryl, an iron-oxygen complex with a resonance structure intermediate between that of $Fe^{2+}\text{-}O_2$ and $Fe^{3+}\text{-}O_2^-$. Ferrous ions and hydrogen peroxide then give the 'Fenton Reaction' to form hydroxyl radicals ($\cdot OH$). Again, it is likely that reaction (7) is an oversimplification of the process of $\cdot OH$ formation. Hydroxyl radical is one of the major products of the radiolysis of water and is responsible for much of the damage to tissue done by radiation. It is highly reactive and able to damage all biological molecules at an almost diffusion-controlled rate.
As can be seen from Equations 5-6 the formation of $\cdot OH$ radicals from O_2^- and H_2O_2 is a metal-catalysed reaction (8-10). The Fenton Reaction in which a ferrous salt, or ferrous complex, reacts with H_2O_2 to yield a highly aggressive oxidant has been the subject of much debate during the past 40 years (for reviews see 11,12). Opinion has often shifted away from the view that the Fenton reaction produces hydroxyl radical (the same species produced by high energy radiolysis of water) and that an alternative iron-oxygen species is formed instead (12). Biochemical interest in alternatives to the $\cdot OH$ radical was stimulated by Elsner and Youngman (13-15) who observed a Fenton reaction in which the action of the oxidant could not be prevented by some $\cdot OH$ radical scavengers, although it was prevented by others. They described the species as a "crypto hydroxyl radical". Recently, others have suggested that

ferrous ions in the presence of EDTA and H_2O_2 do not generate ·OH radicals (16,17) whereas two independent groups have shown that the characteristics of the oxidant produced by iron-EDTA H_2O_2 mixtures are indistinguishable from ·OH radicals formed in free solution (18-22). There is no doubt that when unchelated (eg. $FeSO_4$) or weakly-chelated ferrous ions are added to hydrogen peroxide the damage done by the oxidant cannot be completely prevented by many well known ·OH radical scavengers (18-20). Because of the metal-binding properties of certain ·OH radical detector molecules, such as deoxyribose, damage is restricted to the site of metal binding and scavengers do not compete for ·OH unless they themselves have a higher iron-binding affinity than the detector molecule. For example, deoxyribose, often used as a detector of ·OH, has a weak binding capacity for iron ions (23), as do the ·OH scavengers mannitol and thiourea. By contrast, when EDTA is present the iron ions remain firmly bound to it, so that the damage done to a target molecule appears to be dependent on ·OH radical generated in free solution. This is the 'radiometric' effect on EDTA (20).

An alternative proposal is that the species formed from ferrous ions and H_2O_2 is not the ·OH radical but an iron (IV) species known as ferryl radical ion (FeO^{2+}) (21,24,25). Ferryl species would be expected to be less reactive than ·OH and not to react with certain ·OH radical scavengers. By implication ferryl would also not be expected to react with ·OH spin traps nor to hydroxylate aromatic compounds (as the ·OH radical does). It is, of course, possible that both ferryl and ·OH are produced in Fenton systems and each might cause damage under different circumstances.

1. Lipid peroxidation

The oxygen radicals so far discussed (HO_2·, O_2^-, ·OH) are inorganic oxygen radicals. However, organic oxygen radicals are equally important in biological systems. Probably the most studied organic oxygen radical reactions are the autoxidations of polyunsaturated fatty acids. Lipid peroxidation, as this process is now more widely known, proceeds by a chain reaction. When a hydrogen atom is abstracted from an unsaturated lipid (LH) by a free radical (R·), the resulting carbon-centred radical (L·) reacts rapidly with oxygen to give a peroxy radical (LO_2·) and this is its most likely fate under aerobic conditions. LO_2· has sufficient energy to abstract a hydrogen atom from another lipid molecule to perpetuate the chain, converting itself into a lipid hydroperoxide (LOOH) (Equations 9-11).

$$LH + R\cdot \longrightarrow L\cdot + RH \qquad [9]$$

$$L\cdot + O_2 \longrightarrow LO_2\cdot \qquad [10]$$

$$LO_2\cdot + LH \longrightarrow LOOH + L\cdot \qquad [11]$$

$$L\cdot + O_2 \longrightarrow LO_2\cdot \qquad [10]$$

The reaction will end when the substrate is exhausted or when radicals meet with each other (Equations 12-13).

$$LO_2\cdot + L\cdot \longrightarrow LOOL \qquad [12]$$

$$L\cdot + L\cdot \longrightarrow LL \qquad [13]$$

Lipid hydroperoxides (LOOH) can be decomposed by transition metal ions to give alkoxy (LO·) and peroxy radicals (Equation 14) and eventually they yield numerous carbonyl products (26).

$$2LOOH \xrightarrow[\text{complexes}]{\text{iron}} LO\cdot + LO_2\cdot + H_2O \qquad [14]$$

This decomposition pathway probably accounts for much of the observed stimulation of lipid peroxidation seen when iron salts or iron complexes are added to bulk lipids, isolated subcellular fractions or tissue extracts since they are certain to contain substantial amounts of lipid peroxides (27). Whenever tissues are damaged there is a possibility that metal complexes are released from vacuoles, organelles or other sequestered sites within the cell. Released metal ions can then interact with cell lipids and bioactive lipid peroxides such as products of the cyclooxygenase and lipoxygenase pathways. The result of this is the frequent observation that lipid peroxidation accompanies many disease processes and is associated with cellular injury by several toxins. The role of lipid peroxidation as a cause of many disease processes has recently been questioned (28).

Aldehydes formed by the iron-catalysed decomposition of lipid peroxides are responsible for many of the odours, colour and flavour changes associated with deteriorating foods, damaged plant leaves and fruits. Malondialdehyde is probably the best known of these aldehydes although biologically it is probably one of the least reactive. It appears to be only formed in large amounts when microsomal preparations are peroxidised (29). More-reactive aldehydes have been described and characterised recently (30). Aldehydes are reactive molecules and will thus not get far from their site of formation, and, although they are not free radicals, they must be considered as part of the toxicity of 'oxidant stress' resulting from free radical damage.

B. Are Metal Ions Available for Fenton Chemistry?

The normal adult human contains some 4g of iron and 80 mg of copper with about two-thirds of the iron present in haemoglobin. Iron is retained by the body and constantly re-cycled for use as there are no specific physiological mechanisms for iron excretion. Subtle changes in the delicate balance between iron absorption and iron loss can readily lead to iron-overload or to iron-deficiency diseases. Iron salts never exist free in biological systems; they will be bound to biological molecules such as proteins, lipids, nucleic acids, carbohydrates and low-molecular-mass chelators. Sometimes the binding constants are low (eg. binding of iron to albumin or ADP) and sometimes they are very high (eg. binding of iron to transferrin or lactoferrin). Which then of these metal complexes are able to catalyse radical damage in biological systems? Loosely-bound iron is the most likely promoter of ·OH radical damage in vivo and attempts have been made to detect and measure such forms of iron in biological samples by assays using the anti-tumour drug bleomycin (31,32). The release of the aldehyde (malondialdehyde) from DNA can be measured and related to the amount of iron present (low micromolar range) in the biological sample (31,32). There has been considerable debate as to whether the iron binding proteins transferrin and lactoferrin are Fenton catalysts (see discussion in 11). However, a recent assessment of reaction conditions concludes that they are reactive only when an iron chelator is present and the proteins have been incorrectly loaded with iron (33), i.e. they only work if, under the assay conditions used, iron becomes detached from the proteins. Similarly, the proteins ferritin and haemoglobin, as intact proteins, are unlikely to catalyse ·OH formation detectable in free solution (1). However, they do appear to promote ·OH formation when present in superoxide generating systems. The reason for this appears to be the ability of superoxide or hydrogen peroxide to release iron from the protein-metal centres and it is this low-molecular-mass chelatable iron which is the true catalyst in the Fenton reaction (34-36). In this way, superoxide radicals can provide all the necessary ingredients in a biological system for ·OH radical formation.

The key role of iron in biological radical formation offers the potential for protective intervention by the use of iron-chelation therapy. Unfortunately, very few of the chelators well known to chemists are suitable for clinical use because they usually facilitate stages of radical formation by altering the redox potential of iron and increasing its solubility (1). However, desferrioxamine, an iron-chelator licensed for clinical use in humans, is exceptionally good at inhibiting iron-dependent radical reactions (37). The in vivo use of desferrioxamine has given considerable evidence consistent with the biological importance of iron-dependent radical reactions.

C. Biological Antioxidant Protection

There is no universal biological antioxidant. Specialised mechanisms have evolved to deal with problems of oxidant stress in specific environments. Inside the cell oxygen is reduced to water and, as we have seen, superoxide and hydrogen peroxide are produced. The superoxide dismutase enzymes function to catalytically remove superoxide radicals, bringing about a dismutation (Equation 5) at a considerably faster rate than occurs normally in the absence of enzyme (4-7). The product of this reaction is hydrogen peroxide and two different enzymes exist to remove it, namely catalase and glutathione peroxidase (selenium enzyme). Providing superoxide and hydrogen peroxide are effectively removed, no $\cdot OH$ radicals will be formed and a low molecular mass iron pool (38) can safely exist within the cell to provide iron for the synthesis of iron-containing proteins. The inside of the cell membrane is a lipophilic region in which lipid-soluble radicals are formed. Protection depends to a large extent on lipid-soluble scavenging antioxidants such as vitamin E or β-carotene. Despite their lack of involvement in oxygen metabolism, extracellular fluids are, nevertheless, constantly exposed to O_2^- and H_2O_2 fluxes especially from activated phagocytic cells. Extracellular fluids contain little, if any, of the protective enzymes found inside the cell yet protection is required against oxygen radicals. This is achieved by preventing metal complexes from participating in radical reactions. Iron binding by transferrin and lactoferrin, the binding of haemoglobin and haem by haptoglobins and haemopexin, copper binding by albumin, and ferrous ion oxidation by the copper-containing protein caeruloplasmin (ferroxidase I) contribute a substantial antioxidant potential to extracellular fluids (reviewed in 11). Most of these proteins are part of the body's 'acute-phase' response, increasing in concentration whenever tissue damage occurs. Tissue damage will of course lead to 'mal-placed' iron and copper becoming available for radical reactions as described in Section B.

References

1. Halliwell, B., J.M.C. Gutteridge. 1985. Free Radicals in Biology and Medicine. Oxford University Press, Oxford.
2. Babior, B.M. 1978. N.Engl.J.Med. 298, 659.
3. Babior, B.M. 1978. N.Engl.J.Med. 298, 271.
4. Fridovich, I. 1974. Adv.Enzymol. 41, 35.
5. Fridovich, I. 1975. Annu.Rev.Biochem. 44, 147.
6. Fridovich, I. 1978. Science 209, 875.
7. Fridovich, I. 1986. Arch.Biochem.Biophys. 247, 1.
8. McCord, J., E.D. Day. 1978. FEBS Lett. 86, 139.
9. Fong, K.L., P.B. McCay, J.L. Poyer, H.P. Misra, B.B. Keele. 1976. Chem.Biol.Interact. 15, 77.
10. Halliwell, B. 1978. FEBS Lett. 92, 321.
11. Halliwell, B., J.M.C. Gutteridge. 1985. Molec.Aspects of Med. 8, 89.
12. Halliwell, B., J.M.C. Gutteridge: Biochemistry (in press)
13. Youngman, R.J., E.F. Elstner. 1981. FEBS Lett. 129, 265.
14. Youngman, R.J., W.F. Osswald, E.F. Elstner. 1982. Biochem. Pharmacol. 31, 603.
15. Youngman, R.J. 1984. Trends Biochem.Sci. 9, 280.
16. Koppenol, W.H. 1985. J.Free Rad.Biol.Med. 1, 281.
17. Rush, J.D., W.H. Koppenol. 1986. J.Biol.Chem. 261, 6730.
18. Gutteridge, J.M.C., B. Halliwell. 1986. In: Oxygen and Sulfur Radicals in Chemistry and Medicine (A. Breccia, M.A.J. Rodgers and G. Semerano, eds.). Lo Scarabeo Bologna, p. 47.

19. Gutteridge, J.M.C. 1984. Biochem.J. 224, 761.

20. Gutteridge, J.M.C. 1987. Biochem.J. 243, 709.

21. Winterbourn, C.C. 1987. Free Rad.Biol.Med. 3, 33.

22. Halliwell, B., J.M.C. Gutteridge, O.I. Aruoma. 1987. Analyt.Biochem. 165, 215.

23. Aruoma, O.I., M. Grootveld, B. Halliwell. 1987. J.Inorg. Biochem. 29, 289.

24. Rush, J.D., W.H. Koppenol. 1987. J.Inorg.Biochem. 29, 199.

25. Sutton, H.C., G.F. Vile, C.C. Winterbourn. 1987. Arch. Biochem.Biophys. 256, 462.

26. Schauenstein, E., H. Esterbauer, H. Zollner. 1977. Aldehydes In Biological Systems. Pion, London.

27. Gutteridge, J.M.C., P.J. Kerry. 1982. Brit.J.Pharmac. 76, 459.

28. Halliwell, B., J.M.C. Gutteridge. 1984. Lancet 1, 1396.

29. Esterbauer, H., K.H. Cheeseman, M.U. Dianzani, G. Poli, T.F. Slater. 1982. Biochem.J. 208, 129.

30. Benedetti, A., M. Comporti, H. Esterbauer. 1980. Biochim.Biophys.Acta 620, 281.

31. Gutteridge, J.M.C., D.A. Rowley, B. Halliwell. 1981. Biochem.J. 199, 263.

32. Gutteridge, J.M.C., Y. Hou. 1986. Free Rad.Res.Commun. 2, 143.

33. Aruoma, O.I., B. Halliwell. 1987. Biochem.J. 241, 273.

34. Gutteridge, J.M.C. 1986. FEBS Lett. 201, 291.

35. Gutteridge, J.M.C. 1985. Biochim.Biophys.Acta 834, 144.

36. Puppo, A., B. Halliwell. 1988. Biochem.J. 249, 185.

37. Gutteridge, J.M.C., R. Richmond, B. Halliwell. 1979. Biochem.J. 184, 469.

38. Jacobs, A. 1970. Blood 50, 433.

SELENIUM IN INFLAMMATION AND IMMUNITY

A. Peretz

Free University of Brussels, Saint-Pierre Hospital, Department of Rheumatology and Physical Medicine, rue Haute, 322, B-1000 Brussels, Belgium.

Introduction

The effects of nutrition on the host defense mecanisms against inflammation, infections and on the regulation of the immune system are widely recognized. Important roles for essential nutrients such as iron, zinc and selenium have emerged from clinical and epidemiological studies as well as from experiments in laboratory animals. A drop in serum iron is observed in inflammatory diseases and infections ; however any attempt to reverse it is followed by a worsening of the symptoms. Zinc deficiency produces a reversible atrophy of lymphoid tissues including the thymus, spleen and lymph nodes. Consequently intercurrent infections with opportunistic germs have a greater incidence. Selenium deficiency, in animals, increases the susceptibility to infectious diseases (1).
The aim of this paper is to review the role played by selenium (Se) in the inflammatory response and the immune system. Although they are interrelated, they will be considered as separate chapters. The involvement of Se in the biological response is described, followed by the experimental evidences coming from in vivo and in vitro studies on inflammation and dealing with non specific, humoral and cell-mediated immunity.

Inflammation

Involvement of Se in the biological response

The inflammatory response involves a sequence of reactions among which

Selenium in Medicine and Biology

phagocytosis plays a key role and represents the non specific defense against invading pathogens or foreign particles . Such as indicated in Figure 1, the mechanism of phagocytosis is complex and depends on multiple cellular events. The motile phagocyte reaches its target by chemotaxis , a process involving the liberation of substances (chemotactic factors) in the vicinity of the pathogen; after engulfing the invading organism, the eliminatory or killing process begins by the liberation of destroying enzymes and the activation of the respiratory burst. This process increases the oxygen consumption and liberates highly reactive oxidizing species: the superoxide anion $O_2^{\bar{\cdot}}$ through the action of the membrane enzyme NADPH oxidase, and derived species. $O_2^{\bar{\cdot}}$ is metabolized into H_2O_2 by superoxide dismutase enzyme. The two species give rise to the hydroxyl radical (OH•) by the Haber-Weiss reaction , to singlet oxygen (:ÖÖ), and to chlorite anions (ClO^-). All these aggressive compounds are able to damage biological systems through peroxidation of membrane lipids and attack of various cell components including proteins and nucleic acids. In addition, activation of phagocytes increases the glucose oxidation via the hexose monophosphate shunt. Such as indicated in Figure 1, this metabolic pathways also generates H_2O_2. Finally, the activation of phospholipase A_2 by H_2O_2 and free radicals produced during the respiratory burst, initiates the biosynthesis of prostaglandins and leukotrienes.

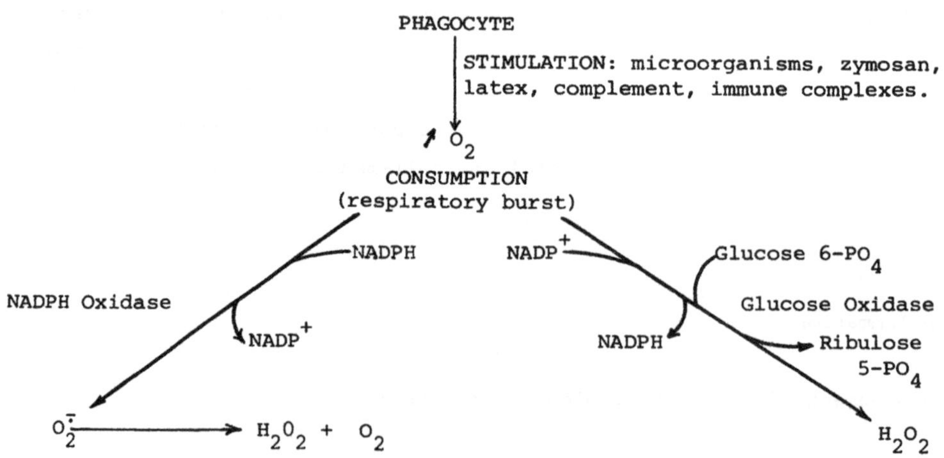

Figure 1: Metabolic pathways leading to peroxide production in activated phagocytes.

Regulation of these reactions occurs in physiological conditions through the action of several enzymes and low molecular weight molecules generally called antioxidants. Among them, Se has to be considered as component of the active site of the enzyme glutathione peroxidase (GSH-Px) as shown in figure 2.

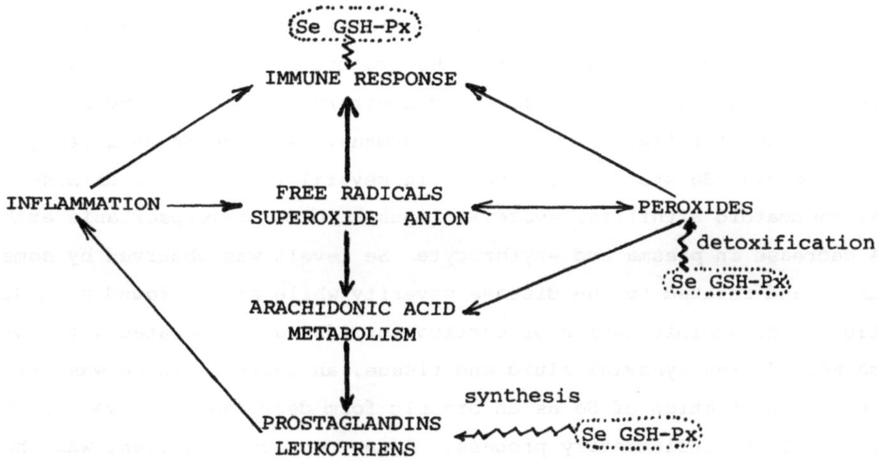

Fig. 2 : Selenium involvement in the Immune and Inflammatory Response.

One of the important functions of this enzyme, in this context, is the reduction of hydrogen peroxide and organic peroxides generated during the respiratory burst, limitating the widespread of the free radical chain reaction. Moreover the hexose monophosphate shunt utilizes the glutathione peroxidase-glutathione-reductase system to detoxify the hydrogen peroxide and generate NADPH (3). Finally, GSH-Px is involved in the multiple steps of the biosynthesis of lipoxigenase and cyclooxigenase pathways (4). In pathological conditions resulting from chronic inflammatory diseases, inappropriate stimulation of phagocytes by immune complexes or complement..., contributes to the tissue damage and the pathogenesis of the disease.

Experimental evidences

In 1963, Roberts reported using the Selye granuloma pouch assay in rats, the antiinflammatory properties of inorganic substances from specific liver fractions which were identified as Se (5). This first investigation prompted studies in diseases characterized by increased levels of peroxides in the plasma and the tissues both in animals and humans.

Animal in vivo studies showed that Se deficiency enhanced acute inflammation in the model of carragenin paw oedema and chronic inflammation in adjuvant arthritis of rats (6). A beneficial effect of the administration of Se was even observed. During the evolution of adjuvant arthritis, the Se profile of rats was not modified contrary to clinical and biological parameters of inflammation (7). In humans, isolated Se-deficiency is rare. However Se status was studied in several inflammatory diseases such as rheumatoid arthritis, systemic lupus erythematosus, psoriasis arthritis. A decrease in plasma and erythrocyte Se levels was observed by some authors and related to the disease severity while others found no modification. The administration of corticosteroids was associated with low plasma Se. In the synovial fluid and tissue, an increase in Se was reported. The administration of Se as an organic form decreased in some selected patients the inflammatory process. A new compound, Ebselen, was shown to possess antiinflammatory properties mainly as leukotriene inhibitor and opens new perspectives (8).

In vitro studies were performed in neutrophils and macrophages from animals which were studied for their ability to produce oxidant species and kill microorganisms. Such as indicated in Table 1, Se deficiency decreased the chemotacxism and the random migration of neutrophils.

Phagocytosis was decreased for zymosan particles but remained normal for Candida and Staphylococcus aureus. The bactericidal activity was normal or decreased while the fung cidal capacity was decreased. Alveolar macrophages of rats exhibited a normal bactericidal capacity while macrophages from peritoneal exsudates presented an enhanced release of hydrogen peroxide. A decrease in intracellular glutathione peroxidase activity was measured and related to the abnormal response of cells. Se supplementation restored the chemotaxis , the migration, the phagocytosis and the bactericidal activity (9). In humans cells, incubation of neutrophils with Se resulted in an enhancement of phagocytosis and bactericidal activity within hours at physiological concentrations (100-200 ng/ml). The rapidity of the effect suggested the incorporation of Se in GSH-Px enzyme with restoration of its activity. At higher dose, an inhibitory effect was observed (10). However a 3 months supplementation with organic Se had no effect on chemotaxism and phagocytosis of neutrophils. Only a slight increase of the intracellular killing of Staphylococcus Aureus was noted (11).

Table 1 : Selenium and Phagocytic Cells

methods	selenium status	effects	references
chemotaxis	deficient	↙	Mc Callister, 1980 Aziz, 1984, 6
	supplemented	↗	Aziz, 1984, 6
neutrophil migration	deficient	N ↙	Mc Callister, 1980 Aziz, 1984
	supplemented	↗	Aziz, 1984
neutrophil ingestion	deficient	N ↙	Serfass, 1975 Boyne, 1979, 81 Gyang , 1984 Aziz, 1984
	supplemented	↗	Arvilomni, 1983 * Urban, 1987 *
neutrophil bactericidal activity	deficient	N ↗	Bass, 1977 Bassford, 1977 Gyang, 1984
neutrophil fungicidal activity	deficient	↙	Boyne, 1979, 81 Serfass, 1975
macrophages bact. act.	deficient	N	Eskew, 1985
macrophages H_2O_2 prod.	deficient	↗	Parnham, 1983

* in humans

In conclusion, Se deficiency enhances the inflammatory reaction in animals while Se supplementation has an opposite effect. From in vitro studies, it appears that the effect may vary following the animal tested and the specific function which was analyzed. Neutrophils are more sensitive to modification of Se in the medium than macrophages. The inflammatory process by itself does not clearly modify the Se status in animals or man.

Immunity

Involvement of Se in the biological response

Inflammation and immunity are interrelated and products resulting from the activation of the inflammatory process can modulate the immune response. On the other hand, the activation of the immune system with production of specific antibodies and stimulation of lymphocytes enhances the reactivity of phagocytic cells. The immune response is a complex system which recognizes and eliminates cellular and non cellular entities through the action of several components : the non specific, the humoral and the cellular responses. The development of a proper response depends on cell-to-cell interactions which themselves depend on the functional integrity of the immunocompetent cells.
As indicated in Figure 3, Se may affect all components of the immune system, mainly by its essential function in the GSH-Px enzyme. The inability of cells deficient in GSH-Px to metabolize H_2O_2 causes an accumulation of reactive oxygen species in the cellular microenvironment with pronounced deleterious effects on cellular membranes by lipid peroxidation. Immunocompetent cells are sensitive to alterations in their structural integrity because they are involved in the recognition of antigens, the binding of mitogens or antigens, the release of soluble mediators, the binding and subsequent lysis of foreign cells. Cytoskeletal structures like microtubules are also sensitive to oxidation; they play an important role in the maintenance of appropriate surface receptors (12). Se is also involved in the ability of cells to proliferate in response to stimuli via non GSH-Px mediated properties.

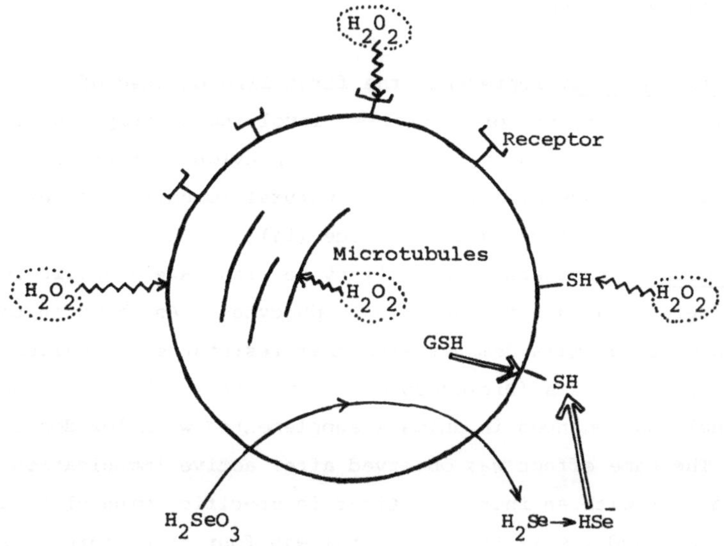

Figure 3: Selenium involvement in immunocompetent cells.

The presence of protein sulfhydril groups in the reducted form at the external membrane is a prerequesite to provide the expression of immunocompetent cell functions. In non activated lymphocytes, the thiol groups exist in a protected form as disulfide bonds. The cleavage of these bonds by reducing agents depends on the GSH content of cells, that in turn is related to the reduction of Se by glutathione reductase. The regulatory role of GSH occurs probably by an exchange of disulfide bonds between GSH and surface thiol proteins. This hypothesis is supported by experimental studies showing the thiol dependency for mitogen activation of lymphocytes and the concomitant increase in intracellular GSH content (13). These studies need to be asserted with immunocompetent cells. Hydrogen selenide (H_2Se) is another derivative that could alter the balance of disulfide bonds at the surface of lymphocytes (Figure 3). Such a role has been demonstrated in erythrocytes and epithelial cells,but remains to be proved in lymphocytes (9).

Experimental evidences

Non specific immunity represents the first line defense of the host against foreign antigens or particles involving neutrophils, mononuclear and macrophages cells whose function was previously reviewed. Complement and lyzozyme are also indices of natural immunity and were reported increased by Se supplementation in mice (14).
Humoral immunity represents the ability to elaborate specific immunoglobulins after clonal activation of B lymphocytes. As shown in table 2, in vivo studies in animals demonstrated that resistance to challenge with several microorganisms (Diplococuccus Pneumoniae, Salmonella, Leptospirosis Plasmodium) was enhanced in animals supplemented with low doses Se (0.5 - 2 ppm). The same effect was observed after active immunization (vaccination) together with an increased titer in specific immunoglobulins. Inorganic Se (SeO_2, selenite, selenate) was four times more potent than organic Se (selenomethionine, selenocysteine) (14). At high doses (5 ppm) an inhibitory effect was observed (15). Se-deficient animals presented an increased susceptibility to infection, low antibody titers with a delayed response, differing when the primary or the secondary response and the class of immunoglobulins were considered (IgM, IgG) (16, 17). In vitro studies showed that Se supplementation in animals increased the number of antibody forming cells what afforded to explain the in vivo increase in antibody titers (18). In humans, very few studies are available. A 3 months supplementation with organic Se in subjects with low Se status had no effect on the in vitro antibody production of lymphocytes (11).
Cell-mediated immunity which concerns the host defense against intracellular pathogens and tumoral cells, is responsible for the graf rejection and autoimmune reactions. Complex interactions exist between T lymphocytes (Helper, suppressor, memory and cytotoxic cells), monocytes and phagocytes. After the recognition of an antigen, sensitized lymphocytes undergo transformation into primitive blasts with increased synthetic activities leading to differentiation and proliferation. Finally the expression of the specific immune response takes place.

Table 2 : Selenium and Humoral Immunity

methods	selenium status	effects on resistance	references
microorganism challenge	deficient supplemented	↙ ↗	Serfass, 1974 Spallholz, 1981 Boyne, 1986
challenge after active immunization	deficient supplemented	↙ ↗	Sheffy 1978 Bernstein, 1972 Norman, 1976 Desowitz, 1980
antibody titers	deficient	↙	Sheffy, 1978-Marsh,1981 Mulhern, 1985
	supplemented	↗	Spallholz, 1973, 4, 5 Koller, 1979 Shackelford, 1980 Aleksondrowiz, 1977 Sheffy, 1970
antibody forming cells	supplemented	↗	Koller, 1979
antibody production in vitro	low	N	Arvilomni, 1983 *

* in humans

As shown in table 3, in vivo Se supplementation was reported to enhance the delayed hypersensitivity reaction (the reaction to subcutaneously injected antigen) in some animals (15) and to decrease the time required for graft rejection. In vitro, the proliferative response of lymphocytes to mitogens and antigens was decreased in most Se-deficient animals (16). The ability to produce lymphokines, the soluble mediators of the immune response, was somewhat modified by the Se status but dependend on the tested lymphokines. Leucocyte migration inhibitory factor was decreased while interleukin-2 production was normal in lymphocytes from Se-deficient goat (19). Cytotoxic and T natural killer (NK) lymphocytes from Se-deficient animals were markedly depressed while antibody-mediated-cytotoxicity (ADCC) gave conflicting results (15). The protection against experimentally transplanted tumors, a lymphocyte-mediated reaction, was increased when Se was dispensed before or at the time of transplantation, suggesting a role for Se in the early phase of tumor growth (20).

Table 3 : Selenium and Cell-mediated Immunity

methods	selenium status	effects	references
delayed-type hypersensitivity	deficient	↗	Martin, 1977
		↙	Koller, 1986
graft rejection	deficient	↗	Aleksondrowiz, 1977
mitogen stimulation	deficient	↙	Sheffy, 1978, 9 Parnham, 1983 Eskew, 1985
		N	Aziz, 1986 Arvilomni, 1983 *
	supplemented	↗	Larsen, 1981 Tuner , 1985
lymphokine production	deficient	↙	Aziz, 1986
		N	Aziz, 1986
	supplemented	↗	Watson, 1986 *
		N	Koller, 1986 Arvilomni, 1983 *
NK , ADCC cytotoxicity	deficient	↙	Mecker, 1985
		↗	Eskew, 1985
	supplemented	↗	Talcott, 1984 Koller, 1986 Dimitrov, 1986
tumor growth	supplemented	↙	Weisberg, 1956 Poirier, 1979, 84 Greeder, 1980 Milner, 1981 Ip, 1981
		N	Schrauzer, 1979 Medina, 1981

* in humans

In humans, as for humoral immunity, few data are available despite the evidence from epidemiological studies that Se plays a role in cancer prevention. Se supplementation had no effect on in vitro proliferation response to mitogens and lymphokines production (11). The addition of Se compounds to cultured human lymphocytes from healthy subjects modulated the response : high doses of inorganic Se had a suppressor effect while low doses had a variable effect which depends on the compound (21). Another study showed an enhancement of NK activity after Se supplementation (22). In vivo studies are preliminar. A group of patients with SIDA had low plasma Se.

In conclusion, these studies showed, at least in the animals, that Se deficiency depresses the humoral and cell-mediated response whereas Se supplementation increases or restores it. The moment, the dose, the form administred are an importance : low doses have a stimulatory effect while high doses an inhibitory effect. Inorganic Se seems more effective than organic but few data are available for the latter. Other studies are needed in human diseases with immune disorders to evaluate the influence of Se on the immune response.

Acknowledgement

The author thank the Fonds National de la Recherche Scientifique de Belgique (contract n° 3.4510.87) and Labcatal S.A., France, for their supports.

References

1. Chandra, R.K., 1985. J. Am. Coll. Nutr. 4, 5-16.

2. Babior, B.M., 1978. New. Engl. J. Med. 298, 659-668.

3. Lands, W.E.M., 1984. Prostagland. Leukotr. Med. 13, 35-46

4. Mc Carty, M., 1984. Med. Hyp. 13, 45-50

5. Roberts, M.E., 1963. Toxicol. and Appl. Pharm. 5, 485-499.

6. Parnham, M.J., Winkelmann, J., Leyck, S. 1983. Int. Immunopharmac. 5, 455-461.

7. Nève, J., Fontaine, J., Peretz, A., Famaey, J.P. 1988. Agents and Actions, in press.

8. Peretz, A., Nève, J., Famaey, J.P., Molle, L., 1987. J. Pharm. Belg., 42, 395-403.

9. Kiremidjian-Schumacher, L., Stozky, S, 1987 . Environ Res., 42, 277-303.

10. Urban, T., Jarstrand, C., 1986. Immunopharmacol. 12, 167-172.

11. Arvillomi, H., Poikonen, K., Jokinen, I., Muukkonen, O., Rasonen, L., Forenan, J., Huttunen, J.K., 1983. Infect. Immunol. 41, 185-189.

12. Nicolson, G.L., 1974. Int. Rev. Cytol. 39, 89-190.

13. Noelle, J.J., Laurence, D.A., 1981. Biochem. J. 198, 571-579.

14. Spallholz , J.E., 1981. Adv. Exp. Med. Biol., 135, 43-62.

15. Koller, L.D., Exon., J.H., Talkatt, P., Osborne, C.A., Henningren, G.M., 1986. Clin. Exp. Immunol., 63, 570-576.

16. Sheffy, B.E., Schultz, R.D., 1979. Fed. Proc. 38, 2139-2143.

17. Mulhern, S.A., Taylor, G.L., Magruder, L.E., Vessey, A.R. 1985. Nutr. Res., 5, 201-210.

18. Koller, L.D., Isaacson-Kervliet, N. Exon, J.H., Brauner, J.A., Patton, N.M., 1979. Arch. Environ Health. 34, 248-252.

19. Aziz, G.Z., Kesius, P.H., Frandsen, J.C., 1984. Am. J. Vet. Res. 45, 1715-1718.

20. Milner, J.A., Greeder, G.A., Poirier, K.A.V., 1981. In : Selenium in Biology and Medicine (J.E. Spallholz J.L., Martin, H.E., Ganther, eds AVI pub Co, Wesport, pp. 146-159.

21. Watson, R.R., Moriguchi, S., Mc. Rae, B., Tobin, L., Mayberry, J.C., Lucas, D., 1986. J. Leuk. Biol. 39, 447-456.

22. Dimitrov, N.V., Charamella, L.J., Meyer, CL., Stowe, H.D., Ku, P.K., Ulrey, D.E., 1986. J. Nutr. Growth and Cancer, 3, 193-198.

SELENIUM AND OTHER TRACE ELEMENTS IN PATIENTS WITH RHEUMATOID ARTHRITIS.

J. Arnaud, V. Imbault-Huart, A. Favier
Laboratoire de biochimie C, CHRUG, BP 217X, 38043 Grenoble, France

A. Zagala, X. Phelip
Rhumatologie et Hydrologie, CHRUG, BP 217X, 38043 Grenoble, France

Introduction

Trace elements play an important part as the active site of differents enzymes involved in oxidation-reduction processes (superoxide dismutase, glutathione peroxidase, catalase). Oxygen free radicals have been implicated in inflammatory process. In patients with Rheumatoid Arthritis (RA), serum selenium (1,2,3,4), serum zinc (5), serum iron (6) levels are abnormally low and serum copper level (7,8) is high. These abnormal trace elements levels may be related to inflammatory process and to the increase of oxygen free radicals.

In the present work, we study serum trace elements levels in RA patients and relationships between selenium and inflammatory proteins.

Patients and Methods:

Blood was collected from two groups of patients:

- Group 1: 21 patients (15 females and 6 males) suffering from RA as defined by the criteria of the American Rheumatism Association (ARA). 16 of these patients had an active RA. The other parameters was listed in table 1.

Selenium in Medicine and Biology

Table 1 : Criteria of RA patients

	range	mean
age	38 - 83 years	60 years
duration of the disease	3 months - 40 years	10 years
duration of morning stiffness	0 - 4 hours	1.4 hours
number of joints with active arthritis	3 - 9	6
articular index	8 - 30	19

All patients received antirheumatic therapy (corticosteroids for 6 patients, D penicillamin for 4 patients, gold for 7 patients and antimalarial drugs for 2 patients.

- Group 2: 21 control patients match for age and sex. 10 patients suffering from sciatica, 5 from back pain, 2 from osteoporosis, one from cervical pain, one from acute lung oedema, one from mechanical arthralgia and one from osteoarthritis.

Serum trace elements were measured by direct electrothermal atomic absorption spectrometry. Immunoglobulin M rheumatoid factor (IgMRF), C reactive protein (CRP), ceruloplasmin, fibrinogen, protein electrophoresis, red blood cells count (RBC), white blood cells count (WBC), haemoglobin (Hb) and erythrocyte sedimentation rate (ESR) were also determined according to standard laboratory methods.

Comparisons between groups were performed by match student t test and correlations were calculated using correlation coefficient.

Results:

The data obtained in the two groups are shown in table 2. Patients in the RA group have a mean selenium, zinc and iron levels significantly lower than that of the control group. Serum copper level is significantly higher in group 1. Increase of manganese level in the RA group is not significant. There is no correlation between trace elements.

Table 2 : Results expressed as mean ± 1 standard deviation

	RA patients	Control patients	
Selenium µmol/l	0.60 ± 0.27	0.91 ± 0.28	*
Zinc µmol/l	10.8 ± 1.8	15.2 ± 2.5	**
Manganese nmol/l	32.1 ± 14.4	29.8 ± 13.3	
Iron µmol/l	9.0 ± 3.5	19.0 ± 7.0	**
Copper µmol/l	30.4 ± 6.8	24.8 ± 6.0	*
Albumin %	45 ± 7	51 ± 5	
Alpha 1 globulin %	5.6 ± 1.5	3.8 ± 1.1	
Alpha 2 globulin %	13.8 ± 2.9	11.8 ± 1.9	
Beta globulin %	12.9 ± 2.3	15.5 ± 1.3	
Gamma globulin %	21.4 ± 5.1	17.7 ± 3.9	
RBC g/l	3.7 ± 0.5	4.5 ± 0.5	
WBC g/l	8.5 ± 1.8	7.7 ± 2.5	
Hb g/l	105 ± 19	136 ± 19	
ESR mm 1st h	67 ± 43	18 ± 17	
Fibrinogen g/l	3.7 ± 2.2	3.4 ± 0.7	
CRP mg/l	72 ± 47	< 10	
Ceruloplasmin g/l	0.67 ± 0.21	0.42 ± 0.13	
IgM RF	670 ± 638		

* $p < 0.01$ ** $p < 0.001$

Serum selenium level in RA patients shows no correlation with anemia markers (RBC, Hb), inflammatory proteins (CRP, fibrinogen, ceruloplasmin and alpha 2 globulin) and ESR. There is a correlation between serum selenium level and IgMRF ($r=0.306$, $P < 0.01$) and between selenium and gamma globulin levels ($r=0.456$, $P < 0.01$). Serum selenium in RA patients shows a correlation with disease duration ($r=0.409$, $P < 0,05$), articular index ($r=-0.561$, $P < 0.05$) but no correlation is found with age, morning stiffness, number of tender joints.

Discussion

Trace elements levels found in this work are in agreement with other studies (1,2,3,4,5,6,7,8,9). As Mottonen (2) and Hannonen (1), we find no correlation between selenium level and inflammatory or anemia biological tests. The correlation between selenium level and disease duration and between selenium and gamma globulin levels suggest that selenium level is lower in the beginning of the illness. This fact is in agreement with the serum selenium normalization after RA classical treatment observed by Mottonen (2). The correlation between selenium level and articular index suggests that selenium is lower in patients with active RA. Tarp (3,4) finds the same results.

In this study, we did not determine glutathione peroxidase, superoxide dismutase, catalase and lipid peroxidation products. This lack of data doesn't allow us to discuss about relationships between trace elements and oxygen free radicals.

References:

1. Hannonen, P., T. Mottonen, M. Oka. 1985. Scand. J. Rheumatol. 14, 440.

2. Mottonen, T., P. Hannonen, O. Seppala, G. Alfthan, M. Oka. 1984, Clin. Rheumatol. 3, 195.

3. Tarp, U., K. Overvad, J.C. Hansen, E.B. Thorling. 1985. Scand. J. Rheumatol. 14, 97.

4. Tarp, U., K. Overvad, E.B. Thorling, H. Grandal, J.C. Hansen. 1986. Acta Pharmacol. Toxicol. 29, suppl. 7, 382.

5. Llorca, G., E. Lejeune, C. Daudens, F. Bienvenu. 1981. Path. Biol. 29, 73.

6. Blake, D.R., N.D. Hall, P.A. Bacon, P.A. Dieppe, B. Halliwell, J.C.M. Gutteridge. 1981. Lancet, 2, 1142.

7. Conforti, A., L. Franco, R. Milanimo, G. Tarolli, G.P. Velo. 1982. Adv. Inflammation Res. 3, 237.

8. Youssef, A.A.R., B. Wood, D.N. Baron. 1983. J. Clin. Pathol. 36, 14.

9. Aaseth, J., E. Munthe, O. Forre, E. Steinnes. 1978. Scand. J. Rheumatol. 7, 237.

SELENIUM AND CANCER

G.N.SCHRAUZER
Department of Chemistry and Biochemistry, University of California, San Diego, La Jolla, California, 92093.

INTRODUCTION

Selenium in appropriate chemical forms and concentrations exhibits anticarcinogenic properties in numerous animal tumor model systems, is antimutagenic, acts as a modulator of cell division, possesses antiproliferating and cytotoxic properties, stimulates immune functions, affects lymphocyte proliferation, activates macrophages, increases cellular respiration, reduces the pathogenicity or tumorigenicity of viral agents, reacts with alkylating agents and toxic heavy metals and has radioprotective properties. Selenium accordingly is receiving increasing attention as a nutritional cancer protecting factor and therapeutic agent (1-3). The present account focuses on some recent developments, with emphasis on interactions of selenium with normal and malignant cells, and the mechanism of its anticarcinogenic actions.

INTERACTIONS OF SELENIUM WITH NORMAL AND MALIGNANT CELLS

Differential Responses

Although selenium at high concentrations is a universal cellular poison, malignant cells are as a rule more sensitive to selenium, giving rise to differential effects of sufficient magnitude to render selenium of interest as a candidate for possible chemotherapeutic applications.
Studies with mammary cells indicated, for example, that Se at levels of $5x10^{-8}$ M inhibits the growth of mammary tumor cells and of several preneoplastic cell lines, while stimulating the growth of normal mammary cells (4).

Selenium in Medicine and Biology

In more recent work, selenite in vitro at 1 ppm similarly was found to inhibit the growth and mitotic activity of human pulmonary adenocarcinoma cells and to arrest the progression of the cancer cell cycle without affecting human embryonic diploid lung cells (5) . The same group of workers also showed that hepatoma cells from different human donors were sensitive to 1 ppm selenite, while human embryonic liver cells remained unaffected (5). Selenite in vitro also caused diminished growth and mitotic activity of human hepatoma cells from several donors without affecting embyronic liver cells; the hepatoma cells in addition developed vacuolated cytoplasms, distorted nuclei, condensed chromatin and even pyknosis (5).

In related studies with hepatoma ascites cells grown in male Kunming mice, selenite was furthermore demonstrated to affect the concentrations of cAMP, cGMP, and of the cAMP-dependent protein kinase isozymes CAK I and II, which act as regulators of cell division and differentiation (6). In the hepatoma cells, cAMP levels were lower and those of cGMP higher than in normal hepatocytes.

Selenite in addition reduced elevated Na^{+},K^{+}-ATPase activities of hepatoma cells but had no effect on the corresponding activities of normal hepatocytes. Cell membrane fluidity, a measure of membrane structure and lability, of hepatoma cells, was higher than that of normal hepatocytes. Whereas selenite had no effect on membrane fluidity of normal liver cells, it increased the membrane fluidity of the hepatoma cells (6). Two human hepatoma cell lines and two human pulmonary adenocarcinoma cell lines were similarly found to respond to selenite, developing increased adhesiveness in vitro and decreased tumorigenicity in vivo after heterotransplantation (7).

Administration of selenite at levels of 1 mg/kg b.w. also increased cAMP and decreased cGMP in hepatoma-bearing mice; in addition, the CAK I/CAKII ratio was reversed to normal (7). In normal mice, the identical treatment with selenite resulted in a reduction of cAMP and an increase of cGMP each by 20%. These experiments demonstrate that selenium in certain instances apparently is able to reverse the progression of malignant cells.

Effects on Mitotic Cycle

Cytokinetic analyses with human pulmonary adenocarcinoma cells revealed that selenite reduced the number of cells entering the S phase, indicating that Se partially inhibited DNA biosynthesis and the mitotic index (5). In relevant studies with rat H-4 minimal deviation hepatomas and of mouse 3T3 fibroblasts, selenite also produced dose-dependent, reversible growth inhibitory effects(8). Cytokinetic analyses in this system revealed that Se increased the durations of the G_1 and S phases, consistent with the inhibition of protein biosynthesis and retardation of DNA replication. Selenium also prolonged the duration of the premitotic resting phase, G_2. By the moderating DNA replication, selenium could reduce the probability of error formation; by prolonging the G_2 phase, Se could in addition provide conditions favorable for DNA repair. While Se did not affect the M phase in this case, retarding effects on S, G_2 and M were observed with mouse mammary cells (9).

Antiproliferative Properties of Se in Liver Regeneration Processes.

In efforts to gain deeper insights into the mechanism of antiprolifera tive action of selenium, two independent groups of workers investigated the effects of selenium on the proliferation rates of hepatocytes in experimental animals. In Se-presupplemented weanling male rats subjected to partial hepatectomy, LeBeouf, Laishes and Hoekstra (10) observed slower liver regeneration rates than in the controls maintained on a low-Se casein diet (10). While this experiment clearly demonstrates the antiproliferative properties of selenium, the supplementary level of Se chosen (6 ppm in the diet) was outside the nutritional range. Hence, the possibility could not be excluded that the diminished proliferation rates were due to a toxic effect of Se. That selenium exerts its antiproliferative effects at nutritional levels, i.e., 0.3 ppm, was demonstrated by Vogl et al. (11) in studies with male albino mice subjected to partial hepatectomy, and in normal mice given injections of partially purified hepatopoietin, a glycoprotein hormone which stimulates liver cell proliferation (12).Significantly higher liver regeneration rates were observed in the Se-deficient mice than in the Se-adequate controls. The paradoxically enhanced proliferative activity of the hepatocytes was plausibly interpreted as a pathological response to stress under conditions of compensated Se-deficiency (11).

Selenium: A Universal Antiproliferative Agent of Metabolically Activated Cells?

The observations with hepatocytes discussed above suggest that selenium could be a universal nutritional antiproliferating agent. Selenium should thus also be effective in modulating the increased proliferation rates of virus-infected or carcinogen-exposed cells. Several experiments may be cited to support this hypothesis. In mice, Se-pretreatment prior to inoculation with Cocksackie B virus significantly reduced myocardial damage (13).

Infection of Se-deficient mice with Cocksackie B virus, in contrast, produced severe myocardial damage with multifocal necrotic lesions typically observed in victims of Keshan disease (13).

In studies with mice infected with bovine leukemia virus Rauscher, selenium pretreatment reduced splenomegaly and improved survival; in vitro, additions of selenite to the growth medium furthermore inhibited reverse transcriptase activity of this RNA virus(14). Since murine mammary tumorvirus (MMTV) is also a RNA virus, a similar mechanism for the anticarcinogenic effects of Se in this system may be postulated, although other mechanisms are possible by which Se prevents the integration of the virus genome with the cellular DNA.

Infection by hepatitis B virus is believed to be one of the main etiological agents responsible for the high incidence of primary liver cancer (PLC) in the population of Qidong county, a low-Se region North of Shanghai, China. The discovery of a fourfold difference in PLC mortality in different sections of Qidong county led to studies which produced statistically significant inverse associations between PLC incidence and the blood Se levels of healthy subjects of Qidong (15). Accordingly, selenium supplementation of the population at risk was suggested as a promising means of PLC prevention. Several pilot human selenium intervention trials are being conducted in Qidong to test this hypothesis. If selenium supplementation lowers the hepatitis incidence as expected, a decline of the PLC mortality rates should eventually become noticeable. Initial results indeed suggest that the incidence of hepatitis B infections in the supplemented groups is declining. Hepatitis B virus infected ducks nesting in the area contribute to the spread of the disease and were for this reason included in the studies. In the Se-supplemented animals the incidence of hepatitis was lower and in their livers fewer preneoplastic alterations were observed than in the unsupplemented controls(15,16).

Since exposure to aflatoxin B_1, in addition to hepatitis B virus infection, is considered to be another major PLC risk factor, lymphocytes were isolated from the blood of Se-supplemented and unsupplemented subjects and exposed to aflatoxin B_1 in vitro. The lymphocytes of the Se-supplemented individuals showed a significantly higher resistance against aflatoxin B_1-induced DNA damage (as measured by the intensity of unscheduled cellular DNA biosynthesis) than those of the unsupplemented controls (17).

In these studies a lower percentage of the subjects receiving supplemental Se developed elevated serum levels α-fetoprotein (AFP) than the placebo controls (17). As serum AFP is elevated in patients with PLC, the observed lowering of the AFP levels may be taken as a clear indication of the protective role of selenium in human liver cancer development.

Immunopotentiating Effects of Selenium

Selenium is indispensable for the maintenance both of humoral and cellular immunity (18-20). Its cancer-protecting properties accordingly may also be linked with its effects on the immune system. However, this protective mechanism will obviously play a role only after the malignant transformation has occurred and the organism is attempting to rid itself of the malignant cells by immunological means. As selenium pretreatments of experimental animals caused significant increases of immunoglobulin titers against several bacterial and cellular antigens (18-20), the joint employment of selenium with cellular antigens is a promising modification of the immunotherapy of malignant diseases. The first model experiments along this line was conducted by Howells et al. (21). Inoculating tumor-bearing mice with Corynebacterium parvum for immunostimulation, a significant improvement of survival was only observed in animals that were also receiving 1 ppm Se in the drinking water. Since the C. parvum plus Se effect was abolished by the potent prostaglandin inhibitor indomethacin, the authors concluded that selenium either modulates the immune response and/or tumor-prostaglandin interactions. They also reported preliminary data indicating that selenium plays a role in lymphocyte activation: Spleen cells from Se-pretreated animals exhibited 2-7 times enhanced proliferative responses in polyclonal mitogen induced blastogenesis assays (21).

In more recent studies with human lymphocyte cultures (HCL), selenium was found to exert dose-dependent regulatory and inhibitory effects (22). Reversible inhibition of the growth of mixed HLCs occurred at Se concentrations of 2-3 $x10^{-6}$ M and up to $5x10^{-6}$ M. Effects of Se toxicity became apparent only at significantly higher Se levels, i.e. $1x10^{-5}$ M, as evidenced by a slight irreversible loss of proliferative activity. Additions of Se to mixed human lymphocyte cultures produced selective growth inhibitions of certain lymphocyte subsets. Specifically, the in vitro effects of Se observed appeared to be equivalent to a reduction of the number of T suppressor cells and/or an elevation of the number of T helper cells in vivo, which in turn would be expected to modulate several humoral and cellular immune functions, notably the production of tumor-specific antigens, the activity of NK cells through the increased production of lymphokines such as interferon or of interleukin-2 (22). This hypothesis is supported by the work of Koller et al. (23), who observed stimulation of NK cell activity, decreased prostaglandin E2 synthesis, decreased delayed hypersensitivity of macrophages from Se-supplemented rats but no enhancement of interleukin-2 production.

Boylan et al. (24), on the other hand, recently found that Se-supplemented Balb/c mice tended to have higher numbers of T cytotoxic suppressor cells than Se-deficient animals but noted no other differences in cell populations. The chemical form of selenium employed in such experiments may be critically important. Johannson and Lindh (25,26) found that selenite accumulated preferentially in neutrophils, followed by erythrocytes and thrombocytes, while L-selenomethionine (or the Se-derived therefrom) accumulated in the opposite order of cells. The dosage level of Se is equally as important. Although the immunopotentiating effects of Se occur at level above the minimal dietary requirements, higher levels of Se are clearly immunosuppressive. In the experiments with Swiss Wistar mice challenged with sheep erythrocytes, Spallholz et al. (27) observed he maximum antibody production at dietary selenite-Se levels of 1.25 ppm. The antibody production declined from Se levels of 1.75 ppm onward, which is close to the lower limit of beginning chronic Se toxicity.

The anticarcinogenic effects of selenium also reach a maximum at intermediate supplementary levels of Se. In life-term experiments with MMTV-carrying mice, the incidence of spontaneous mammary tumors reached a minimum at supplementary levels of 1-2 ppm Se (28). The fact that the incidence of mammary tumors in these experiments was higher in the mice exposed to chronically toxic Se levels of 5 ppm Se (as selenite in the drinking water) shows the difference between the function of selenium as a nutritional cancer preventive agent and its applications as a cytotoxic drug in cancer therapy.

CHEMICAL MECHANISMS OF ANTIPROLIFERATIVE AND ANTICARCINOGENIC ACTIVITY

Although glutathione peroxidase (GSH-Px) is not considered to be responsible for the antiproliferative effects of Se, the protection of cells and tissues against oxidative free radical damage accorded by this enzyme may increase their resistance against environmental agents or drugs which stimulate peroxide metabolism. This may increase the resistance against certain carcinogenic stress factors but is not a general mechanism of anticarcinogenic action of selenium. In the majority of systems studied selenium exerts its antiproliferative and anticarcinogenic properties at levels above those required for maximum GSH-Px activity. Thus, in experiments with rats treated with DMBA and fed different levels and types of fat, no correlation between the anticarcinogenic effect of selenium and its ability to suppress lipid peroxidation was observed (29). Selenium instead appears to act as a catalyst of cellular oxidation-reduction reactions, especially reactions involving biogenic thiols, notably GSH, as the electron donors and ultimately oxygen as the electron acceptor. GSH is present in all mammalian cells and participates in a variety of reactions, including the detoxification of drugs and xenobiotics. Although the oxidation of GSH to GSSG occurs during normal cellular respiration, the cellular oxidation state as measured by the GSSG:GSH ratios depends on cell type, is subject to adaptive responses, and can be influenced by catalytic amounts of reactive forms of selenium such as sodium selenite or selenocystine. For example, additions of selenite to the culture media increased the GSSG:GSH ratios in vitro in experiments with minimum deviation hepatoma cells and caused an increase of the levels of oxidized pyridine nucleotides (8). Similar oxidation-reduction processes are also catalyzed by selenium in vivo, as evidenced by the increased GSSG:GSH ratios in the livers of Se-supplemented rats as compared to unsupplemented controls (8). Selenium thus may be regarded as a biocatalyst of cellular respiration capable of shifting cells into more oxidized metabolic states. As many workers in the past have linked cancer development to a "disturbance of cellular respiration" and it is known that the proliferation rates of malignant cells are in general inversely proportional to their respiratory activity (30), a physiological means of increasing the oxygen consumption of normal and transformed cells is at hand which provides a logical basis for the explanation of the anticarcinogenic and cytotoxic activity of selenium.

In the catalysis of the oxidation of thiols to disulfides, selenite is reduced to selenotrisulfides and selenodisulfides such as RSSeSR, $RS\text{-}Se_2H$ and RSSeH (31,32). Based on kinetic studies, the ions $RSSe^-$ and $RSSe_2^-$ were identified as the catalytically active species (33). In the corresponding reaction of SeO_3^{--} with GSH the selenotrisulfide GSSeSG is sufficiently stable to be isolated. The compound was found to be highly cytotoxic against a variety of malignant cells, both in vitro and in vivo (34,35). However, the recognition of selenium as a catalytic cytotoxic agent is important in this context (36,37). It suggests that the efficacy of selenium in cancer therapy could be enhanced through joint administration with GSH (37). The catalytic properties of selenium could also be utilized against drug resistant malignant cells to induce a depletion of GSH, preferably in conjunction with a suitable inhibitor of GSH synthesis. This may lead to a resensitization of cells whose drug resistance is due to elevated levels of cellular GSH (37).

The Se-promoted activation of cellular respiration will not only affect the GSSG/GSH ratio. It may be expected that other biologically active peptides with SH or -S-S- bonds are also affected. Such peptides could be growth factors similar to "Elongation Factor 2" (EF_2), which regulates protein biosynthesis in polyribosomes of rat liver (38). EF_2 contains SH groups in the active form and -S-S- residues in the oxidized inactive form Antiproliferative effects could thus be induced by promoting the oxidation of EF_2 in vivo with selenium as the catalyst. It may be significant that human tumornecrosis factor (TNF) is a protein consisting of 157 amino acids which contains one cys-S-S-cys residue (39). It could be hypothesized, therefore, that TNF is inactivated by reduction of the cys-S-S-cys bond and that an enhancement of its efficacy could be achieved by Se-mediated GSH oxidation in the tumor environment.

Selenium in the -2 oxidation state has a high affinity for a variety of heavy metals. Exposure to such metals is known to diminish the uptake and physiological activity of selenium. It is therefore not surprizing that the anticarcinogenic properties of Se are abolished by a variety of Se-antagonistic elements (40,41).

Se^{-2} is also a powerful nucleophile and is readily methylated under physiological conditions. While the methylation of Se is primarily a mechanism of detoxification, the anticancer properties of Se could be weakened by methylation under special nutritional or metabolic conditions (methionine- or choline-rich diets?). Reactions with other alkylating agents are alsoopossible and potentially detrimental. This should be particularly true for reactions involving Se^{-2} and cytostatic alkylating drugs. These as yet little studied reactions could further compromise the selenium status of cancer patients. .

CONCLUDING REMARKS

Research during the past ten years has reaffirmed the value of selenium as an anticancer agent in numerous animal tumor model systems. Based on these studies and epidemiological evidence, several large-scale human intervention trials utilizing supplemental selenium have been initiated in China and in the U.S.A. which should provide the ultimate proof of the efficacy of selenium in human cancer prevention. Large-scale clinical trials have yet to be performed to test the value of selenium in human cancer therapy. While selenium should not be expected to replace conventional cancer therapy, its use as an adjuvant may be justified and should be encouraged.

References

1. Schrauzer, G.N.(Ed.). 1988: Selenium: Present Status and Perspectives in Biology and Medicine. Biol.Trace El.Res. 15 and references therein.

2. Combs, Jr., G.F., O.A.Levander, J.E.Spallholz, J.E.Oldfield (Eds): Selenium in Biol. and Med. Parts A and B. Avi-Van Nostrand Reinhold Co New York, and references therein.

3. Ip, C. 1986. J.Amer.Coll.Toxicol 5: 7.

4. Medina, D. and C.Osborn. 1981. Cancer Lett. 13, 333.

5. Pung, A., Z.Mei, S.-Y. Yu. 1987. Biol. Trace El. Res. 14, 1, 29.

6. Yu , S.-Y., A.Pung, L.M.Wang, S.L.Huang, H.C.Chen, X.P.Lu, Q.Y.Liu. 1988. Biol. Trace El. Res. 15. 243.

7. Pung A., Z.Mei, S.-Y.Yu. 1988. Biol. Trace El. Res. 15, 19.

8. LeBoeuf, R.A., B.A.Laishes, W.G.Hoekstra. 1985. Cancer Res. 45, 5496.

9. Medina, D., C.J.Oborn. 1983. Cancer Res. 43(Suppl.), 2460.

10. LeBoeuf,R.A., B.A.Laishes, W.G.Hoekstra. 1985. Cancer Res. 45, 5489.

11. Vogl,S. M.Goldberg, G.Ruhenstroth-Bauer, R.Otter, A.Wendel. 1987. J.Hepatol. 4, 212.

12. Ruhenstroth-Bauer, G., M.Goldberg, S.Vogl. 1984. Naturwiss. 71, 404.

13. Bai, J. et al. 1982. YingYang Xuebao 4, 235.

14. Balansky, R.M., R.M.Argirova. 1981. Experientia 37, 1194.

15. Yu,S.-Y., Y.-J.Chu, X.-L.Gong, C.Hou, W.-G.Li, H.-M.Gong, J.-R.Xie. 1985. Biol. Trace El. Res. 7, 21.

16. Yu, S.-Y., personal communication, March 1988.

17. Yu,S.-Y., Y.J.Chu, W.G.Li, 1988. Biol. Trace El. Res. 15, 231.

18. Spallholz,J.E., Selenium in Biology and Medicine. (Spallholz, J.E., J.L.Martin, H.E.Ganther, Eds.), Avi Publ. Co., Westport CT. 1981., 103.

19. Kirmidjian-Schumacher, L., G.Stotzky. 1987. Environ. Res. 42. 277.

20. Shimura,J., C.C.Chang, J.H.Chen, G.F.Combs,Jr., I.C.Campbell, V.Utermohlen. 1982. Fed.Proc. 41, 1707.

21. Howells, J.M., R.G.Crounse, J.M.Thomas, T.K.Whitley, M.Finn, J.T.Bray, A.M.Smith, C.Jones. 1982. Abstr. Conf. Soc. Envir. Geochem Health., 36.

22. Petrie, H.T., L.W.Klassen, M.A.Tempero, D.Kay. 1986. Biol. Trace El. Res. 11, 129.

23. Koller, L.D., J.H.Exon, P.A.Talcott, C.A.Osborne, G.M.Henningsen. 1986. Clin.Exp.Immunol. 63, 570.

24. Boylan, M., H.Larsen, L.Lemon, J.E.Spallholz. 1988. Fed.Proc. 2, A1621.

25. Johansson, E., U.Lindh. 1983. Health Effects and Interactions of Essential and Toxic Elements. Studentlitteratur, Lund, 1983. Abstr. 109.

26. Johansson, E., U.Lindh., 1987. see op.loc.cit.ref. 2, p.236.

27. Spallholz, J.H., J.L.Martin, M.L.Gerlach, R.H.Heinzerling. 1973. Proc. Soc. Exp. Biol. Med. 143, 685.

28. Schrauzer, G.N., D.A.White, C.J.Schneider. 1976. Bioinorg. Chem. 6, 265.

29. Ip,C., Sinha,D. 1981. Carcinogenesis 2, 435.

30. Bauer,K.H. 1963. In: Das Krebsproblem. 2.Ed. Springer Verl.Berlin, p.164.

31. Ganther, H.E. 1971. Biochem. 10, 4089.

32. Hu,M.-L., A.L.Tappel. 1987. J.Inorg.Biochem. 30, 239.

33. Rhead, W.J., G.N.Schrauzer. 1974. Bioinorg. Chem. 3, 225.

34. Poirier,K.A., J.A.Milner. 1983. J.Nutr. 113, 2147.

35. Milner, J.A., M.Fico (1987) in op. loc cit. ref. 2, p.1034.

36. Schrauzer, G.N. 1983. In: Selen: Neuere Entwicklungen Biologie, Biochemie u. Medizin. Verl. f. Medizin Dr.Ewald Fischer, Heidelberg, p. 72.

37. Batist, G. 1988. Biol. Trace Element Res. 15, 233.

38. Vernie, L.N., J.G.Collard, A.P.M.Eker, .A. De Wildt, I.T.Wilders. 1979. Biochem.J. 180, 213.

39. Old,L.J. 1985. Science 230, 630.

40. Schrauzer, G.N. 1979. In: Advan. in Nutritional Res. (Draper, H.H., Ed.), p. 219.

41. Schrauzer, G.N. 1987. JUEH, Kitakyushu, Japan, 9, 208.

SELENIUM AND OTHER ANTI-OXIDANTS IN BREAST CANCER

M. Gerber

INSERM, Centre Paul Lamarque, 34094 Montpellier Cedex, France

S. Richardson

INSERM U.170, 16,av. P. Vaillant-Couturier, 94807 Villejuif Cedex, France

P. Chappuis

Laboratoire Central de Biochimie, Hôpital Lariboisière, Paris, France.

Introduction

A case-control study on host related risk factors in breast cancer (BC) has been conducted in Montpellier (Southern France) and Milan (Italy), based on questionnaires and blood measurements. The analysis of 250 cases and controls in Milan and 120 in Montpellier demonstrated a statistically significant difference with regards to plasma level of vitamin E, higher in cases than in controls ; the significance of the difference in vitamin E levels remained after adjustment on cholesterol levels. Moreover increasing levels of vitamin E were associated with increasing odds ratios for BC : 4.2 ; 95% CI : (1.9-9) in the highest quintile. There was no difference in vitamin E intake (1).

Vitamin E, associated to vitamin C, is the first line of defense of plasma membrane against free radicals. Selenium (Se), through the seleno-dependant glutathione-peroxidase, zinc(Zn) and copper (Cu), co-factors of the enzyme superoxi-dismutase (SOD), participates in this defense. The question of the role of anti-oxidants in cancer is very controversial. Some animal studies support a preventive role of anti-oxidants in cancer (2,3). In human prostective studies also, epidemiological indications exist for a protective role for Se or vitamin E (4), but other studies deny this finding (5,6). In view of our first results on vitamin E, an investigation of the status of other anti-oxidants was considered in a sub-group of our study.

Selenium in Medicine and Biology

Subjects and Methods

The subjects were French women, aged 25 to 65 years, recruited in Languedoc-Roussillon. The cases were hospitalized in Cancer Center Paul Lamarque, for the first time, with a diagnosed BC at stages T1, N0, N1 or N2, without metastasis. The controls were hospitalized in Neurology or Neuro-Surgery for the first symptoms of diseases other than cancer or cardio-vascular conditions.

Vitamin E was assessed by high pressure liquid chromatography (Hoffmann la Roche, Bâle, Switzerland). Serum Se, Zn and Cu were measured by atomic absorption spectrophotometry (Dr. Philippe Chappuis, Hôpital Lariboisière, Paris, France). Nail clipping Se was assessed by the same method (Paul van Noord, Preventicon, Utrecht, and M. de Bruin, Interuniversity Reactor Institute, Delft, The Netherlands).

Results

The levels of anti-oxidants are overall higher in cases than in controls.(Table 1)

Table 1 Anti-oxidant levels in various samples of cases and controls.

	CASES			CONTROLS			
	number	mean	SD	number	mean	SD	p[c]
Serum Se (µmole/l)	46	1.19	0.21	50	1.13	0.27	NS
NailSe (mg/g)	25	0.69	0.18	41	0.68	0.13	NS
Serum Zn (µmole/l)	47	13.90	1.58	49	13.55	1.52	<0.01
SerumCu (µmole/l)	47	17.47	3.61	49	15.83	2.54	=0.01
Plasma vit.C (mg/l)	48	11.98	18.86	52	8.98	13.54	NS
Leukoc. Vit. C ($\mu g/10^8$ leuk)	45	16.24	19.51	51	9.85	4.09	<0.05
Leukoc. Vit. E. ($\mu g/10^8$ leuk)	32	7.23	4.61	35	4.74	3.25	<0.01
Erythr Vit.E (µg/ml erythr. pellet)	29	1.07	0.63	33	0.83	0.63	<0.1

However, the difference is statistically significant only for serum Zn, serum Cu, plasma vitamin E, leukocyte vitamin E and C levels. Moreover, when case and control mean Cu levels were adjusted for age, total cholesterol and oral contraceptives, which are variables generally associated with serum Cu levels, the difference between case and control mean levels was no longer significant. On the contrary, when case and control mean Zn levels were adjusted for age, HDL cholesterol and oral contraceptives, the difference between case and control mean levels remains significant. There exists a good correlation among serum Zn, serum Se and plasma vitamin E, all 3 factors implicated in anti-oxidants functions. This is observed either in each sample, or in the addition of both samples, indicating that it is a general characteristic of human female populations. In the same way, plasma vitamin E is correlated with erythrocyte vitamin E, whereas serum Se is correlated with nail Se, as expected, only in controls. It cannot be said wether the absence of correlation in cancer patients is fortuitous or related to a metabolic alteration of Se associated with cancer, (table 2)

Table 2. Significant correlations among anti-oxidants

	CASES			CONTROLS			CASES + CONTROLS		
	number	r	p	number	r	p	number	r	p
Serum Zn/ serum Se	47	0.31	<0.05	48	0.33	<0.05	95	0.34	<0.001
Plasma vit. E / serum Zn	46	0.22	NS	46	0.13	NS	92	0.25	<0.05
Plasma vit. E / serum Se	46	0.27	<0.1	47	0.15	NS	93	0.23	<0.05
Nail Se / serum S	24	0.02	NS	40	0.33	<0.05	64	0.29	NS
Plasma vit. E / erythr. Vit. E	29	0.38	<0.05	31	0.30	NS	60	0.37	<0.01

Discussion

It has been observed in this subgroup of a case-control study on BC that the levels of humoral and cellular anti-oxidants are overall higher in cases than in controls. The difference is statistically significant for serum Zn, leukocyte vitamin E and C levels. This is in agreement with our previous result showing a significantly higher level of plasma vitamin E in cases than in controls(1). A recent study on fat soluble vitamin and cancer has shown significant higher levels of plasma vitamin E in BC (7). Higher levels of Se (8) have been found in BC patients compared to control samples. We did not look for anti-oxidants in tumor tissue but higher levels of Se, Zn and Cu have also been described in BC (9). Our results showing that a statistically significant difference of vitamin E and Zn levels is associated with increased risk of cancer support the hypothesis of an eventual lower peroxidation in tumor tissues protecting cancer cells against deleterious effects of free radicals, and enhancing tumor growth through this protection.

Acknowledgement

This work has been supported by the "Ligue Nationale contre le Cancer" ans PIREN-CNRS.

References

1. Gerber, M ., Cavallo, F., Marubini, E., Richardson, S., Barbieri, A., Capitelli, E., Costa, A., Crastes de Paulet, A., Crastes de Paulet, P., Decarli, A., Pastorino, U., Pujol, H., Int. J. Cancer (in press).

2. Ip, C., 1986. JNCI 77, 299.

3. Burke, J.P., Frenton, M.R., Proc. Soc. Exp. Biol. and Med. 1985, 179, 187.

4. Willet, W.C., Polk, B.F., Morris, J.S., Srampfer, M.J., Pressel, S., Rosner, B., Taylor, J.O., Schneider, K., Hames, C.G. 1983 Lancert, 2, 130.

5. Menkes, M.S., Comstock, G.W., Vuilleumier, J.P., Helsing, K.J., Rider, A.A., Brookmeyer, R. 1986 New Engl. J. Med. 315, 1250.

6. Virtamo, J., Valkeila, E., Alfthan, G., Punsar, S., Huttunen, J.K., Karvonen, M.J. 1987 Cancer, 60, 145.

7. Rougereau, A., Person, O., Rougereau, G., 1987 Internat J.Vit. Nutr. Res. 57, 367.

8. Meyer, F., Verreault, R. 1987. Amer, J. Epidemiol. 125, 917.

9. Rizk, S.L., Sky-Peck, H.H. 1984 Cancer Res., 44, 5390.

SELENIUM AND CANCER IN CHILDREN (PRELIMINARY DATA)

J.M.D. Malvy, D. Amédée-Manesmes

Unité de Recherche en Hématologie Pédiatrique INSERM U056 et Département de Pédiatrie, Université Paris-Sud, Centre Hospitalier de Bicêtre, 78, rue du Général Leclerc, 94275 Le Kremlin-Bicêtre, France

J. Arnaud, A. Favier

Laboratoire de Biochimie C, CHRU de Grenoble, 38043 Grenoble Cedex, France

O. Houot

Centre de Médecine Préventive, 2 avenue du Doyen Jacques Parisot, 54501 Vandoeuvre Lès Nancy Cedex, France

B. Montagnon

Unité de Recherche sur la Nutrition et l'Alimentation INSERM U01, Hôpital Bichat, 170, boulevard Ney, 75877 Paris Cedex 18, France

Introduction

The role of dietary factors is suggested in the etiology of cancer. Several human studies suggest that low intake and low serum levels of selenium may be associated with cancer (1, 2, 3). Selenium has not been studied in children population with cancer. Therefore, we measured selenium in serum collected in 1986 and 1987 in 71 children from 1 to 16 years old with recently diagnosed cancer, at diagnosis time and before treatment. These measurements were compared with those from 71 controls who were matched for age, sex, geographical origin and time of blood collection, and who remained free of cancer. The pilot study was conducted in France, in the cities and countrysides of Paris, Lille, Bordeaux, Tours and Nancy. Selenium was analysed by electrothermic atomic absorption method.

Selenium in Medicine and Biology

Results

There was no difference between sex and age for the measurements of selenium levels. Serum selenium was significantly lower in children with cancer than in sex- and age- matched healthy controls at diagnosis time. (serum selenium : 0.82 ± 0.33 μmol/l (mean ± S.D.) versus 1.05 ± 0.31 μmol/l, $p < 0.001$). Significant differences between cases and controls were also found, according to site of cancer. Selenium levels were significantly lower in patients with leukemia (N = 33; 0.67 ± 0.21 μmol / l versus 0.93 ± 0.19 μmol/l, $p < 0.001$); lymphoma (N = 10; 0.99 ± 0.38 μmol / l versus 1.39 ± 0.48 μmol/l, $p < 0.05$); and bone tumours (N = 6; 0.85 ± 0.14 μmol/l versus 1.00 ± 0.20 μmol/l, $p < 0.05$) than in matched healthy controls.

Discussion

The non-prospective nature of the study prevents any conclusion to know whether low selenium levels predispose to cancer or are the consequence of the active inflammation induced by the malignant disease. However, selenium status may be reliable predictor of malignant disease prognosis and supplementation in these patients is questionable.

References

1. Salonen, J.F., G. Alfthan, J.L. Huttunen, P. Puska. 1984. Am. J. Epidemiol., 120, 342-349.

2. Salonen, J.F., R. Salonen, R. Lappetelainen, Ph. Maenpaa, G. Alfthan, P. Puska. 1985. Br. Med. J. 290, 417-420.

3. Salonen, J.F. 1986. Ann. Clin. Res. 18, 18-21.

BLOOD SELENIUM CONCENTRATION AND GLUTATHIONE PEROXIDASE ACTIVITY IN MALE PATIENTS WITH ALCOHOL DEPENDENCE

B.A. Zachara[x], J. Rybakowski[xx], K. Borowska[xxx], E. Pilaczyńska[xx], R. Zamorski[xxx] and B. Stefaniak[x]

Departments of Biochemistry, Academy of Medicine[x] and Institute of Agriculture[xxx], Department of Psychiatry, Academy of Medicine[xx], Bydgoszcz, Poland

Introduction

The biological role of selenium /Se/ is related to glutathione peroxidase /GSH-Px/, the enzyme which protects cellular membranes against lipid peroxidation /1/. Chronic ethanol consumption is frequently associated with depletion of Se /2,3,4,5/. Both acute and chronic ethanol ingestion increases lipid peroxidation which may be associated with liver disease /6/. This suggests that the role of Se in alcoholic patients may be related to the activity of GSH-Px and to the prevention of lipid peroxidation. This study was undertaken to examine the Se concentration as well as to measure the GSH-Px activity in blood of male patients with alcohol dependence.

Materials and Methods

Male patients /n=26/ with alcohol dependence were studied after the admission to the Addiction Treatment Unit. The duration of alcohol abuse ranged 3-30 yrs /mean, 12.4 yrs/. Control group comprised 22 male healthy persons. Whole blood and plasma Se levels were determined by the fluorometric method of Watkinson /7/. The coefficients of variation for whole blood and plasma /n=8/ were less than 3%. Animal blood standard samples of the IAEA /A-13/ were used to monitor the accuracy of Se analysis /8/. The result was 0.25$\pm$ 0.015 ppm Se, compared to 0.24 ppm /IAEA assay value/. The GSH-Px activities of red cell haemolysates and plasma were measured by the coupled method of Paglia and Valentine /9/ with t-butylhydroperoxide as substrate. The coefficient of variation for haemolysate was 2.8% and for plasma 1.5%. A detailed description has been published earlier /10,11/. The significance was determined using Student's t-test at the 95% confidence interval.

Selenium in Medicine and Biology

Results

The whole blood and plasma Se levels of alcoholic patients were 63.8±7.9 and 49.6±8.4 ng/ml, respectively, and in healthy subjects, 72.0±5.9 and 57.0±7.9 ng/ml, respectively /Fig.1/. Se concentrations both in whole blood and plasma were significantly lower in alcoholic patients compared to control subjects /$p < 0.001$ for whole blood and $p < 0.05$ for plasma/. The mean GSH-Px activity in alcoholics was 14.8±3.1 U/g Hb in red cells and 174±39 U/l plasma. In the control group the enzyme activities were 16.8±3.5 U/g Hb and 195±48 U/l Plasma /Fig. 2/. These activities were not significantly different between groups studied. In eight alcoholic patients the Se levels

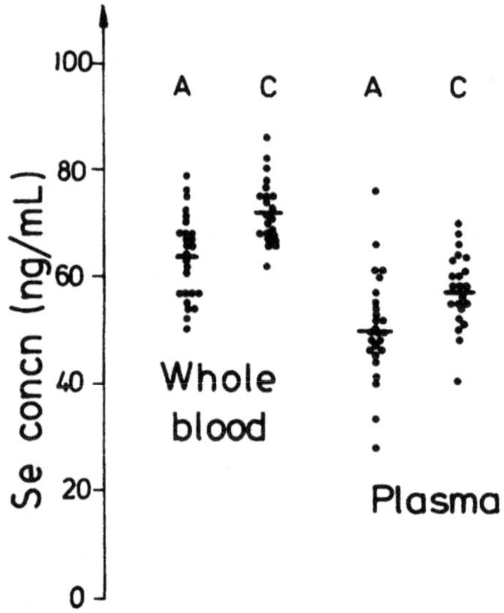

Fig. 1. Se concn in whole blood and plasma of alcoholic patients /A/ and control subjects /C/.

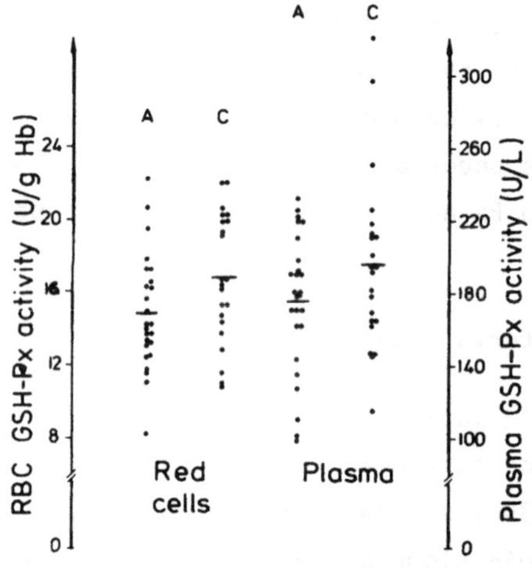

Fig. 2. Red blood cell and plasma GSH-Px activity in alcoholic patients /A/ and control group /C/.

and GSH-Px activities were re-measured at 7-8 weeks after admission to the unit. Both Se levels in whole blood and plasma as well as the GSH-Px activties in red cells and plasma remained unchanged after the treatment period.

Discussion

The present study has shown that chronic alcoholic patients have lower whole

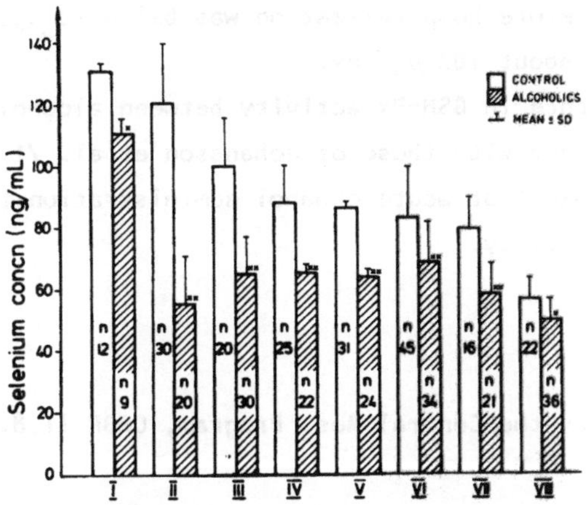

Fig. 3. Estimated plasma /serum/ Se concn in healthy subjects and alcoholic patients in some countries presented by other authors. Legend: I, Dutta et al. /12/, Baltimore, USA; II, Aaseth et al. /2/, Oslo, Norway; III, Dworkin et al. /3/, Valhalla, USA; IV, Korpela et al. /4/, Oulu, Finland; V, Thuong et al. /15/, Paris, France; VI, Välimäki et al. /16/, Helsinki, Finland; VII, Johansson et al. /17/, Lund, Sweden; VIII, Our results. Statistical significances: x, $p < 0.05$, xx, $p < 0.001$.

blood and plasma Se concentrations as compared to healthy subjects. We also found that our healthy persons had very low Se levels in blood. These values are markedly lower than in the USA /3,12,13/, Canada /14/, Norway /2/ and some other European countries /4,15,16,17; Fig. 3/. Since there is a positive correlation between the Se intake and blood Se concentration /18/ it seems most likely that low blood Se levels in the groups examined by us depend on poor dietary Se intake. The calculated daily dietary Se intakes by our healthy subjects and alcoholics were 24 and 15 µg Se, respectively. In all available studies on plasma /serum/ Se levels in alcoholic patients published in the last few years, the authors consistently found significantly low Se concentration /Fig. 3/. The reduced serum Se levels were found not only in alcoholics with liver cirrhosis /2,4,16,17/ but also in those without liver injury /3,4,12,15,16/. Although Dworkin et al. /3/ suggested several possibilities responsible for lowered Se levels in alcoholics, the true mechanism is still unknown. It is very likely that the reduced Se level in alcoholics reflects a persistent abnormality of Se metabolism. This supposition is in line with our findings of unchanged levels of this element in alcoholic patients following 7-8 weeks of hospitalization. The last mentioned observations are in accordance with Välimäki et al. /16/ data. Ours and Finnish /16/ findings are in contrast to Dutta et al. /12/ observations who have shown that after hospitalization of alcoholics their plasma Se level increased very rapidly. Moreover, they proved that the depletion of Se in alcoholics was related to diminished dietary Se intake. The

estimated daily dietary Se intake before hospitalization was below 20 μg, while the hospital diets contained about 102 μg/day.
Our observations on lack of difference in GSH-Px activity between alcoholics and control group are in agreement with those of Johansson et al. /17/. Other authors /19,20/ have also shown that acute ethanol administration to rats did not affect liver GSH-Px activity.

Acknowledgements

This work was partially supported by the Central Res. Program, CPBR 11.8.

References

1. Rotruck, J.T., A.L. Pope, H.E. Ganther, A.B. Swanson, D. Hafeman, W.G. Hoekstra. 1973. Science 179, 588-590.
2. Aaseth, J., A. Smith-Kielland, Y. Thomassen. 1986. Ann. Clin. Res. 18, 43-47.
3. Dworkin, B.M., W.S. Rosenthal, G.G. Gordon, R.H. Jankowski. 1984. Alcoholism: Clin. Exp. Res. 8, 535-538.
4. Korpela, H., J.T. Kumpulainen, E.A. Sotanieni. 1985. Nutr. Res. Suppl. I, 424-425.
5. Zarski, J.-P., J. Arnaud, L. Dumolard, A. Favier, M. Rachail. 1985. Gastroenterol. Clin. Biol. 9, 664-669.
6. MacDonald, C.M. 1973. FEBS Lett. 35, 227-230.
7. Watkinson, J.H. 1966. Anal. Chem. 38, 92-97.
8. Muramatsu, Y., R.M. Parr. 1985. Internat. Atomic Energy Agency /RL/128.
9. Paglia, D.E., W.N. Valentine. 1967. J. Lab. Clin. Med. 70, 158-169.
10. Zachara, B.A., W. Wąsowicz, J. Gromadzińska, M. Skłodowska, G. Krasomski. 1986. Biol. Trace Element Res. 10, 175-187.
11. Wąsowicz, W., B.A. Zachara. 1987. J. Clin. Chem. Clin. Biochem. 25, 409-412.
12. Dutta, S.K., P.A. Miller, L.B. Greenberg, O.A. Levander. 1983. Am. J. Clin. Nutr. 38, 713-718.
13. Levander, O.A., V.C. Morris. 1984. Am. J. Clin. Nutr. 39, 809-815.
14. Dickson, R.C., R.H. Tomlison. 1967. Clin. Chim. Acta 16, 311-321.
15. Thuong, T., P. Auzepy, A.M. Blondeau, C. Richard. 1986. Presse Med. 15, 693-696.
16. Välimäki, M.J., K.J. Harju, R.H. Ylikahri. 1983. Clin. Chim. Acta 130, 291-296.
17. Johansson, U., F. Johnsson, B. Joelsson, M. Berglund, B. Åkesson. 1986. Brit. J. Nutr. 55, 227-233.
18. Thomson, C.D., M.F. Robinson. 1980. Am. J. Clin. Nutr. 33, 303-323.
19. Kocak-Toker, N., M. Uysal, G. Aykac, A. Sivas, S. Yalcin, H. Öz. 1985. Pharmacol. Res. Commun. 17, 233-239.
20. Baker, A.L., A.P. Autor. 1983. Fed. Proc. 42, 1309.

SERUM SELENIUM IN ALCOHOLIC DISEASES

J. Arnaud, A. Favier
Laboratoire de Biochimie C, CHRUG, BP 217X, 38043 Grenoble, France

J.P. Zarski, M. Rachail
Clinique d'hépatogastroentérologie, CHRUG, BP 217X, 38043 Grenoble, France

H. Labadie, M. Beaugrand
Service d'hépatogastroentérologie, Hôpital Jean Verdier, Avenue du 14 juillet, 93140 Bondy, France

Introduction

Several studies demonstrated that alcoholic patients with liver disease have lower serum selenium level compared to control subjects (1,2,3,4). Selenium depletion in alcoholic subjects may be related to diminished dietary intake (1,2,3,4), alcohol consumption which impaired selenium intestinal absorption (2), increase of lipid peroxidation (1,2,3,4), increase of urinary and fecal loss of selenium (1) or decrease of alpha 2 globulin which have been proposed to be linked with serum selenium (1).
The purpose of this study is to determine serum selenium concentration in chronic alcoholic patients with or without cirrhosis and to study the relationships between serum selenium level and liver biological tests.

Patients and Methods:

Blood was collected in three groups of patients. All the alcoholic subjects (groups 1 and 2) had an alcohol consumption of more than 80 g/d for at least five years.

Selenium in Medicine and Biology

- Group 1: 32 alcoholic patients (27 males, 5 females). The mean age was 52 years (range: 33-69 years). Histological diagnosis was steatosis (21 patients) or alcoholic hepatitis (11 patients).
- Group 2: 37 alcoholic cirrhosis patients (26 males, 9 females) with a mean age of 56 years (range: 41-77 years). Cirrhosis diagnosis was histologically documented and 6 patients had also an acute alcoholic hepatitis.
- Group 3: 30 healthy subjects without renal or liver failure.

Serum selenium concentration was determined by electrothermal atomic absorption spectrometry after dilution 1:1 with nitric acid 0,1 M in triton X 100 0,1% (w/V). Nickel solution was then injected in the graphite furnace as a matrix modifier. The instrument employed was a Perkin Elmer 560 atomic absorption spectrometer equiped with a HGA 500 graphite furnace and an AS 40 autosampler.
Total bilirubin, plasma prothrombin index (PI), alanine aminotransferase (ALAT), alkalin phosphatase (Alk P), gamma glutamyl transferase (GGT), albumin and transferrine (TRF) were also determined according to standard laboratory methods.
A "t" student test was used to compare the results. Correlations were calculated using Pearson's correlation coefficient.

Results:

Serum selenium level and results of liver function tests are provided in table 1. Serum selenium level is lower in groups 1 and 2 than in group 3. The difference is statistically significant only in group 2 ($P < 0.01$ versus group 3, $P < 0.05$ versus group 1). In chronic alcoholic patients serum selenium decrease is not statistically significant.
In group 1, serum selenium level is not correlated with liver function tests (PI, ALAT, Alk P, GGT, total bilirubin). On the other hand, a significant correlation is found between serum selenium and albumin ($r=0.313$, $P < 0.05$).
In group 2, serum selenium level is negatively correlated with PI

Table 1 : results expressed as mean ± 1 standard deviation

	chronic alcoholic	cirrhosis	controls
Selenium µmol/l	0.92 ± 0.25	0.79 ± 0.26	0.97 ± 0.17
Total bilirubin µmol/l	26 ± 14	40 ± 33	< 10
PI %	91 ± 13	70 ± 18	100
ALAT UI/l	45 ± 25	33 ± 19	17 ± 5
Alk P UI/l	104 (14-480)	169 (17-740)	22 ± 4
GGT UI/l	280 ± 251	214 ± 213	17 ± 5
Albumin g/l	37 ± 6	29 ± 6	43 ± 2
TRF g/l	1.97 ± 0.55	1.97 ± 0.69	2.8 ± 0.3

($r = -0.343$, $P < 0.02$), Alk P ($r = -0.343$, $P < 0.02$), GGT ($r = -0.315$, $P < 0.05$); a positive correlation is found with total bilirubin ($r=0.531$, $P < 0.001$). In this group, no correlation is observed between serum selenium and protein (albumin, TRF).

Discussion:

This study confirms very low serum selenium level in alcoholic cirrhosis. These results are in agreement with previous studies (1,2,3,4). Although reduced serum selenium concentration are noted in several studies (2,3,4), the serum selenium in our chronic alcoholic patients (group 1) is not significantly depressed, but our patients have a normal nutritional status. This observation suggests that selenium reflects the liver injury and that selenium level is not related to alcohol consumption. Contrary to Korpela (3), in our cirrhosis patients, we found negative correlations between serum selenium and some of liver function tests (Alk P, ALAT, PI, GGT). Yet, we note a surprise and unexplain positive correlation between selenium and total bilirubin. On the other hand, liver function tests measured in our patients are not very abnormal, except GGT and Alk P.

To understand the relation between selenium and liver damage, we need other studies, particularly lipid peroxidation products determination.

References:

1. Aaseth J., Y. Thomassen, J. Alexander, G. Norheim. 1980. N. Engl. J. Med. 303, 944.
2. Dutta S.K., P.A. Miller, L.B. Greenberg, O.A. Levander. 1983. Am. J. Clin. Nutr. 38, 713.
3. Korpela H., J. Kumpulainen, P.V. Luoma, A.J. Arranto, E.A. Sotaniemi. 1985. Am. J. Clin. Nutr. 42, 147.
4. Tanner A.R., I. Bantock, L. Hinks, B. Lloyd, N.R. Turner, R. Wright. 1986. Digestive Dis. Sci. 31, 1307.

PLASMA SELENIUM IN CARDIOMYOPATHY

P. Thérond, M. Marchand, D. Biou, J.F. Demelier

Laboratory of Biochemistry, Bretonneau Hospital, 2 rue Carpeaux, 75018 Paris, France.

Ph. Auzépy, C. Richard

Units of Cardiology and Intensive care, Bicêtre Hospital, 94270 Paris, France.

Introduction

Congestive cardiomyopathy was reported as associated with selenium deficiency. The first diagnosis of selenium deficiency in congestive cardiomyopathy was found in China under the name of "Keshan disease"(1). On the other hand, in order to explain the high incidence of coronary atherosclerosis (2) in Finland,the selenium deficiency has been advanced. We report here the measure of plasma selenium, glutathione peroxidase and vitamin E concentrations in french adults patients suffering from congestive cardiomyopathy.

Materials and methods

Plasma selenium concentration (µg/L) was measured by thermoelectric atomic absorption spectrophotometry (Philips SP9);plasma glutathione peroxidase (U/L) by the enzymatic method of Paglia and Valentine and Vitamin E by high pressureliquid chromatography (HPLC), in 29 patients distributed in two groups as follows :

-Group I (control group): 18 healthy volunteers from medical staff with a mean of age of 34 years.

-Group II : 11 patients suffering from congestive cardiomyopathy with a mean of age of 56 years.

None out of the 29 patients suffered from renal failure,anemia,or malnutrition. Results were expressed as mean ± standard error on the mean. Statistical analysis used was a non parametric test: test U of Mann and Whitney.

Selenium in Medicine and Biology

Results :

Mean plasma selenium level is 123 $\pm$ 4 µg/L in the control group and 79 $\pm$ 5 µg/L in Group II,significantly lower than Group I ($p<0.001$). Plasma glutathione peroxidase activity shows similar variations (Group I: 450 U/L $\pm$ 7 ; Group II: 280 U/L $\pm$ 10; $p<0.001$).
On the other hand we don't find difference between plasma vitamin E concentrations in GroupI and II (Group I: 11.5 $\pm$ 0.5 mg/L; Group II: 13.4 $\pm$ 1 mg/L).

Discussion

Low concentrations of selenium have been reported to increase risk for both cancer (3) and ischemic heart disease (4). Previous data had demonstrated the relationship between low selenium status and fatal cardiomyopathy during long term parenteral nutrition (5). Keshan disease was a cardiomyopathy with a high case mortality that usually affected Chinese children and women of child-bearing age ingesting food very deficient in selenium. We suspected in our patients the association of chronic alcohol abuse and selenium deficiency because alcohol consumption is known to increase urine elimination and decrease blood concentration of selenium. This association might explain the severity of congestive cardiomyopathy. In fact,we haven't found significant difference between plasma selenium level in patients with alcoholic or idiopathic congestive cardiomyopathies. Contrast investigations on plasma vitamin E have been conducted by several researchers. Furthermore, no significant difference in vitamin E levels was observed between control group and patients with congestive cardiomyopathy. In conclusion, insufficient dietary intake of selenium might account for a moderate decrease of plasma selenium without depletion in vitamin E. The nutrition status of our patients was within normal range that seemed to exclude a possibility of denutrition. Thus, the isolated selenium deficiency might be an other factor of risk of congestive cardiomyopathy.

References

1. Chen, X., G. Yang, J. Chen, Z. Wen, K. Ge. 1980. Biol. Tr. El. Res. 2, 91.
2. Virtamo, J., E. Valkeila, G. Alfthan. Am. J. Epidemiol. (in press)
3. Salonen, J.T., G. Alfthan, J.K. Huttunen, P. Puska. 1984. Am. J. Epidemiol. 120, 342.
4. Salonen, J.T., G. Alfthan, J.K. Huttunen, J. Pikkarainen, P. Puska. 1982. Lancet 2, 175.
5. Fleming, C.R., J.T. Lie, J.T. Mc Call, J.F. O'Brien, E.E. Baillie, J.L. Thistle. 1982. Gastroenterology 83, 689.

IS NON OBSTRUCTIVE MYOCARDIOPATHY(NOMC)IN AIDS SELENIUM-DEFICIENCY-RELATED?

J.F. Zazzo, A. Lafont, H. Darwiche, F. Sayegh, F. Camus
Infections Diseases Unit, Hôpital Claude Bernard, Paris, France

P. Chappuis
Department of Biochemistry, Hôpital Lariboisière, Paris, France

J. Chalas, C. Benattar
Department of Biochemistry, Hôpital Antoine-Béclère, Clamart, France

Introduction

Despite little clinical suspicion, cardiac abnormalities are frequent in Acquired Immunodeficiency Syndrom (AIDS). Myocarditis is the most common finding (1). Opportunistic myocardial pathogens are seen in only third of cases (2). In the two third remaining patients the non obstructive cardiomyopathy (NOCM) is idiopathic. AIDS patients are undernourished and trace elements deficiency, especially Selenium (Se) may be implicated in the pathogenesis of NOCM. Low blood Se (3) and low cardiac levels (4) have been reported in AIDS. We submitted 10 AIDS patients with NOCM (left ventricular shortening fraction-LVSF < 28 % by echocardiography) to sodium selenite p.o. (800 µg/day during 15 days followed by 400 µg/day during 8 days). Plasma Se and glutathion-peroxidase (GSHpx) in red blood cells were assessed before and during supplementation. Simultaneously echocardiography was performed once a week.

Results

Plasma Se, before treatment, was significantly reduced in 8 out of 10 patients ($0.65 \pm 0.12 \mu mol.l^{-1}$ VS $1.10 \pm 0.15 \mu mol.l^{-1}$ $p < 0.001$). GSHpx was in the normal range. Among these 8 patients, 6 normalized simultaneously LVSF and biology within 15 days, one died, and one had a thiamine deficiency (fig 1).

Selenium in Medicine and Biology

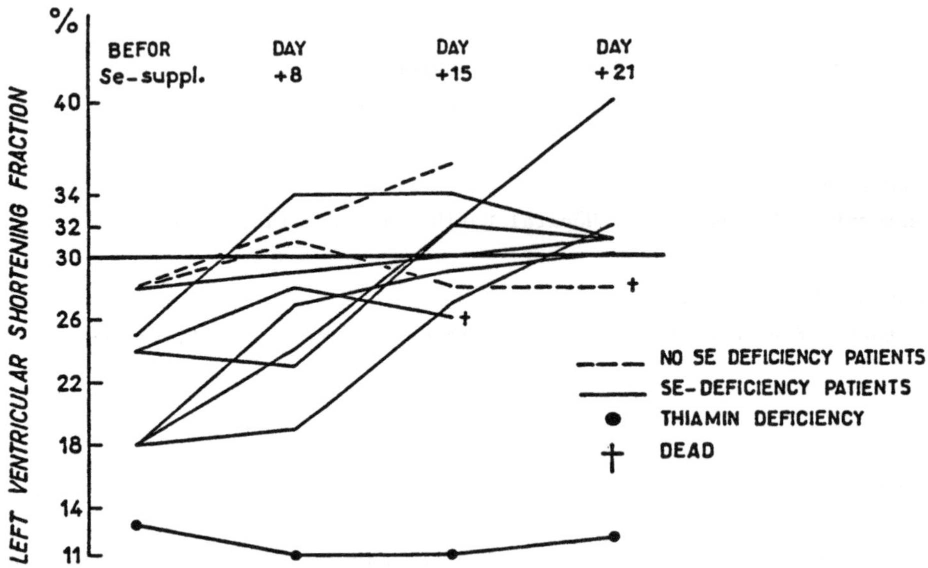

Figure 1 : Evolution of LVSF in AIDS patients supplemented with sodium selenite

Discussion

Immune depression provides a clinical environment fostering the development of cardiotrophic viral myocarditis which has not yet been proved. Nutritional factors may be advocated especially Se deficiency.
A prospective double blind trial is performing to establish causation relationship.

References

1. Roldan, E.O., L. Moskowitz, G.T. Hensley. 1987. Arch. Pathol. Lab. Med., 111, 943.

2. Anderson, D., R. Virmani, Macher. 1987. IIIrd International conference on AIDS. M. 11. 6 (Abstr.).

3. Dworkin, B., W. Rosenthal, G. Wormser. 1986. J.P.E.N. 10, 405.

4. Antonecchia, P., B. Dworkin, F. Smith. 1988. J.P.E.N. 12, (suppl.), 78 (abstr.).

LIPID PEROXIDATION, RELATIONSHIP WITH SELENIUM AND GLUTATHIONE PEROXIDASE ACTIVITIES IN PATIENTS WITH CHRONIC RENAL FAILURE.

M.J. Richard, J. Arnaud, N. Koukay, A. Favier
Laboratoire de Biochimie C, CHRUG, BP 217X, 38043 Grenoble, France

C. Jurkovitz, E. Dechelette
Hémodialyse, CHRUG, BP 217X, 38043 Grenoble.

T. Hachache, H. Meftahi, M. Foret
A.G.D.U.C., CHRUG, BP 217X, 38043 Grenoble.

F. Laporte
Laboratoire d'Enzymologie, CHRUG, BP 217X, 38043 Grenoble.

Introduction

The survival of most patients with chronic renal disease is now ensured as a result of improved dialysis methods. However, this treatment is not yet optimal. Alterations of the trace elements concentrations in plasma are known to occur in these patients on or without intermittent hemodialysis (HD) treatment (1,2,3). Recent investigations found a selenium (Se) deficiency in these patients (4,5). Moreover, in the course of HD on cellulosic membranes, it has been suggested that polymorphonuclear activation can release oxygen radicals which damage endothelial cells (6).
The present investigation was undertaken to clarify the relationship of lipid peroxidation to circulating selenium (Se), zinc (Zn) and copper (Cu) in chronic uremia hemodialysis patients.

Selenium in Medicine and Biology

Patients and Methods

Patients : 17 uremia patients (aged 20-61 years) on maintenance HD (2 or 3/weekly) since fewer than five years were investigated. 18 healthy subjects (aged 20-62 years) served as controls. Were excluded: alcoholic, diabetic and patients with respiratory failure. Before of the patient connection to the dialyser, samples (10 ml) of venous blood were collected into heparinized, trace element free, polypropylene tubes. Plasma was assayed immediatly for glutathione peroxidase (GPX) and frozen for trace elements and lipid peroxidation products determination. The red blood cells (RBC) were washed three times and the washed packed RBC were assayed for GPX, superoxide dismutase (SOD), and catalase determination.

Methods : Trace elements (Se, Zn, Cu and Mn) were determined by flameless atomic absorption spectrometry (PERKIN ELMER 560 with HGA 500). GPX activity in plasma and hemolysates were assayed by a modified method of Gunzler and al (7). For SOD determination we used two different methods : a luminol assay according to Bensiger et al (8) measuring SOD activity and an immunologic assay (9) for the SOD protein determination. For monitoring lipid peroxidation in plasma, three parameters were studied. Malonaldehyde (MDA) was determined by a modified fluorometric micromethod with thiobarbituric acid reagent (unpublished). The measurement of lipid hydroperoxides (HPO) was performed by an enzymatic method (10). Conjugated dienes were measured by a method based on those described by Lunec and Dormandy (11), and the results were expressed as a relative absorbance by ml of plasma.

A student's t.test was used to compare the results. Correlations were calculated using Pearson's correlation coefficient.

Results

Table 1 : Trace element status and enzymatic antioxidant system in plasma and red blood cells in renal failure patients on dialysis

	HD patients n = 17	control subjects n = 14	
Plasma			
. Se $\mu mol.L^{-1}$	0.69 ± 0.12	1.05 ± 0.13	*
. Zn $\mu mol.L^{-1}$	13.4 ± 1.1	16.7 ± 2.0	*
. Cu $\mu mol.L^{-1}$	16.0 ± 1.75	15.5 ± 1.5	NS
. Fe $\mu mol.L^{-1}$	15.5 ± 4.1	19 ± 12	NS
. Mn $\mu mol.L^{-1}$	20.6 ± 2.4	19.9 ± 2.3	NS
. GPX UI/L	117.8 ± 16.7	334.6 ± 19.2	*
Erythrocyte			
. RBC-GPX U/gHb	17.2 ± 1.65	20.0 ± 1.7	**
. Cu-Zn SOD :			
Apoenzym µg/gHb	746 ± 49	850 ± 50	*
Activity µg/gHb	500.35 ± 116.7	720 ± 150	*
. Catalase UI/mgHb	202 ± 24	140 ± 40	*

ND = Non significant *p < 0.001 ** p < 0.050

Trace element levels and antioxidant enzymatic activity are provided respectively in table I. The mean plasma selenium level is significantly lower in HD patients than in control group ($p < 0.0001$). Plasma GPX activity is markedly decreased as compared to the controls. But, in these subjects, plasma Se does not correlate with plasma GPX. On the contrary, RBC GPX is well correlated with plasma Se ($r=0.63$, $p<0.0001$). Plasma Zn in HD patients is significantly below normal. Statistically, copper and manganese status are not perturbated. We noted a correlation between Zn status as well as Cu status and erythrocyte CuZn-SOD activity (respectively $r=0.51$ $p<0.030$; $r=0.60$ $p<0.01$). A higher catalase activity is reported in the erythrocytes of HD patients than in healthy subjects.

The results of the plasma lipoperoxide products are given in Fig.1.

In all the HD patients, elevated levels of MDA are found. Increase of conjugated dienes and lipidic hydroperoxides confirm this hyper lipidoperoxidation. MDA is negatively correlated with plasma Se ($r=-0.59$ $p<0.001$) (Fig. 2).

Fig. 1 - Lipoperoxide levels : malonaldehyde (MDA), organic hydroperoxides (HPO) and conjugated dienes (DC) in blood plasma of HD uremia patients (HD) and control subjects (C).

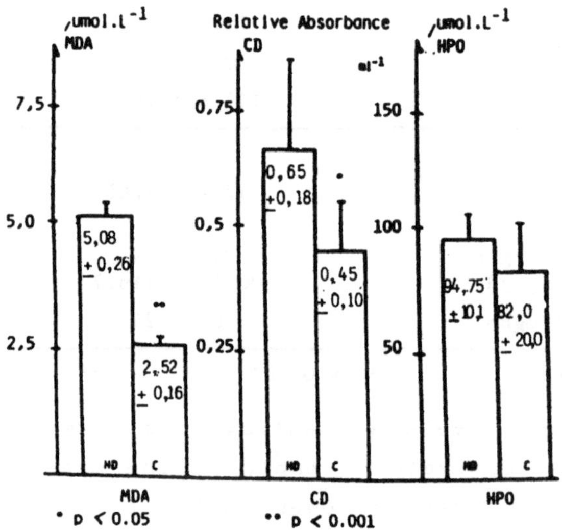

Fig. 2 - Plasma MDA vs plasma selenium levels in control (▪) and dialysis renal failure patients (●) ($p<0.001$).

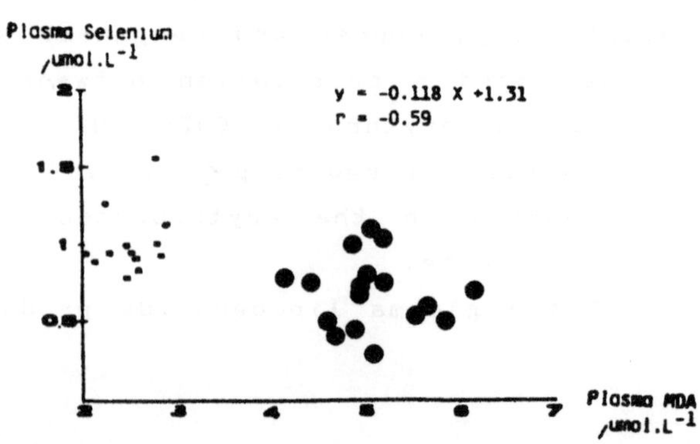

Discussion

This study confirms abnormalities of plasma trace elements in uremia patients. Low selenium level are in agreement with previous studies (4,5) and well correlated with RBC GPX activity decrease. Little is known about Selenium homeostasis but the kidney does play a regulatory role. Plasma GPX breakdown cannot be explained by selenium level. Takahashi (12) demonstrated that purified plasma GPX is chemically, immunologically, electrophoretically and in some ways fonctionally different from purified RBC GPX, but this glycoprotein is still unknown. Is it a kidney synthetized protein ?

Our results are in agreement with the essential roles of Zn and Cu in the structural part of SOD (13). This protein contains Zn as well as Cu. The catalytic activity is intimately associated with its copper prosthetic group and the role of Zn is important in stabilizing the native configuration of the apoenzyme.

The findings presented here suggest, in this pathology, a decreased enzymatic intracellular protection against oxygen-derived species such as $O_2^{\cdot -}$, hydrogen peroxide (H_2O_2) produced by polymorphonuclear leucocytes. So, in the presence of transition metals as iron, the powerful hydroxyl radical ($OH^{\cdot}$) can be formed. In these HD uremia patients, a superoxide generating system has been suggested. The complement activation occuring during dialysis induces the phenomenon of neutrophils activation which can subsequently damage endothelial cells by an oxydative mechanism (6). MDA and lipidoperoxides products found in this work support the hypothesis that peroxidative membrane damages may be an important factor in the secondary pathogenesis of HD : eg. aging, atherosclerosis, cardiopathy, rhumatologic complications,... Although decreased activities of two antioxydant enzymes : (SeGPX and ZnCuSOD) have been found it is not clear whether increased lipid peroxidation arise from deficiencies of protective mechanism or from raised free radicals levels due to an oxidative stress or from a synergic combination of these two mechanisms.

References

1. Sprenger, K.B.G., V. Krivan, H. Geiger, H.E. Franz. 1985. Nutrition Research, suppl, 1, 350.
2. Favier, A., J.L. Lafond, B. Desrousseaux, J. Arnaud. 1984. In: Trace Element Analytical Chemistry in Medicine and Biology (P. Brätter, P. Schramel, eds.). Walter de Gruyter Berlin NY, 3, p. 711.
3. Cornelis, R., L. Mees, S. Rengori, J. Hoste. 1979. Mineral Electrolyte Metabolism 2, 88.
4. Foote, J.W., L.J. Hinks, B. Lloyd. 1987. Clin.Chim.-Acta. 164, 323.
5. Dworkin, B., S. Weseley, W. Rosenthal, E. Schwartz, L. Weiss. 1987. Am.J.Med.Sci. 293, 6.
6. Market, N., C. Heierli, T. Kuwahara, J. Frei, J.P. Wranters. 1988. Clin.Nephrology. 29, 129.
7. Gunzler, W.A., H. Kramers, L. Flohe. 1974. Z.Klin.Chem. Klin.Biochem. 12, 444.
8. Bensiger, R.E., C.M. Johnson. 1981. Anal.Biochem. 116, 145.
9. Flohé, L., F. Otting. 1984. In: Methods in Enzymology (L. Pocher, eds.) Academic Press INC. 105, p. 93.
10. Heath, R.L., A.L. Tapel. 1976. Anal.Biochem. 76, 184.
11. Hunter, M.I.S., J.B. Muhameo. 1986. Clin.Chim.Acta. 155, 123.
12. Takahashi, K., P. Wewburger, H.J. Cohen. 1986. J.Clin. Invest. 77, 1402.
13. Forman, H.J., I. Fridovich. 1973. J.Biol.Chem. 248, 2645.

BLOOD SELENIUM, GLUTATHION PEROXIDASE AND CADMIUM IN CHRONIC RENAL FAILURE PATIENTS

F. Dubois, B. Nicolas, R. Hoffman, F. Belleville, P. Paysant
Department of Biochemistry, Brabois University-Hospital, route de Neufchateau F 54511 Vandoeuvre Cedex, France

M. Kessler, B. Jonon, T. Cao-Huu
Department of Nephrology, Brabois University-Hospital, Route de Neufchateau, F 54511 Vandoeuvre Cedex, France

Introduction

Patients maintained on regular haemodialysis present frequent and severe complications : ischemic heart diseases, acute or chronic infections... Aetiology of these complications has yet to be fully elucidated. Elsewhere epidemiological studies reveal the importance of selenium (Se) in pathogenesis of cardiovascular diseases (1). Accumulated evidences also indicated that Se deficiency in animals is associated with an impairment of immune responsiveness (2).
Little is known about Se metabolism, however the Se urinary excretion represents the principal known process of Se regulation (3). Because Se deficiency could be involved in pathologic states associated with chronic renal failure and hemodialysis, we studied plasma Se levels in patients with chronic renal failure before dialysis and in patients undergoing long term hemodialysis. Glutathion peroxidase GSH-Px and cadmium (known to accumulate in chronic renal failure) were also assayed in hemodialysis patients.

Selenium in Medicine and Biology

Material and methods

Plasma selenium was tested in 3 groups : group I included control subjects. Group I A was composed of 125 subjects aged 20 to 59 years (mean = 40.6), group I B was composed of 110 subjects aged 60 to 95 years (mean = 80). Group II : 10 patients aged 31 to 69 years (mean 44.5) with end stage renal disease, before the first dialysis. Group III : 46 subjects treated with chronic hemodialysis. Group III A : 23 patients aged 25 to 59 years (mean = 43) and group III B : 23 patients over 60 years (mean 72.7). These patients were hemodialysed from 1 month to 16 years.
Plasma Se and Cd levels were determined by electrothermal atomic absorption spectrometry. GSH-Px was assayed by Paglia's method (4).

Results and discussion

Selenium : plasma Se of groups I A, I B, II, III A, III B are reported in table 1. Plasma Se was significantly ($P < 0.0001$) decreased in chronic renal failure and hemodialysis patients compared with normals. Post-dialysis plasma Se levels were not significantly different from pre-dialysis values.
Our results confirmed those obtained by Dworkin (5). The mechanism of this abnormality is not understood. We can hypothesize : (i) a protein malnutrition, because Se in foodstuffs is linked to protein content. But we did not find a correlation between plasma Se and serum total proteins ($r = -0.182$). (ii) Loss of Se during hemodialysis. We observed that the renal chronic failure patients have already very low Se levels. Pre and post dialysis plasma Se levels showed no changes. Furthermore there was no correlation between duration of dialysis sessions (in hours) and plasma Se levels ($r = 0.277$). (iii) Interactions with heavy metals. Se

interacts in the body with heavy metals and it is effective in preventing their toxic effects (6, 7).
The plasma Cd levels measured in patients of group II was significantly increased ($P < 0.0001$) versus smoker controls.

	Se µg/l	GSH-Px activity U/gHb	Cd µg/l
Group I A	71.7 ± 11 n = 125	8.3 ± 1.9 n = 30	1.18 ± 0.5 n = 51 no smokers 2.17 ± 1.09 n = 35 smokers
Group I B	65.5 ± 13.3 n = 110		
Group II	36 ± 10.5* n = 10		
Group III A			
pre-dialysis	40.9 ± 11.6* n = 23	8.5 ± 1.5 n = 17	5 ± 2.2* n = 19
post-dialysis	40.4 ± 12.6* n = 22	7.86 ± 1.5 n = 17	
Group III B			
pre-dialysis	36.3 ± 12.7* n = 23	7.7 ± 2.10 n = 14	6.1 ± 2.8* n = 13
post-dialysis	36.2 ± 12.6* n = 20	6.77 ± 1.4 n = 14	

Table I : Plasma Se, Cd levels and erythrocyte GSH-Px activity in controls (group IA, IB) chronic renal failure patients (group II) hemodialysis patients (group III). The results are compared to age matched controls. * $P < 0.0001$. Plasma cadmium (Cd) levels are compared to smoker control.

Glutathion peroxidase has been determined in patients of group II. When the results were expressed in U/gHb, there was no significant difference between the results obtained in patients of group II and in controls. But as a constant

feature of chronic renal failure is the presence of anemia (7) the whole GSH-Px activity was decreased. The GSH-Px activity in post-dialysis was not significantly diminished compared to GSH-Px activity in pre-dialysis, the light difference can be to assign to hemoconcentration.
Low plasma Se levels in chronic renal failure and in hemodialysis patients raised the problem of Se supplementation.

References

1. Salonen, J.T., Alfthan, G., Huttunen, J.K., Pikkarainen, J., Puska, P. 1982. Lancet 2, 175.

2. Kiremidjian-Schumacher, L., Stotzky, G. 1987. Environm. Res. 42, 277.

3. Robberecht, H.J., Deelstra, H.A. 1984. Clin. Chim. Acta 136, 107.

4. Paglia, D.E., Valentine, W.N. 1967. J. Lab. Clin. Metab. 70, 158.

5. Dworkin, B., Wesely, S., Rosenthal, W.S., Schwartz, E.M., Weiss, L. 1987. Am. J. Med. Sci. 30, 6.

6. Babich, H., Devenas, M.A., Stotzky, G. 1985. Environm. Res. 37, 253.

7. Nève, J., Vertongen, F., Molle, L. 1985. Clin. Endocrinol. Metab. 14, 629.

8. Foote, J.W., Hinks, L.J., Lloyd B. 1987. Clin. Chim. Acta. 164, 323.

IMPLICATION OF SELENIUM AND GLUTATHIONE PEROXIDASE DEFICIENCY IN THE PATHOLOGY OF HEMODIALYZED PATIENTS.

M. SAINT-GEORGES, C.L. d'AUZAC
Néphrologie-Dialyse : Hôpital National de Saint-Maurice - 94410 - FRANCE

D. BONNEFONT, B. BOURELY, M.C. JAUDON, A. GALLI
Laboratoire de Biochimie (Dr. GALLI) - Hôpital Salpétrière - PARIS - FRANCE

Introduction

The trace element Selenium, essential for its role in glutathione peroxidase activity, is found in lowered level in hemodialyzed patients plasma (4). The pathological consequences of this deficiency, particularly the muscular and cardiovascular consequences, are well know. (2-3-5). The purpose of our study was :

1. to evaluate Selenium deficiency and the lowering of plasma and erythrocyte glutathione peroxidase activity in 39 hemodialysis patients.
2. to investigate the implication of these deficiencies on the patients pathology in order to justify correcting the Selenium deficiency.

Patients and Methods

39 patients (23 men and 16 women) from 29 to 74 years of age (mean age 55 ± 12), having been dialyzed for from 1 to 17 years (mean 8 ± 5). 19 of them have had 10 years or more of hemodialysis treatment. They were dialyzed for four or five hours, three times a week ; 18 on cuprophan or cellulose acetate hollow-fiber dialyzers, 21 on polysulfon fibers or polyacrilonitrile. Plasma Selenium (Se), plasma (GPxp) and erythrocyte (GPxe) glutathione peroxidase activities were determined for the 39 patients and for 15 controls. Selenium was determined by flameless atomic absorption

Selenium in Medicine and Biology

spectrophotometry at 196 nm (Perkin Elmer spectrophotometer 5 000 with graphite furnace HGA 400). Glutathione peroxidase activity was measured by spectrophotometry at 340 nm using the Beutler technique (substrate : t-butyl hydroperoxide). Vitamin E was assayed bu high-performance liquid chromatography. A dietary review done by a dietitian for one week helped to determine the protein average (g/kg/d) and caloric (kcal/kg/d) intake for each patient.

In addition to the clinical examination, the effects of these deficiencies on pathology was invertigated through :

- measuring total muscle mass (TMM) by triceps skinfold thickness method (corrected formula of Heymsfiels). The measurement was taken after dialysis on the arm opposite the arteriovenous fistula. The TMM was adjusted for body weight in order to facilitate comparisons between patients.
- A Merieux multitest [R] (multiple/single cutaneous test of cellular immunity), was done on 34 patients.
- An echocardiographic study was performed on 29 of the 39 hemodialyzed patients, with uni- and bi-dimensional ultrasonography. All the ultrasonograms were done by the same operator.

Results

The results are presented in Tables 1 through 3.

Biological Aspects

Plasma and erythrocyte glutathione peroxidase and Selenium levels were significantly lower in the hemodialyzed patients than in the control group (Table 1).

Table 1 : Results

	Se - μg/l	GPxp - IU/l	GPxe - IU/g of hemoglobin
HD (39)	38 ± 14 *	153 ± 32 ***	19 ± 4 **
Controls (15)	88 ± 17 *	334 ± 41 ***	26 ± 4 **

* $p < 0.001$ ** $p < 0.05$ *** $p < 0.001$

We found a strong correlation between Se and GPxp ($p < 0.01$) but no between Se and GPxe. Levels were not influenced by age, sex, protein or even caloric intake, hemoglobin, or vitamin E levels. Serum Vit. E level was found normal or increased and correlated with the age of patients. There was a strong correlation between Se or GPxp and duration on dialysis ($p < 0.01$). The dialysis membrane type was important in regard to Selenium deficiency : Se was lower in patients dialysed with highly permeable membrane (polyacrilonitrile or polysulfon) than those dialysed with cuprophan or cellulose acetate membrane ($p < 0.002$).

Clinical Aspects

Cellular immunity : Neither Se nor GPx levels were statistically correlated to multitest scores. The TMM was significantly lower in subjects having a GPxp less than 150 IU/ml (Table 2).

Table 2 : Muscle Mass - Selenium and Glutathione Peroxidase

	Se (17) ⩽ 35	Se > 35 (22)	GPxp (20) ⩽ 150	GPxp (19) > 150
MMT/Wt	33,6 ± 5,8 *	33,8 ± 7,1 *	32,32 ± 4,1 **	35,2 ± 8,2 **

MMT/Wt : Total muscle mass adjusted to body weight (percentage).
Se : Plasma Selenium in µg/l.
GPxp : Plasma glutathione peroxidase in IU/l.
* NS ** $p < 0.01$ (n) : number of patients

Cardiomyopathy : we did not find lower levels of Se or GPx in hypertensive subjects or those suffering from a more severe coronaropathy. The echocardiographic study showed that interventricular septal hypertrophy was more marked in subjects having lower Se and GPxp (Table 3).

Table 3 : Thickness of the Septum - Selenium and Plasma Glutathione Peroxidase

	Se (15) ⩽ 35	Se (14) > 35	GPxp (15) ⩽ 150	GPxp (14) > 150
I.V.S.T	12,3 ± 2,4 *	9,50 ± 3,8 *	12,33 ± 2,4 **	10,9 ± 1,2 **

I.V.S.T. : Thickness of the Interventricular Septum in mm.
Se : Plasma Selenium in µg/l
GPxp : Plasma Glutathione peroxidase in IU/l.
* $p < 0.01$ ** $p < 0.05$ (n):number of patients.

The parameters of contractability did not correlate significantly either with Se or GPx, contrary to expectations. But our sample is small; however the results, for this reason, should be quoted. There is, in particular, a positive correlation between the fractional shortening and the GPxp ($r = 0.33$; $0.1 < p < 0.05$) but none with the duration on dialysis ; but on the other hand, the velocity of circumferencial fiber shortening is positively correlated with the GPx ($r = 0.31$) and negatively with the duration on dialysis ($r = -0.31$).

Discussion :

The involvement of Selenium and glutathione peroxidase deficiencies in the pathology of hemodialysis is difficult to establish due to the multiple pathogenic factors due to the heavy pathologic background of patients and the small number of patients in our study. Nevertheless the long-term survival requires a good muscular-trophic state, and the cardiovascular problems stay the first cause of death in hemodialysis. Consequently, based on the importance of Selenium deficiency and the lowering of glutathione peroxidase activity on one hand, and on ultrasound cardiography and total muscle mass results, on the other hand, it appeared desirable to try to normalize Selenium and consequently glutathione peroxidase activity. We therefore gave our patients Selenium supplements. They received 500 µg per os of sodium Selenite, following each dialysis, with a monthly monitoring of Se, GPxp and of GPxe.

References :

1. Nève J., Vertongen F., Molle L. : Clinics in Endocrinology and Metabolism 1985 14 : 629 - 657.

2. Oster O., Prellwitz W., Kasper W. and Meinretz T. : Clin. Chim. Acta. 1983 : 128 : 125 - 132.

3. S. Kenneth Pehrsson, Lars-Eric Lins : Nephron 1983. 34 : 93 - 98

4. Salonen J.T., Puska P. : Lancet 1982 ii : 175 - 179.

Section V

Selenium supplementation and toxicity

SELENIUM SUPPLEMENTATION IN HUMANS

A. Aro and J.K. Huttunen

National Public Health Institute, Mannerheimintie 166, SF-00280 Helsinki, FINLAND

Introduction

The possible links between inadequate dietary intake of selenium and human disease have stimulated interest in the nutritional bioavailability of various selenium compounds. Numerous models have been examined to assess selenium availability in animals (1). This paper reviews the data on selenium supplementation in humans and describes the effects on foods and on plasma selenium of supplementation of all agricultural multimineral fertilizers with sodium selenate, effective in Finland since 1984.

Bioavailability studies

Most of human studies on the bioavailability of selenium compounds have been conducted in populations with low selenium intake. Under such conditions, both the concentration of selenium and the activity of glutathione peroxidase can be used as indices of selenium status. On the contrary, no change in the activity of glutathione peroxidase in response to selenium supplementation is observed in populations with high dietary intake of selenium (2,3).

In a study conducted in Finland in the early 1980's, the effects of supplementation with 200 µg of selenium from

Selenium in Medicine and Biology

wheat, yeast and sodium selenate were compared with placebo in healthy men with low dietary selenium intake (serum selenium level <70 µg/l) (4). Organic selenium from wheat and yeast produced a rapid and pronounced increase in serum selenium, while the effect of inorganic selenate was significantly smaller. Serum selenium values did not reach a plateau during the 11-week experiment. Red blood cell selenium concentration increased progressively during supplementation with selenium-rich wheat and yeast, whereas no change was seen in subjects who received sodium selenate. Although there was a highly significant difference between the effects of wheat and yeast selenium and inorganic selenate on plasma and erythrocyte selenium concentrations, the availability of selenium for platelet glutathione peroxidase was similar for all three forms used in the study. However, 10 weeks after the supplements were discontinued, platelet glutathione peroxidase activity stayed at a higher level in the groups that had received wheat or yeast selenium as compared with those who had received sodium selenate.

Several other investigators have reported similar observations. A Chinese study (5) found differences in plasma and erythrocyte selenium concentrations in response to supplementation with various forms of selenium. In this study the plasma and red blood cell selenium levels reached a plateau after 4 months of supplementation. Studies conducted in New Zealand indicated better bioavailability of selenomethionine as compared with sodium selenite when plasma selenium concentration was used as the index (6,7). On the other hand, no difference was observed in the glutathione peroxidase activity of whole blood. Kumpulainen et al. (8) studied the effects of various selenium supplements (100 µg/day) in lactating mothers. The highest serum selenium concentrations both in the serum of mothers and infants and in mothers' milk were found during supplementation with yeast selenium, suggesting better availability and more efficient incorporation into tissues and milk. Better absorption of organic selenium as compared with inorganic forms has previously been documented by Young et al. (9).

Prevention of manifestations of selenium deficiency

Chinese investigators have carried out extensive prevention trials of Keshan disease and Kashin-Beck disease, which are thought to be due to dietary selenium deficiency. Supplementation with sodium selenite in the form of tablets or in table salt lowered the incidence of Keshan disease (10,11,12). In Kashin-Beck disease, selenium supplementation of children for one year with inorganic selenium compounds reduced skeletal changes as assessed by radiography (13,14). Both Keshan disease and Kashin-Beck disease are manifested in childhood. Based on our knowledge about the bioavailability of different selenium compounds, a better effect might be anticipated by supplementation with organic selenium, especially if supplemention is started during pregnancy and lactation and is continued through childhood.

Selenium fertilization in Finland

In studies performed in the 1970's, the selenium intake in Finland was found to be low, 20-30 µg per day (15). The reason for the low intake was low selenium concentration in domestic agricultural products. During periods when high-selenium wheat was imported from North America, the selenium intake increased considerably (16). This resulted in large variations of serum selenium concentrations in the population (17).

Because of concerns expressed about the nutritional adequacy of domestic agricultural products, an official decision was made to increase their selenium concentration by supplementing multimineral fertilizers with sodium selenate. The decision to use selenate was based on Finnish studies showing that selenate is incorporated into plants, while more reduced forms are unavailable in the soil (18). As selenium is incorporated more effectively into the vegetative parts of plants than into grains (18), two different levels of supplementation were chosen: 6 mg selenium per kg fertilizer for fodder and hay production and 16 mg per kg for cereal production. The supplementation was started in 1984 and has affected the selenium concentration of foods from 1985 on.

The results of selenium fertilization will be published in detail elsewhere (19). The primary goal was to increase the selenium concentration of cereals tenfold. The result was a 20-fold increase of selenium concentration in spring wheat, while the effect on winter cereals (rye, wheat) remained smaller (3-4 -fold). Considerable increases were also observed in the selenium concentrations of milk and cheese (7-8 -fold), meat (4-9 -fold) and eggs (2-3 -fold). The selenium in plants is largely in organic form and is effectively incorporated into most foods produced in the country. No measurable effect on the selenium content of soil and water has been observed. The lack of environmental effects is also indicated by an unchanged selenium concentration in fish (19).

The intervention has had a pronounced effect on the mean selenium intake of people. The calculated intake was about 90 µg per day in 1986 as compared with 40 µg per day in 1984 and the even lower values of 1970's. A corresponding increase in serum selenium concentrations has been observed in both urban and rural people (19). Selenium fertilization has also affected the relative contributions of different foods to total dietary selenium intake. Fish accounted for about one-third of the mean selenium intake of Finns in the 1970's. In 1986 meat, milk products, eggs and cereals were the most important dietary sources of selenium, and fish accounted for only approximately 10 % of the total selenium intake (19). The effects of the intervention are carefully monitored by an expert group appointed by the government.

Summary and conclusions

Marked differences in nutritional availability for various forms of selenium have been reported in man. Organic selenium (selenomethionine) is absorbed better than inorganic selenium salts and is superior to selenate in terms of elevating plasma and erythrocyte selenium levels. Both sources of selenium increase the activity of platelet glutathione peroxidase activity in acute experiments, but organic selenium is better than selenate in sustaining the enzyme activity once the supplementation is discontinued. All these observations suggest that organic forms of selenium are effectively deposited in tissue stores. Continuous accumulation of selenium in tissues might create a risk for health in subjects who ingest high amounts of organic selenium for long periods of time.

Experience in Finland since 1984 suggests that enrichment of agricultural fertilizers with selenate is an effective, quite predictable, and apparently safe way of increasing the selenium content of food and the selenium intake in populations with inadequate intake.

References

1. Levander, O.A. 1983. Fed.Proc. 42.

2. Schrauzer, G.N., D.A. White. 1978. Bioinorg.Chem. 8, 303.

3. Steiner, G., H. Menzel, I. Lombeck, F.K. Ohnesorge, H.J. Bremer. 1982. Eur.J.Pediat. 138, 138.

4. Levander, O.A., G. Alfthan, H. Arvilommi, C-G. Gref, J.K. Huttunen, M. Kataja, P. Koivistoinen, J. Pikkarainen. 1983. Am.J.Clin.Nutr. 37, 887.

5. Yang, G., L. Zhu, S. Liu, L. Gu, P. Qian, J. Huang, M. Lu. 1987. In: Selenium in Biology and Medicine (G.F. Combs Jr., L.E. Spallholz, O.A. Levander and J.E. Oldfield, eds.) AVI, p. 589.

6. Thomson, C.D., M.F. Robinson, D.R. Campbell, H.M. Rea. 1982. Am.J.Clin.Nutr. 36, 24.

7. Robinson, M.F., C.D. Thompson. 1983. Nutr.Abstr.Rev. Clin.Nutr. 53, 3.

8. Kumpulainen, J., L. Salmenperä, M.A. Siimes, P. Koivistoinen, J. Perheentupa. 1985. Am.J.Clin.Nutr. 42, 829.

9. Young, V.R., A. Nahapetian, M. Janghorbabi. 1982. Am.J.Clin.Nutr. 35, 1076.

10. Chen, X., X. Chen, G. Yang, Z. Wen, J. Chen, K. Ge. 1981. In: Selenium in Biology and Medicine (J.E. Spallholz, J.L. Martin and H.E. Ganther, eds.) AVI, p. 171.

11. Xu, G., S. Hong, H. Song, J. Xie. 1985. Nutr.Res. Suppl. I, 187.

12. Wen, Z., X. Chen, F. Ping, P. Qian, R. Liu, J. Huang. 1987. In: Selenium in Biology and Medicine (G.F. Combs Jr., J.E. Spallholz, O.A. Levander and J.E. Oldfield, eds.) AVI, p. 902.

13. Li, C., J. Huang, C. Li. 1987. In: Selenium in Biology and Medicine (G.F. Combs Jr., J.E. Spallholz, O.A. Levander and J.E. Oldfield, eds.) AVI, p. 934.

14. Liang, S., J. Zhang, X. Shang, S. Mu, F. Zhang. 1987. In: Selenium in Biology and Medicine (G.F. Combs Jr., J.E. Spallholz, O.A. Levander and J.E. Oldfield, eds.) AVI, p. 938.

15. Varo, P., P. Koivistoinen. 1980. Acta Agric.Scand.Suppl. 22, 165.

16. Mutanen, M., P. Koivistoinen. 1983. Int.J.Vitam.Nutr. Res. 53, 102.

17. Alfthan, G. Nutr.Res. (in press)

18. Yläranta, T. 1985. Increasing the selenium content of cereal and grass crops in Finland. Academic dissertation, University of Helsinki, Helsinki.

19. Varo, P., G. Alfthan, P. Ekholm, A. Aro, P. Koivistoinen. Am.J.Clin.Nutr. (in press)

A COMPARISON OF TEN SELENIUM SUPPLEMENTATION PRODUCTS.

J. Clausen and S.A. Nielsen

Institute of Life Sciences and Chemistry, University of Roskilde, DK-4000, Denmark.

Introduction

Selenium is recognized as an essential nutrient for humans and animals (1,2,3). Recent studies have brought evidence that Se may prevent cancer and stimulate the immune apparatus (6,7). Many populations have Se intakes in what is now considered the suboptimal range (8,9). Se in diets and dietary supplementation products occur in different forms: inorganic Se compounds and organic forms such as those occurring in animal tissue, plants and yeast, i.e. D,L and L-Se-methionine, Se-cystine and Se-cysteine. These compounds are known to have quite different properties in respect to bio-availability and biochemical pathways in animals (10,11,12). Since a widespread use of Se supplementation seems to be desirable, it is important to explore the effects of various form of Se supplementation.

In this study, the effects on Se-status of humans, resulting from a daily Se supplementation containing 200 µg of Se, were compared for 10 Se-preparations, reflecting the range of supplementation types available to consumers.

Organic Se-preparations, especially those containing pure L-Se-methionine, had the strongest and most stable effect on Se status and GSH-Px activity. In addition, our results indicate that Se assimilation and retention is improved in the combined "antioxidant cocktail" product "Bio-Selenium + Zinc".

Selenium in Medicine and Biology

Material and Methods

Subjects: 135 normal healthy individuals aged from 20 to 62 y, median of 36y, of which 115 were females. Normal values of whole blood selenium (bSe) and glutathione peroxidase activities (GSH-Px) were measured in all subjects. Apart from age and sex, information on smoking habits was recorded. Out of this group, the ten experimental groups were selected randomly. During 3 months period each subject received 200 µg of Se per day. Then during 4 months withdrawal followed. bSe-and GSH-Px were measured at onset, after 1 and 3 months of supplementation, and finally at the end of the withdrawal period. bSe and GSH-Px in erythrocytes were assayed as previously described (13,14).

Supplementation preparations used in this study:

Product name:		Manufacturer:	Declared as:
Pure Selenite		-	-
Pure Selenate		-	-
Al-Selena	*)	"Alko", SF	Seleno-yeast with vitamin E
Multi-vitamin	*)	"Brugsen", DK	Vitamin/mineral cocktail with selenite
Multi-vitamin	*)	"Ferrosan", DK	Vitamin/mineral cocktail with selenite
Selenium + ABDE	*)	"Nutana", DK	Vitamin/mineral cocktail with seleno-yeast
Nutrition 21	*)	"Nutrition 21", USA	Seleno-yeast
Selenium ACE	*)	"Wassen", GB	Vitamin/mineral cocktail with seleno-yeast
Pure L-Se-methionine		-	-
Bio-Selenium + Zinc		"Pharma Nord", DK	Antioxidant cocktail with L-selenomethionine

***) The 6 products marked with an asterisk will be referred to in the following as p.3 through p.8.**

Supplementation preparations: 3 of the preparations were reference preparations containing 3 pure Se compounds: selenite, selenate and L-Se-methionine. The other 7 preparations were the commercially available supplementation products mentioned in the table above (in random order).
The 6 commercial products marked with an asterisk will be referred to in the following as p.3 through p.8. The manufacturers of the seventh, "Bio-Selenium + Zinc" (Pharma

Nord, Vojens, Denmark) have supported this study, and have agreed to the identification of their product in the graphs and tables in the following.
The study was basically non-biased and double blinded concerning preparations. The dosage of 200 µg was given according to the manufacturers' declarations of Se contents. All preparations were analyzed for their actual Se content (table 2) by atomic absorption(hydride method).

Results

Smokers had significantly lower whole bSe values than non-smokers but identical GSH-Px activities (apart from a slightly lower value found in females with the t-BHP assay method) (Table 1). Fig.1 compares the developments in bSe throughout the experiment for ten experimental groups. Preparations containing organic Se gave rise to a stronger response (apparently, with a rather different time dependence) than did preparations with inorganic compounds. The change after one month in whole bSe for selenate (average went from 100 to 123 µg/l) and L-Se-methionine (from 120 to 200), was significant (p.05) The difference is even greater at 3 months (p.01). Similar differences are found, e.g., between the effects of selenite and L-Se-methionine (1 month: p.01, 3 months: p.01), between selenite and Bio-Selenium (p.01, p.01), between selenate and Bio-Selenium (p.01, p.01), and, even more pronounced, between the combined groups receiving L-Se-methionine preparations, and the combined selenite/selenate groups (p.002, p.002). Within organic Se preparations, pure L-Se-methionine showed a stronger effect on bSe than Se-yeast. This difference was non-significant after one month;after 3 months, however, the difference was significant (p.01). Whereas values for organic Se continue to rise during the whole 3 months' period of supplementation, the values for inorganic Se stagnated after 1 month (even with an slight decline). Furthermore, inorganic preparations did not induce a long term increase in bSe. bSE levels of the organic

Table 1: Whole blood selenium levels and the GSH-Px activities assayed either with t-BHP or with H_2O_2 as peroxide donors (numbersin parantheses are standard deviations).

	selenium, µg/L whole blood	GSH-Px t-BHP	GSH-Px H_2O_2
Whole group, n=135	109 (27)	56 (33)	81 (31)
Females, n=115	109 (28)	58 (35)	80 (29)
Males, n=20	110 (21)	42 (17)	90 (44)
Smokers, n=73	102 (16)	56 (38)	79 (26)
Non-smokers, n=62	118 (34)	56 (24)	83 (36)

GSH-Pxactivities are expressed in kat/mole and measured in erythrocytes.

Table 2: Actual selenium contents of experimented daily dosage forms (corresponding to a declared content of 200 µg Se)

Pure L-selenomethionine	209.8 µg
Bioselenium + Zinc	188.0
p.3	185.6
p.4	196.0
p.5	166.8
p.6	198.0
p.7	172.6
p.8	172.6
pure selenate	188.0
pure selenite	166.0

The mean error of this assay is estimated to 10 µg or 5 %. The largest deviation of actual daily dosage is about 35 µg or 20 %.

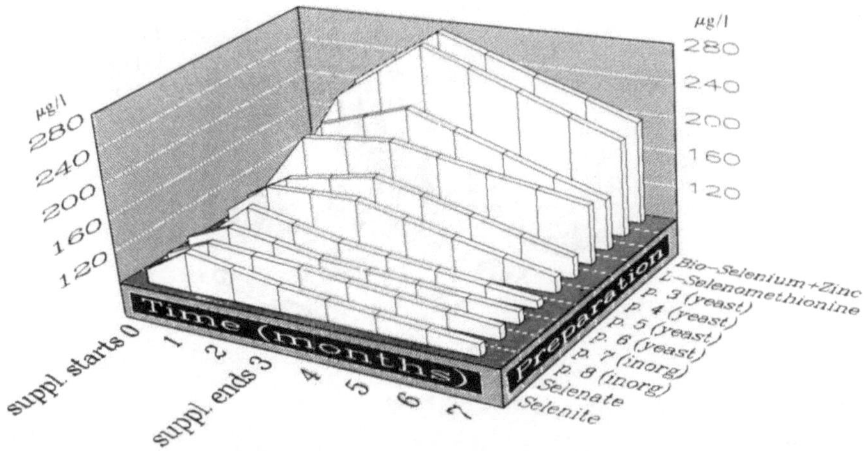

Figure 1: Whole blood selenium levels.

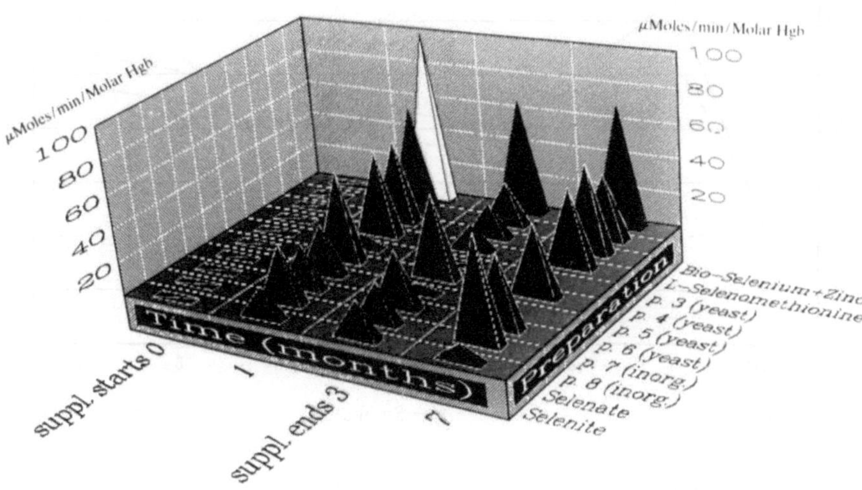

Figure 2: Changes in glutathione peroxidase activity.

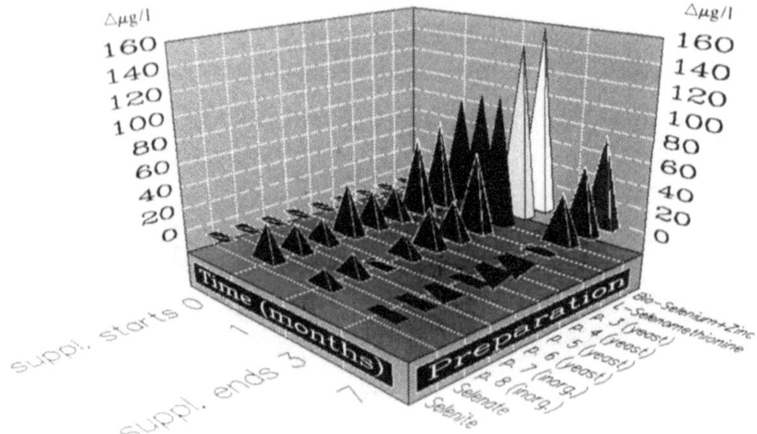

Figure 3: Changes in whole blood selenium content.

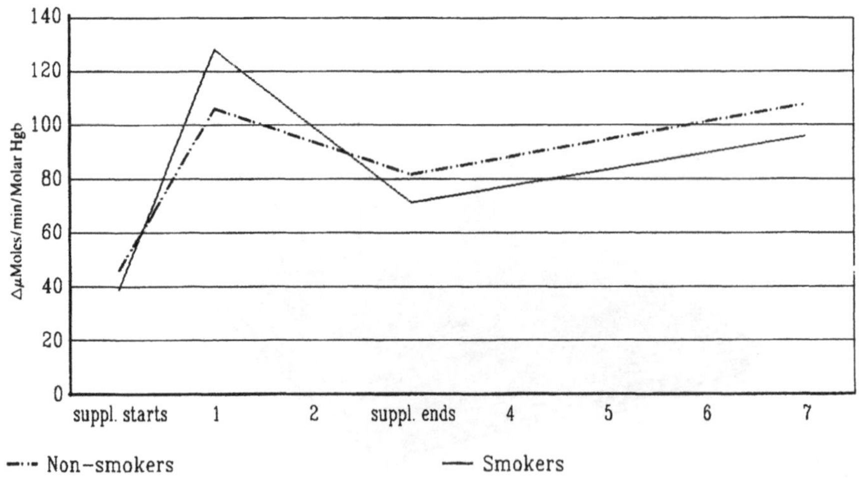

Figure 4: Glutathione peroxidase activity and smoking.

groups were much higher after the withdrawal period than at outset.

Fig.2 compares GSH-Px activities for 4 of 10 preparations. Inorganic forms gave rise to a protracted rise, and to a stable increased level lasting at least for 4 months of the withdrawal period. Again, the organic forms show a stronger effect with a different time factor: a fast peak after one month (although the peak values are not significantly higher than inorganic values), followed by a decline to values close to the inorganic 3 months values; and finally a steady level (with an insignificant increase) after the end of supplementation. Here, too, the differences between response to preparations based on L-Se-methionine and on inorganic Se, respectively, are statistically significant (p.02 in all cases).

Splitting data into subgroups of smokers and non-smokers (table 1 and fig. 4) reveals a tendency of higher bSe in non-smokers. After 3 months of supplementation, there is a clear trend of higher bSe and GSH-Px in GSH-Px bSe in smokers, than in non-smokers. (For the selenite group p.008; p0.08 for selenate). In smokers a fast rise in GSH-Px and bSe during the first month of supplementation was followed by a marked decline in GSH-Px.

Taking both parameters - bSe and GSH-Px - into consideration, obviously organic Se are seen in this study to have the strongest effect in changing selenium status. Within the preparations declared to contain organic Se the preparations with pure L-Se-methionine had a significantly stronger effect than yeast-based preparations. Finally the effects of the L-Se-methionine based antioxidant cocktail Bio-Selenium + Zinc were consistently even stronger than the effects of pure L-Se-methionine. However, it is interesting to note that this latest difference is observed in spite of an appr. 10% higher Se content in the pure L-Se-methionine preparation used in this study (see table 2).

Discussion

In accordance with earlier works (16,17,18,19) the present study indicates that both organic and inorganic Se compounds are absorbed from the intestinal tract, and that organic Se gives rise to higher serum or bSe values than the inorganic compounds. However, in this study it is found that erythrocyte GSH-Px does not quite follow the patterns of GSH-Px observed previously in lymphocytes. Furthermore, the decline of bSe towards normal values, observed in the 4 months' withdrawal period, was not reflected in a similar decline of erythrocyte GSH-Px values - probably due to the low turn-over rate of these cells. The differences in bSe between individuals receiving organic and inorganic Se may be related to differences in the tissue compartmentalization (20) - thus, organic Se shows a preference for binding to hemoglobin (11,12).

Although smokers showed significantly lower bSe at outset, this was not reflected in GSH-Px. The lower Se level in smokers than in non-smokers may be related to the body burden of heavy metals - e.g., cadmium (21) - since Se is able to inhibit the toxicity of heavy metals (22). Related to this the present study also showed selenium and GSH-Px to rise significantly faster in smokers than in non-smokers, during the first month of supplementation (although long-term effect of supplementation was similar in smokers and non-smokers). This finding should be elucidated further. However, Arduser et al. (10) have shown that in certain animals selenite and selenate are actively transported over the brush border of the intestinal tract in a Na-linked process; and it may well be that in smokers the heavy metals complex anions like sulfate, chromate and thiosulfate - all anions inhibiting the selenate transport. This may then result in a higher intestinal absorption and thereby a higher GSH-Px activity. However, in smokers also the mixed function oxygenase is activated (23), and since Se may stimulate the P-450 activation (24), the increase in the P-450 dependent peroxide generation may also

cause an increased synthesis of GSH-Px.
Generally, the different initial Se status, as well as the difference in supplementation response patterns, seems to indicate that Se requirement of smokers can be greater than that of non-smokers.
Like Levander et al. (25), we found that bSe reached a steady state after two months, and the GSH-Px activity reached a plateau after 3 months.
The present data clearly indicate the significant differences between the response from organic and inorganic Se, respectively. L-Se-methionine had a stronger and more stable effect on Se status than Se-yeast and inorganic Se. Se-yeast tended to produce a mixed response, somewhere in the range between the L-Se-methionine and the inorganic Se - probably reflecting an actual composition of these preparations as a combination of organic and inorganic Se compounds. Finally, the material indicates that L-Se-methionine coctailed with antioxidants i.e. Bio-Selenium + Zinc is the most efficient compound among the Se products tested - probably because of synergistic interactions between the antioxidant nutrients, in assimilation and metabolism.

References

1. H.E. Oksanen,.1980.Proc. Mineral Elements -80, part II, Helsinki, The Academy of Finland, p.445 (ISBN 951-99301-2-4).
2. E.J. Underwood.1977. Trace Elements in Human and Animal Nutrition (IV Edition), Academic Press Inc., N.Y., p.302.
3. Keshan Disease Research Group of the Chinese Acad.Sci.1979. Chinese Med.J., 92, 477.(1979)
4. A.M. van Rij, C.D. Thomson, J.M. McKenzie & M.F. Robinson.1979. Am.J.Clin.Nutr. 32, p.2076.
5. R.J. Stead, L.J. Hinks, M.E. Hodson, A.N. Redington, B.E. Clayton & J.C. Batten.1985. Lancet 19 oct., p.862.
6. L.C. Clark.1985.Fed.Proc. 44, p.2584.
7. S.Y. Yu, P. Ao, L.M. Wang, S.L. Huang, H.C. Cheng, Q.Y. Liu & X.P. Lu.1987.Proc. Present Status and Perspectives of Selenium in Biology and Medicine. Eup.Acad, p.18.
8. E.B. Thorling, K. Overvad & J. Geboers.1986.Ann. Clin. Res. 18,3.
9. J. Clausen & S.A. Nielsen.1988.Clin.Res, in press.
10. F. Arduser, S. Wolfram, E. Scharrer & B. Schneider.1986.Biol. Trace Element Res., 9, p.281 .
11. M.A. Beilenstein & P.D. Whanger.1986.J.Nutr.116, 1701.
12. M.A. Beilenstein & P.D. Whanger.1986.Nutr.116, 1711 .

13. G.E. Jensen, K.S. Shukla, G. Gissel-Nielsen & J. Clausen.1978. Scand.J.Clin.Lab.Invest., 38, p.309.
14. E. Beutler, K.G. Blume, J.C. Kaplan, G.W. Loehr, B. Ramot & W.N. Valentine.1977.Brit. J. Haematol. 35, p.331.
15. D.E. Paglia & W.N. Valentine.1970.J.Lab.Clin.Med., 70, p.158 .
16. G.N. Schrauzer & J.E. McGiness.1980.Proc. 2nd Internat. Symp. on Selenium in Biology and Med., Texas Tech. Univ.
17. M.F. Robinson, H.M.F. Rea, G.M. Friend, R.D.H. Stewart, P.C. Snow & C.D. Thomson.1978.Br.J.Nutr., 39, p.589.
18. O.A. Levander, G. Alfthan, H. Arvilommi, C.G. Gref, J.K. Huuttunen, M. Kataja, P. Koivistoinnen & J. Pikkarainen 1983 Am.J.Clin.Nutr 37, p.887
19. K. Jaakola, J. Tummavuori, A. Pirinen, P. Kurkela, M. Tolonen & A.U. Arstila1983.Scand.J.Clin.Lab.Invest., 43, p.473.
20. S.O. Jacobsson & E. Hansson.1965.Acta Veter.Scand., 6, p.287 .
21. G. Scherer & H. Barkmeyer.1983.Ecotox, Environm.Safety, 7, p.71 .
22. S.C. Rastogi, J. Clausen & K.C. Srivastava.1976.Toxicol. 6, p.377.
23. N. Singh & J. Clausen.1983.Cancer Letters 13, p.53.
24. R.F. Burk & B.S.S. Masters.1973.Arch.Biochem.Biophys. 170, p.124 .

SELENIUM STATUS IN HUMANS AS INVESTIGATED BY THE EFFECTS OF SUPPLEMENTATION WITH Se-ENRICHED YEAST TABLETS.

J. Nève, S. Chamart

Free University of Brussels, Toxicology and Bioanalytical Chemistry Unit, Campus Plaine, 205-1, B-1050 Brussels, Belgium

S. Van Erum, F. Vertongen

Free University of Brussels, Saint-Pierre Hospital, Department of Medical Chemistry, rue Haute, 322, B-1000 Brussels, Belgium

M. Dramaix

Free University of Brussels, School of Public Health, Laboratory for Epidemiology, route de Lennick, 808, B-1070 Brussels, Belgium

Introduction

Although selenium (Se) supplementation has received considerable attention in several countries (1), the Se impact on health remains insufficiently known in other parts of the world. People from Belgium or France appeared to have an "intermediate" Se status, with plasma Se levels in the range 60-100 µg Se/l, and their determined dietary Se intake is in the lower part of the estimated safe and adequate dietary intake of 50 to 200 µg Se per day (1-3). Parameters such as blood and intake levels are however not sufficient to assess the adequacy of Se status. Considerable variations in Se intake are indeed observed, and mechanisms for Se homeostasis are not completely understood (4).
Measuring responses in trace element dependent enzyme activities to supplementation has been suggested as a promising technique to this purpose. Indeed, in most cases, a response could only been observed if the trace element requirement for the enzyme is not met. Moreover, saturation of activity can be observed in some tissues at dietary intake that could be defined as satisfying or adequate nutritional requirement (2, 5).

Selenium in Medicine and Biology

In one of our previous investigations devoted to the characterization of sensitive indices for assessing changing Se status of man (6-8), we observed an increase in platelet glutathione peroxidase activity (a selenodependent enzyme, GPx) as well as a plateau of activity between 15 and 60 days supplementation of a group of Belgian subjects with DL-selenomethionine, 100 µg Se/day. Such a response, that occured for plasma Se levels in the range 110-120 µg Se/l, was only previously reported in subjects with rather low Se status (9-11). We then concluded that Se status of investigated group was insufficient for an optimal function of GPx.
These investigations are extended in the present study by determining the biological effects of another Se form, the Se-enriched brewer's yeast, which is consecutively dispensed to healthy subjects at two intake levels, 100 and 200 µg Se/day for a total of 60 days.

Subjects and Methods

Twenty healthy Belgian adults participated in the study whose protocol was approved by the Ethical Committee. A group of 10 subjects, 6 males and 4 females, 28 ± 5 years old, took the Se supplement (Se enriched brewer's yeast capsules, CELNIUM 50, manufactured by Expansion Aromatique Française, Sanofi, France), two capsules/day during the first month (100 µg), and 4 during the second one (200 µg). The second group of 10 subjects, 4 males and 6 females, 33 ± 6 years, received a placebo. Subjects consumed their usual diets.
Blood was taken by venipuncture before breakfast by days 0, 5, 15, 30, 35, 45 and 60 of the study, and urine was simultaneously collected, for analysis of Se in plasma, erythrocytes and urine and for determination of GPx activity in plasma, erythrocytes, and platelets, using H_2O_2 as substrate (8, 12, 13).
The data were analyzed for differences between values of the two groups at any given day by analysis of variance (IBM computer with SPSS software), and for time and treatment significance by ANOVA with repeated measures (ANOVA-RM, CDC computer with BMDP 2V software).

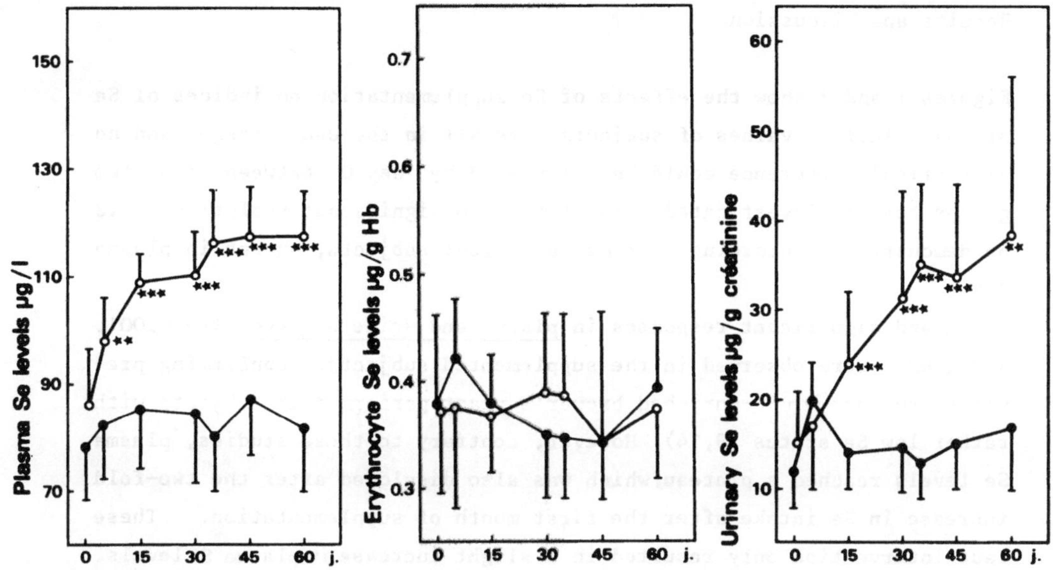

<u>Figure 1:</u> Effects of Se supplementation on plasma, erythrocyte and urine Se (Se-rich yeast, 100 µg/d during 30 days, then 200 µg/d the next 30 days). Each point represents the mean value and horizontal bars the standard deviation. ● : controls; O: Se-supplemented. The asterisks indicate the statistical significance for differences between values for controls and Se supplemented subjects (* $p < 0.05$; ** $p < 0.01$; *** $p < 0.001$, ANOVA).

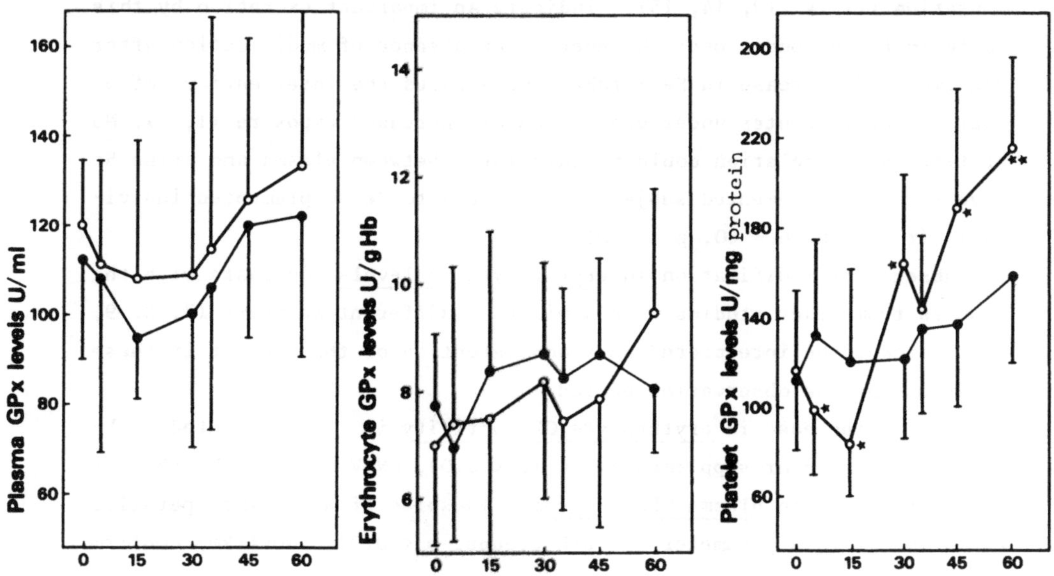

<u>Figure 2:</u> Effects on plasma, erythrocyte and platelet GPx (see also Fig. 1).

Results and Discussion

Figures 1 and 2 show the effects of Se supplementation on indices of Se status. Initial values of subjects were all in the usual range, and no statistical difference could be calculated by day 0 between the two groups for all investigated parameters. No significant evolution could be demonstrated according to time in control subjects, except in plasma GPx.

Rapid and significant responses in plasma and urine Se levels($p < 0.001$, ANOVA-RM) were observed in the supplemented subjects, confirming previous studies with Se-enriched brewer's yeast, performed in subjects with rather low Se status (9,14). However, contrary to these studies, plasma Se levels reached a plateau, which was also disclosed after the two-fold increase in Se intake after the first month of supplementation. These last intervention only resulted in a slight increase in plasma Se levels. Differences in the homeostatic regulation of plasma Se levels seem not only to depend on the chemical form of ingested Se (7-10, 14, 15), but also on the usual Se intake. Urine Se levels progressively increased during the first 30 days of the trial, but plateaued thereafter. Early changes in urine Se levels, which have been neglected in other supplementation trials (9, 14, 15), indicate an important excretion by this route in these conditions; however, the absence of modification after the two fold increase in Se intake demonstrates the intervention of other excretory routes under conditions of increased exposure (1, 2). No statistical correlation could be calculated between plasma and urine Se levels in unsupplemented subjects, contrary to Se supplemented individuals ($r = 0.36$, $n = 60$, $p < 0.01$).

The absence of modification in erythrocyte Se levels contrasts with the results from other studies with similar or different Se forms (7, 8, 9, 15), and could be interpreted as a non retention of the element in these cells during the observation period.

The slight increase in erythrocyte GPx activity in supplemented subjects at the end of supplementation ($p < 0.01$, ANOVA-RM) and the absence of modification in plasma GPx activity (despite of a slight parallel evolution of this parameter in both groups, $p < 0.01$, ANOVA-RM) confirm several studies showing the poor sensitivity of these parameters to changes in Se status, unless in subjects with marginal Se status (8). For the first indice, this has been related to the long life span of red blood cells.

In accordance with our previous study in similar subjects, but with a different Se form (8), platelet GPx activity clearly responded to supplementation ($p < 0.001$ for time-treatment interaction, ANOVA-RM). Such a response, that was only observed in subjects with rather low Se status (50-70 μg Se/l plasma), is therefore disclosed for the second time in subjects with usual plasma Se levels in the range : 70-100 μg Se/l. Moreover, the increase again reached significance when Se levels were approximatively 110-120 μg Se/l. This time however, contrary to our last study showing a plateau of activity after 15 days supplementation, platelet GPx activity continued to increase till the end of the study (ANOVA-RM), probably as a result of the two-fold increase in Se intake after 30 days supplementation. The value of platelet GPx in the assessment of Se status, that has been related to the very short life span of the cells, is confirmed in this study. Significant correlations could be calculated between plasma Se and platelet GPx, both in unsupplemented subjects ($r = 0.29$, $n = 80$, $p < 0.01$), and in supplemented individuals ($r = 0.43$, $n = 60$, $p < 0.001$), supporting the use of these parameters as functional indices of Se status (8).

In conclusion, the present trial again suggests that usual Se status in Belgium is insufficient for an optimal function of GPx, but efforts remain to be made to more precisely identity the adequate intake level. Finally, the Se form tested in this study was adequate to correct the biochemical deficiency.

Acknowledgements

This research is supported by grants from the "Fonds de la Recherche Scientifique et Médicale de Belgique", the Ministères de l'Emploi, du Travail et du Budget de Belgique, and by Expansion Aromatique Française and Laboratoires Millot-Solac, Sanofi, France.

References

1. Nève, J., M. Henry, A. Peretz, J.P. Mareschi. 1987. Cah. Nutr. Diet. 12, 145.

2. Nève, J., F. Vertongen, L. Molle. 1985. Clin. Endocrinol. Metab. 14, 629.

3. National Academy of Sciences, RDA. 1980. Washington, DC, 9th ed.

4. Robinson, J.R., M.F. Robinson, O.A. Levander, C.D. Thomson. 1985. Am. J. Clin. Nutr. 41, 1023.

5. Levander, O.A. 1986. J. Am. Coll. Nutr. 5, 37.

6. Nève, J., F. Vertongen, N. Thonnart, Y.A. Carpentier. 1984. In : Trace Element Analytical Chemistry in Medicine and Biology (P. Brätter, P. Schramel, eds), W. de Gruyter, Berlin, 3, 139.

7. Nève, J., F. Vertongen, N. Thonnart et al. 1986. Acta Pharmacol. Toxicol. 59, 166.

8. Nève, J., F. Vertongen, P. Capel. 1988. Am. J. Clin. Nutr., 48, 139.

9. Levander, O.A., G. Alfthan, H. Arvilomni et al. 1983. Am. J. Clin. Nutr. 37, 887.

10. Thomson, C.D., L.G. Ong, M.F. Robinson. 1985. Am. J. Clin. Nutr. 41, 1015.

11. Menzel, H., G. Steiner, I. Lombeck,F.K. Ohnesorge. 1983. Eur.J. Pediatr. 140, 224.

12. Nève, J., S. Chamart, L. Molle. 1987, ibid ref 6, 4, 349.

13. Nève, J., M. Hanocq, L. Molle, G. Lefebvre. 1983. Analyst, 107, 934.

14. Kumpulainen, O.A., L. Salmenpera, M. Siimes et al. 1985. Am. J.Clin. Nutr. 42, 829.

15. Luo, X., W. Wei, C. Yang et al. 1985. Am. J. Clin. Nutr. 42, 439.

EFFECTS OF SELENIUM SUPPLEMENTATION WITH Se ENRICHED YEAST TABLETS ON HEPATIC, MUSCULAR, RENAL AND HEMATOLOGICAL PARAMETERS IN HUMANS.

S. Van Erum, P. Capel, F. Vertongen

Free University of Brussels, Saint-Pierre Hospital, Department of Medical Chemistry and Hematology, rue Haute, 322, B 1000 Brussels

J. Nève

Free University of Brussels, Toxicology and Bioanalytical Chemistry Unit, Campus Plaine, 205-1, B 1050 Brussels

Introduction

The effects of a 60 days Se supplementation on hepatic, muscular, renal and hematological parameters were investigated in a group of 10 healthy adults aged from 25 to 43 years. Results were compared with those obtained from 10 controls receiving placebo (brewers yeast). For the conditions in which this study was performed we refer to "Selenium status in humans as investigated by the effects of supplementation with Se - enriched yeast tablets by J. Nève and al.

Methods

Hepatic function was assessed by aspartate aminotransferase (AST), alanine aminotransferase (ALT), γ-glutamyltransferase (γ-GT) and alkaline phosphatase (ALP) activities determination. Muscular status was tested by aldolase, creatine kinase (CK) and lactate dehydrogenase (LDH) activities. All these tests were performed on a RA 1000 (Technicon Inc.). Renal function was assessed by serum creatinine, urea and uric acid measurements (UA) performed on AAII (Technicon).

Selenium in Medicine and Biology

Hematological parameters, red blood cell-, leucocyte- and platelet counts as well as hemoglobin (Hb) level were performed on a ELT 800 (Ortho).

Results

No abnormal values were observed in the supplemented nor in the control group and we did not find any significant differences between mean values of supplemented and control populations for any parameter tested.

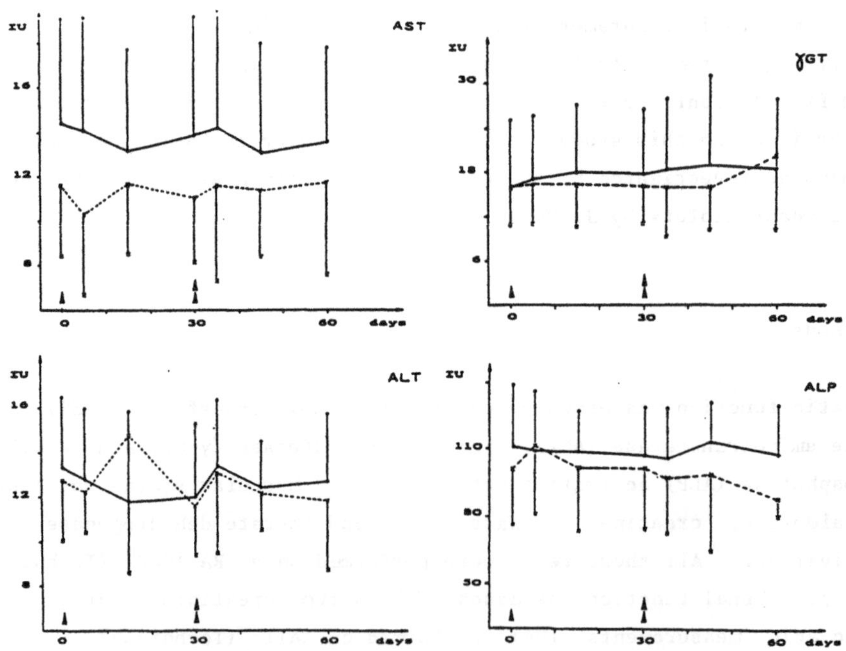

Fig 1. Effects of Se supplementation on hepatic enzymes.

. supplemented subjects x controls

▲ 100 µg Se/d ▲▲ 200 µg Se/d

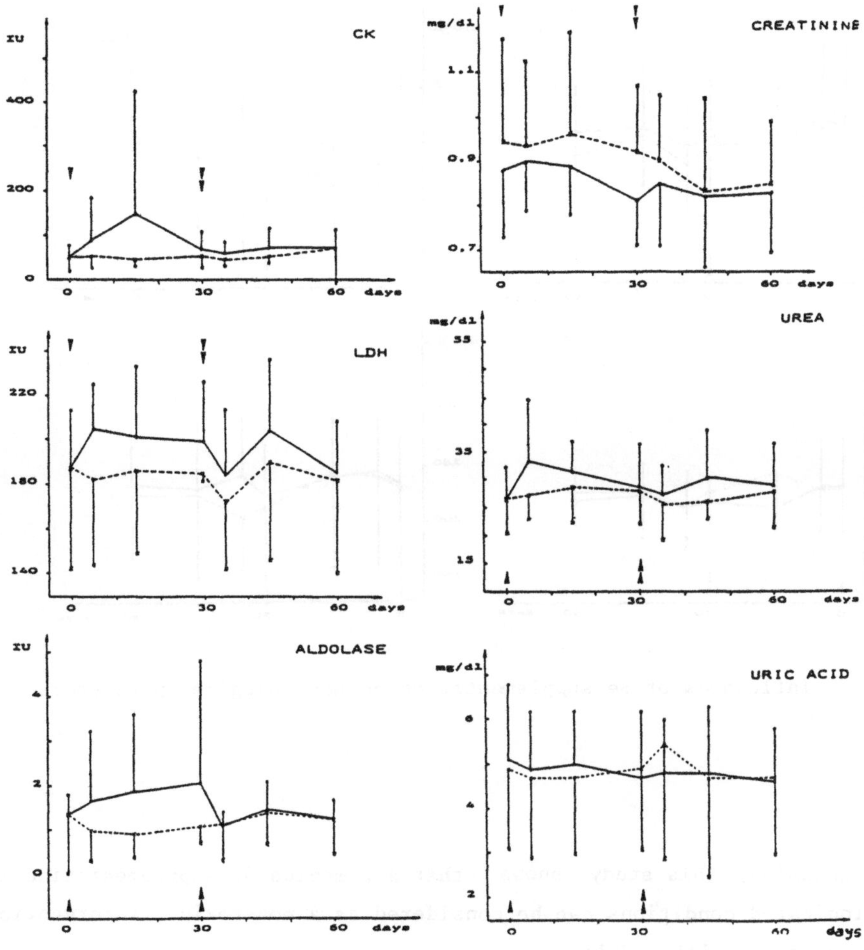

Fig 2. Effects of Se supplementation on muscular and renal parameters.

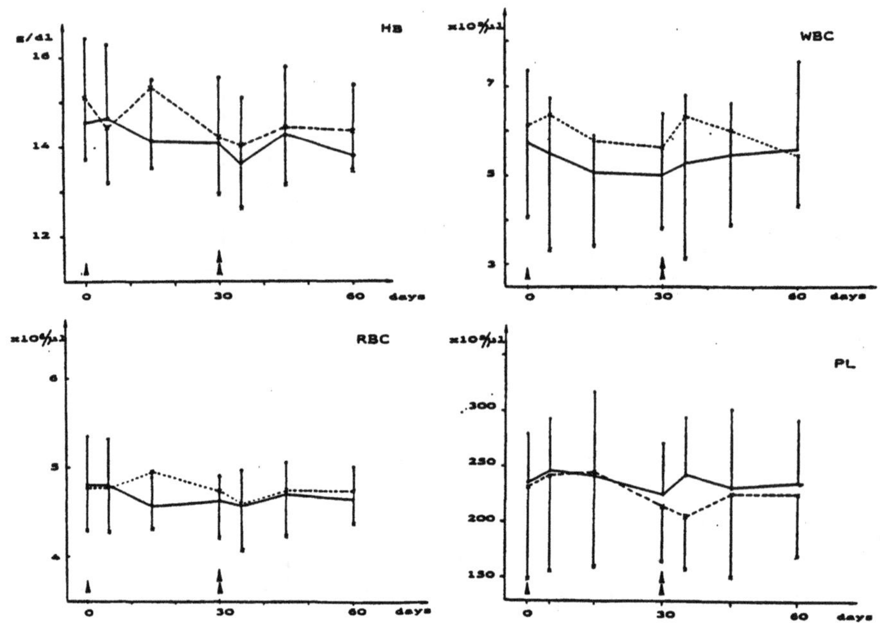

Fig 3. Influences of Se supplementation on hematological parameters

In conclusion, this study shows that a 2 months Se supplementation in the indicated conditions can be considered as a non-toxic intervention for a normal healthy adult.

SELENIUM CONTENT OF THE MINERAL WATER OF LA ROCHE POSAY. INTEREST FOR DIETARY SUPPLEMENTATION

S. Ducos-Fonfrède, F. Clanet

Bio-inorganic Chemistry and Hydrology Laboratory, UER of Pharmaceutical Science, University of Tours. F 37042 Tours

E. Delrez

Dermatologist; 39, avenue des Fontaines, F 86270 La Roche Posay

Introduction

La Roche Posay spa town is located near Poitiers at about 300 km in the South of Paris. For a long time, this spa has been known for treatment of patients with various skin disorders. The treatment consists in topical hydrotherapeutical procedures as filiform douches, baths and sprayings; and in drinking mineral waters. These waters have a selenium content at level reaching 120 µg/l; the mean value being 55 µg/l. A daily intake of about 75 µg of selenium (for drinking 1,5 l per day) can be estimated. The skin diseases more frequently treated by La Roche Posay spa water are psoriasis and eczema. The clinical effects are encouraging and it seems that selenium and glutathione-peroxidase deficiency might influence the clinical symptoms (1).

The purpose of this study was to evaluate hair selenium level, before and after a water ingestion period, utilising only new hair growing during the experimental period. Although blood values are the most widely used indicator of selenium status, we choosed hair as a diagnostic tool in this study (2). Indeed, hair collection is a less invasive procedure, can be stored without loss and is a potential indicator of long term accumulation (3). This investigation is however only preliminary, as hair analysis is a limited procedure to assess nutritional status of individuals (4).

Selenium in Medicine and Biology

Materials and Methods

The mineral springs of La Roche-Posay : mineral content :

The two main springs are CONNETABLE and MELUSINE. They provide drinking water. Table 1 shows their general characters and table 2 their major minerals content.

Table 1 : General characters

	CONNETABLE	MELUSINE
Temperature(•C)	13	12.7
Resistivity (Ω.cm 20°C)	1040	1531
Total residue dried at 180°C (mg/l)	582	407
pH	7.06	7.43

Table 2 : Most abundant minerals (mg/l)

	CONNETABLE	MELUSINE
Bicarbonate	408.7	390
Sulphate	141.6	51.6
Chloride	47.7	16.8
Silicate	53.5	53
Calcium	199.2	152
Sodium	17	9.5
Magnesium	9.72	4.13
Potassium	1.8	2.65

These tables show that major minerals are calcium, sulphate and silicate. Table 3 summarizes trace minerals content. The most abundant are selenium and zinc. Selenium is mainly in the form of selenate, while selenite is less concentrated. Arduser et al. (5) demonstrated that selenate is actively absorbed by the small intestine through the sulphate carrier and that the inhibition by sulphate of selenate intake occurs when the sulfate concentration exceeds the selenate by 1000-fold. As seen in table 3 the best sulphate-selenate ratio is for water from MELUSINE spring.

Table 3 : Trace Minerals in La Roche Posay Waters (µg/l)

	CONNETABLE	MELUSINE
Selenate	75.1	111
Selenite	0.8	1.1
Copper	20	13
Zinc	310	58
Sulphate/Selenate	1885	464.8

Patients and hair sampling

The study was carried out in ten patients with skin disorders; five psoriasis, four eczema, and one acne. They drank mineral water, 1,5 l per day, during three weeks. Their age ranged from 20 to 50 years. No subject used selenium containing shampoos for at least six months prior to beginning the experiment. They were instructed to do the same during the experimental period. Hair samples were collected from the sub-occipital area of the head close to the scalp and stored in paper envelopes until analysis. On the first sampling, before drinking mineral water (day = 1), only two cm hair, in immediate proximity of the head was kept for analysis. At the last sampling (day = 21), all the hair clipped was analysed. Prior to analysis, hair samples (= 100 mg) were washed with bidistilled water and dried; then samples were digested with nitric, perchloric acids and hydrogen peroxide. The pH was adjusted to 1-2 and the chloro-diaminobenzene solution added. The piaz-selenol formed is then extracted by shaking with toluene, and selenium measurement performed by atomic absorption spectrophotometry (6) .

Results and Discussion

Table 4 : Recovery of Selenium from Hair : (ppm)

Psoriasis		Day=1	Day=21	Eczema	Day=1	Day=21
N°	A	0.14	0.19	F	0.01	0.15
	B	0.04	0.30	G	0.01	0.15
	C	0.09	0.20	H	0.04	0.12
	D	0.03	0.30	I	0.04	0.14
				J	0.08	0.11
Acne						
	E	0.27	0.45			

These results indicate that daily ingestion of mineral waters from La Roche Posay increased hair selenium level. However this study is only a preliminary step and should be controlled by other measurements, including plasma and erythrocyte selenium levels and glutathione-peroxidase activity. Mollin and Wester found that 7 µg per day can be lost via desquamation, that is increased in psoriasis and eczema. Therefore, in these skin diseases, change in selenium metabolism is possible. The clinical effects observed in the patients treated in La Roche Posay spa are sufficient reasons to continue this study.

References

1. Juhlin, L., L.E. Edqvist, L.G. Ekman, K. Ljunghall, M. Olsson. 1982. Acta Dermatovener. (Stockholm) 62, 211-214.

2. Bowen, H.J. 1972. Sci. Tot. Environ, 1, 75-78.

3. Maugh, T.H. 1978. Science 202, 1272.

4. Barret S. 1985. Jama 254, 8, 1041-1045.

5. Arduser, F., S. Wolffram, E. Scharrer. 1985. J. Nutr. 115, 9, 1203-1208.

6. Nève, J., M. Hanocq, L. Molle, G. Lefebvre. 1982. Analyst 107,934-941.

ANTIOXIDANT SUPPLEMENTATION REDUCED SERUM LIPID PEROXIDES IN ELDERLY. A Randomized double-blind one-year study.

M. Tolonen, S. Sarna, M. Keinonen
University of Helsinki, SF-00290 Helsinki, Finland

M. Halme
Hartola Health centre, SF 19600 Hartola, Finland

U-R. Nordberg
Killinmäki, SF 02400 Kirkkonummi, Finland

Introduction

In our recent preliminary study geriatric symptoms decreased and the general condition improved among elderly who were supplemented with selenium and vitamin E (1). Here we wish to report biochemical and psychological improvements in a further study on elderly in a Finnish nursing home.

Material and methods

Seventy elderly were selected for the study. They were matched and randomly allocated into the propositae or placebo groups. The mean age was 80 years (range 65-96). In all 45 subjects (21 propositea, 24 placebo) completed all the tests under investigation.
Whole blood selenium (B-Se) was analyzed by means of hydride AAS, erythrocyte glutathion peroxidase (E-GPx) by a modified method (2), TBA reactants (lipid peroxides) in serum according to (3), serum alpha-tocopherol and retinol by means of HPLC. Psychological test included Wechsler memory scale (WAIS), WAIS number series, clock recognition and clock time drawing, logical learning, Luria 10 words learning test, and other cognitive tests.

Selenium in Medicine and Biology

The significance of the difference between the propositae and subjects on placebo were analyzed by means of Wilcoxon's test for independent samples. The changes within the groups were tested using Wilcoxon's signed sanks test (Wilcoxon's pair test), using normal approximation (two-tail).

Supplements

The daily dietary supplements for the elderly included one tablet (Bio-Selen + zinc[R]) containing 9 mg beta-carotene, 2 mg pyridoxal-HCL, 90 mg ascorbate, 60 mg d- α -tocopheryl acetate, 15 mg zinc as gluconate, and 100 µg organic selenium (N-21), one capsule of sodium selenate decahydrate (800 µg inorganic selenium), and 400 mg d-α--tocopheryl acetate (Ido-E[R]). The control group took corresponding placebos.

Results

The Mean B-Se in the propositae tripled (from a level of 85 to 245 µg/l), and the E-GPx increased from 17-20 to 27-28 U/gHb. Plasma α -tocopherol also tripled from 14 to 50 µmol/l, whereas retinol did not change.
The mean levels of serum TBA reactants (µmol/l) were initially relatively high as compared to an <u>ad hoc</u> group of 31 younger controls (2.7 ± 0.7 vs 2.3 ± 0.6, $p < 0.01$). The subjects on placebo showed no change, but the level amongst the propositae decreased in three months to 2.2 ± 0.6, and stayed at this level until the end of the study year (Figure 1). Serum TBA reactant concentration had a significant inverse correlation ($R = -0.428$, $p < 0.01$) with the B-Se (Figure 2).

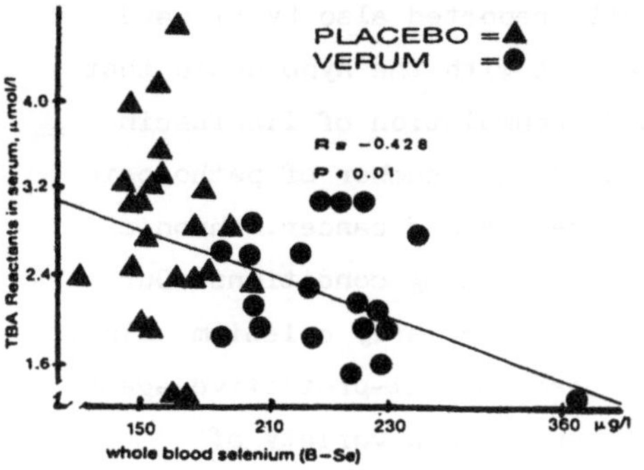

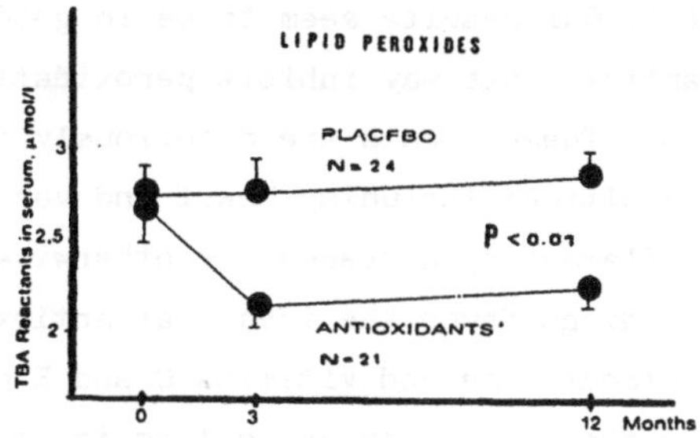

Psycological tests indicated that the propositae improved slightly in several tests, but only the clock drawing task showed a statistically significant improvement ($p < 0.01$) as compared to their own initial values as well as with the placebo group. The subjects on placebo deteriorated slightly in several tests and in the other tests their results remained unchanged. Clinical improvement was reported by the nurses in individuals of the antioxidant group. No adverse side effects were reported.

Discussion

Significant reduction of serum lipid peroxides was attributable to administration of an antioxidant cocktail. The relatively high initial levels of TBA reactants levelled down to the niveau observed in healthy controls. A similar decrease has been obtained with a combination ascorbic acid and α-tocopheryl (4). A significant negative correlation existed between serum TBA reactants and B-Se; a corresponding correlation was more recently reported also by Richard (5). Our results seem to be in good agreement with the hypothesis that antioxidants may inhibit peroxidation and accumulation of lipofuscin (6). These agents are notoriously associated in a number of pathologic conditions including heart and vascular diseases and cancer, chronic inflammatory diseases, or otherwise life-threatening conditions. Our findings favor the idea that antioxidants, particularly selenium, zinc, betacarotene and vitamins C and E may be used as cell-protective agents in the prevention as well as in adjunct therapy in a variety of pathological conditions.

References

1. Tolonen, M., Halme, M., Sarna, S., 1985 Biol. Tr. El Res. 7, p. 161-168.

2. Beutler, E., Blume, K., Kaplan, C., Löhr, G., Ramot, W., Valentine, W., 1988, Br. J. Naematol., 35, 331-340.

3. Yagi, K., 1982, In: Lipid peroxides in Biology and Medicide. Academic Press, New York, p. 223-242.

4. Wartanowicz, M., Panczenko-Kresowska, B., Ziemlanski, S. et al. 1984, Ann Nutr. Metab., 28, pp. 186-191.

5. Richard, M-J., Jurkovitz, C., Arnaud, J. et al., 1988, Selenium in Medicine and Biology. Second congress on trace elements in medicine and Biology. Avoriaz (France). March 15-18th, in press.

6. Tolonen, M., Halme, M., Sarna, S., Clausen, J. 1988. The fourth Biennial General Meeting of the society for free Radical Research International conference on Medical, biochemical and chemical aspects of free redicals, April 9-13, 1988 Kyoto, Japan, in press.

SELENIUM AND VITAMIN E SUPPLEMENTATION IN CYSTIC FIBROSIS.

P. Thérond, M. Marchand, D. Biou, J.F. Demelier, F. Brion
Laboratory of Biochemistry and Pharmacy, Bretonneau Hospital, 2, rue Carpeaux, 75018 Paris, France.

P. Foucaud, J. Navarro
Unit of Gastroenterology, Bretonneau Hospital

Introduction

In spite of 40 years of research it is only to day that a sensitive approach of the genetic localisation has been realised (1). A number of abnormalities in the membrane of CF patients have been observed in vitro concerning the increase of fatty acid turnover in the phospholipids of the red blood cells (2). The purpose of this study was to determine the main antioxidant agents that may affect the ability of the organism to protect biomembranes against oxidative attack : selenium and vitamin E. The other questions are whether children with CF are at risk to develop selenium deficiency and if the supplementation in vitamin E proposed is sufficient (100 mg/day). We assessed the selenium status of our patients in two phases : short and long term phases by the determinations of selenium in plasma and erythrocytes and by a functional phase : red blood cell glutathione peroxidase activity (RBC-GPx). We also measured tocopherol concentrations in plasma reported to total lipids.

Material and methods

26 patients with CF participated in the study. The subjects ranged in age from 1 month to 18 years (mean of age : 5 years). All have severe exocrine pancreas dysfunction with steatorrhea. For 6 to 26 patients (Group 3) the diagnosis of CF was just established at the beginning of the study.

Selenium in Medicine and Biology

They were not supplemented in vitamin E and did not receive pancreatic replacement therapy. The 20 others (Group 2) with normal fat intake received 100 mg/day of oral acetate tocopherol. The control group (Group 1) was obtained from 19 healthy fasting children (mean of age : 8 years). After an overnight fast, venous blood was collected and was separated for measurement of plasma and RBC selenium by atomic absorption spectrophotometry, RBC-GPx activity using the coupled test with cumene hydroperoxide as substrate as described by Paglia and Valentine and vitamin E by high pressure liquid chromatography.

Results

There is a moderate deficiency in selenium in CF patients. In fact plasma selenium values are more significantly reduced ($p < 0.001$) as compared to the RBC selenium values ($p < 0.01$). This low decrease in RBC is confirmed by RBC-GPx activity which is near normal range ($p < 0.05$). Vitamin E follows the same evolution than RBC selenium with a moderate decrease. Furthermore,reported to total lipids the deficiency in vitamin E appears more pronounced(Table 1). In the CF patients from group 3 (without therapy) the deficiency in selenium and vitamin E are, on the other hand, more clear. The decrease in RBC selenium is then sufficient to reduce RBC-GPx.

Table 1 : Selenium and vitamin E status of CF children as compared to a control group (mean ± SD).

	Plasma selenium µg/l	RBC selenium µg/g Hb	RBC GPx UI/g Hb	Vit. E/ to. lipids mg/g
Group 1	136	0.84	45	2.60
	40	0.20	8	0.40
Group 2	80	0.63	36	1.73
	20	0.20	6	0.70
Group 3	47	0.32	27	1.34
	30	0.10	10	0.70

Discussion

The importance of selenium in nutrition is now well established. Deficiency has been involved in the aetiology of CF (3). This hypothesis is to day rejected. Furthermore, a significant decrease in selenium has been observed in many studies (4). Our study confirm these results with the difference that the CF patients without digestive therapy have a strong decrease of all parameters. Children with exocrine pancreatic insufficiency might therefore have impaired absorption and secondary deficiency of this essential micronutrient (5). The double deficiency in selenium and vitamin E during CF might be in part responsible of membrane abnormalities by increasing peroxidation of fatty acids. Our results suggest (particularly in Group 2) that an elective supplementation in selenium is not necessary. In fact the selenium content of the usual pancreatic preparations appears to be sufficient in providing children 10 to 20 µg/day of selenium (6). A contraindication for selenium supplementation exists in animals deficient both vitamin E and selenium, because toxicity occurs at a lower level of selenium administration when vitamin E is low (7). In spite of this toxicity it seemed suitable to supplement children at the beginning of treatment (Group 3) when RBC-selenium is lower than 0.40 µg/g of hemoglobin. This supplementation should occur after correction of vitamin E status to avoid a toxicity of selenium. On the other hand, the proposed supplementation in vitamin E (100 mg/day) could be increased to 200 mg/day like recommended by different authors (8).

References

1. Frezal, J., O. Haguenauer-Cohen, N. Van Cong. 1986. Arch. Fr. Pediatr. 43, 225.

2. Rogiers, V., I. Dab, Y. Michotte. 1984. Pediatr. Res. 18, 704.

3. Wallach, J.D., B. Garmaise. 1979. In : Traces substances in environmental health (Hemphill DD, eds.). University of Columbia Press, Missouri, p. 469.

4. Castillo, R., C. Landon, K. Eckhardt. 1981. J. Pediatr. 99, 583.

5. Heinrich, H.C., E.E. Gabbe, H. Bartels. 1977. Klin. Wschr. 55, 595.

6. Van Caillie-Bertrand, M., F. De Bieville, H. Neijens. 1982. 71, 203.

7. Van Vleet, J.F., K.B. Meyer, H.J. Olander. 1974. 165, 543.

8. Spock, A., D. Ludlow, C. Shoaf. 1986. In : 14th annual meeting of the european working group for cystic fibrosis Budapest, Hungary , p. 33.

PLASMA AND ERYTHROCYTE SELENIUM (Se), GLUTATHIONE PEROXYDASE (GSH-Px), MALONDIALDEHYDE (MDA) AND PLASMA LIPID HYDROPEROXIDES (LH) AS A FUNCTION OF Se SUPPLEMENTATION IN 12 TREATED PHENYLKETONURIC (PKU) CHILDREN.

B. Wilke, M. Vidailhet, C. Guillemin.
Faculté de Médecine de Nancy, France.

A.Favier, Y. Ducros, J. Arnaud, M.J. Richard.
Faculté de Médecine de Grenoble, France.

Introduction

Phenylketonuric (PKU) children are treated with diets containing only minute amounts of natural proteins such as meat, fish, cereals or dairy products which afford usually 93-96 % of total ingested Se. Their main nitrogen intake consists of special mixtures (AMINOGRAN, P.K.U. 2, MAXAMAID X.P.) deprived of L-phenylalanine. Those products are supplemented by vitamin and trace elements, but in most of the cases, not by Se. Then, nutritive Se deficiency can be anticipated and effectively often occurs in treated PKU.
The present study was undertaken to evaluate more precisely this Se deficiency and the effect of Se supplementation on plasma and red blood cells Se and GSH-Px activities as well as possible biochemical effects of an abnormal lipid peroxidation.

Patients and Methods

The study population consists of 12 PKU children (10 females). Their mean age is 4 years with a range from 11 months to 9 years and 3 months. Their protein intake mainly consists of L-amino-acid mixtures : 10 children receive MAXAMAID X.P. (1) (product Se free) and 2 receive a new product :

Selenium in Medicine and Biology

PCU II - F.D. 8 (2) (1.47 µg of Se per g of amino acid corresponding to 2.7 µg/kg/day). Control values are obtained from children paired for age and unaffected with inflammatory, digestive, or nutritional disease. Blood is collected at 9 a.m. after 12 h of fasting and kept into heparinized, trace element free tubes. After centrifugation, the plasma and erythrocytes GSH-Px activities are immediately measured. Plasma and red blood cells (RBC) are frozen at - 20°C until Se, malondialdehyde (MDA) and lipid hydroperoxides (L.H.) analyses. Plasma electrolyte values are controlled at every time of the study and blood cells count and vitamin E at D 0. Plasma Se is measured by atomic absorption spectroscopy (3), erythrocyte Se by stable isotope dilution GC/MS (4), plasma and erythrocyte MDA by fluorimetric assay (5), plasma LH by the enzymatic method of Health R.L. and Tappel AL (6) and plasma and RBC GSH-Px by the method of Paglia and Valentine (7). Wilcoxon's test is utilized for the statistical analysis.

Results

The mean values and standard deviations for Se and GSH-Px in plasma and RBC before and after Se supplementation are indicated in table 1.

Table 1 : Effect of Se Supplementation as Sodium Selenite (10 µg/kg/day) on Plasma, Red Blood Cells (RBC) Se Levels and GSH-Px Activities

	PLASMA				RBC			
	Selenium		GSH-Px		Selenium		GSH-Px	
	µmol/l	n	U/g prot/mn	n	µmol/l	n	U/ml/mn	n
Control values	1.11 ± 0.29	12	2.7 ± 0.4	12	0.90± 0.26	12	10.4 ± 1.7	12
Day 0	0.27 ± 0.19	12	1.0 ± 0.7	12	0.42± 0.19	12	5.1 ± 2.2	12
Day 2	0.46 ± 0.24	11	1.3 ± 0.6	11	0.54± 0.22	10	5.9 ± 1.7	11
Day 10	0.82 ± 0.36	12	1.9 ± 0.5	12	0.61± 0.30	12	6.3 ± 2.1	12
Day 28	1.07 ± 0.37	12	2.7 ± 0.4	12	0.77± 0.20	10	6.9 ± 2.1	12
Day 56	1.17 ± 0.24	10	2.4 ± 0.6	12	0.95± 0.29	10	8.8 ± 1.8	12
Day 120	0.90 ± 0.29	5	2.5 ± 0.6	11	0.79± 0.16	5	9.3 ± 2.3	11

Plasma and erythrocyte Se deficiencies are confirmed as well as decreased plasma and red blood cells GSH-Px activities. The specific oral supplementation with sodium selenite results in an increase of plasma Se and GSH-Px. The values of plasma Se and GSH-Px are already significantly (respectively $p < 0.01$ and $p < 0.05$) higher after 2 days. Despite the small number of cases studied until now at 120 days, plasma Se seems to reach a plateau after 1 month of treatment. Erythrocyte Se levels and GSH-Px activity increase more slowly; within 10 days however the values are already significantly ($p < 0.01$) higher; erythrocyte GSH-Px activity reaches normal levels only after 2 months. Plasma products of polyinsatured fatty acids peroxidation assayed as LH as well as plasma and red blood cells MDA showed no significant change after supplementing with Se during 2 months (table 2).

Table 2 : Effect of Se Supplementation on Plasma Lipid Hydroperoxides (HL µmol/l) plasma and RBC Malondialdehyde (MDA µmol/l)

		Day 0	Day 2	Day 10	Day 28	Day 56	normal adult values
PLASMA	HL	143 ± 66	120 ± 53	154 ± 43	149 ± 53	153 ± 64	62 ±102
	MDA	2.74±0.51	2.92±0.41	2.61±0.34	2.76±0.75	2.63±0.45	2.36±0.45
RBC	MDA	6.28±2.29	5.3 ±1.5	6.22±2.62	6.64±1.99	6.03±2.93	

Highly significant ($p < 0.01$) positive correlations are found between : 1-plasma Se levels and plasma GSH-Px activity ($r = 0.66$); 2-red blood cell Se levels and RBC GSH-Px activity ($r = 0.59$); 3-RBC and plasma Se ($r = 0.62$); 4-RBC and plasma GSH-Px activities ($r = 0.52$).

Discussion

PKU children on L-amino-acid mixture or protein hydrolysate diets have low plasma and R.B.C. Se levels and GSH-Px activities without any apparent clinical consequence. Their mean plasma and R.B.C. Se values correspond respectively to 56 % and 49 % of mean control values. The lowest values of plasma and RBC Se are respectively at 2.2 % and 17 % of normal means. Two children receiving a Se supplemented mixture (PCU 2 - FD 8, affording 2.7 µg/kg/day) since 15 months, have the same low plasma Se levels (0.53 and 0.23 µmol/l), erythrocyte Se levels (0.44 and 0.46 µmol/l) plasma GSH-Px (2.9 and 1.6 U/g prot/mn) and RBC GSH-Px (6.9 and 8.8 U/ml/mn) proving the insufficiency of this level of Se supply. These facts contrast with the normal estimated Se needs (2-5 µg/kg/day) perhaps because of a poor bioavailability of selenium salt added to amino-acid mixture. Comparatively to normal plasma adult values, M.D.A. and L.H. are high in these Se deficient children and don't vary 2 months after the beginning of Se supplementation.

References

1. Maxamaid X.P. : Liverpool Scientific supplies limited (England)

2. PCU 2 - FD 8 : Gallia Laboratory (France)

3. Nève, J., L. Molle. 1986. Acta Pharmacol. Toxicol. 59, 606-609.

4. Reamer, D.C., C. Vellon. 1981. Anal. Chem. 53, 2166-2169.

5. Dousset, J.C., M. Trouilh, M.J. Foglietti. 1983. Clin. Chim. Acta, 129, 319-322.

6. Heath, R.L., A.L. Tappel. 1976. Ann. Biochem. 76, 184-191.

7. Paglia, D.E., W.N. Valentine. 1967. J. Lab. Clin. Med. 70, 158-169.

SELENIUM FERTILIZATION IN PRACTICE IN FINLAND

T. Yläranta
Agricultural Research Centre of Finland
SF-31600 Jokioinen

The selenium content of Finnish agricultural soil usually varies between 0.1 mg/kg and 0.6 mg/kg. The highest selenium content, in some cases exceeding 1 mg/kg, has been found in organogenic soil. The selenium content of mineral soils is closely dependent on the clay fraction of the soil. On average, only 4 % of the total selenium present in soil is extractable in hot water (1).

The natural selenium content of Finnish crops is very low, averaging only 0.01-0.02 mg/kg of dry matter. Attempts have been made in Finland since the 1960's to ensure an adequate supply of selenium for animals by adding selenium to commercial feeds and through selenium medication.

The addition of selenium as sodium selenate to the most common multinutrient fertilizers used in agriculture and horticulture was started in Finland in 1984. Selenium has been added to the fertilizers used mainly for cereal crops at the rate of 16 mg/kg, and those used mainly for grassland cultivation at the rate of 6 mg/kg. As the average fertilizer application is about 500 kg/ha, each cereal crop receives about 8 g/ha of selenium and each grass crop 3 g/ha. The target selenium content of plants has been 0.1 mg/kg of dry matter.

The aim of raising the selenium content of plants is mainly to increase the selenium intake of man. The Finns have not been shown to suffer from any disorders caused by selenium deficien-

Selenium in Medicine and Biology

cy. Nutritionists have been so concerned about the low intake of selenium among the Finns that there has even been talk of a risk to the nation's health. The figure of 50-200 µg put forward by the United States National Research Council (2) has therefore been taken as the target daily intake of man.

The selenium fertilization was at the first time in use in whole farming area in 1985. The selenium content of different crops since that year has been 10-20 times higher than those before the beginning of selenium fertilization (Table 1).

Table 1. The mean selenium content of some Finnish cultivated plants before the beginning of selenium fertilization and using selenate fertilizers in 1985 and 1986.

Crop	Se mg/kg of dry matter		
	Before Se fert.	1985	1986
Spring wheat	0.01	0.23	0.23
Barley	0.01	0.16	0.21
Oat	0.01	0.15	0.20
Timothy grass (silage)	0.02	-	0.17
Timothy hay	0.02	0.19	0.13
Potato	0.01	0.06	-

A typical example of the range of selenium contents in the grains of cereals is presented in Table 2. The top concentrations of grass selenium have been higher, up to 1 mg/kg of dry matter. The main reason for those cases is the use of selenated fertilizers intended for cereal crops containing the higher addition of selenium.

The effect of selenium fertilization on the selenium levels of animal products has been very clear. For instance, the selenium concentration of cow milk is ranging now from 20 µg/l to 30 µg/l which level is 3-4 times of that before the use of selenated fertilizers. The selenium levels of beef and pork are rised more sharply staying now at the 9 and 4 times, respectively, higher level than in the 1970's.

Table 2. The selenium content of barley and oat grain 1984-1986.

Crop		No. of samples	Se mg/kg of dry matter	
			Mean	Range
Barley	1984	50	0.008	0.004-0.016
	1985	240	0.16	0.007-0.56
	1986	211	0.21	0.013-0.58
Oat	1984	49	0.009	0.004-0.015
	1985	200	0.15	0.008-0.41
	1986	216	0.20	0.005-0.52

At the moment the Finns receive on average some over 100 µg/day of selenium from their diet (Fig. 1). This calculation is based on the energy intake of 10 MJ/day. That selenium intake is 3-4 times of that in the 1970's.

The effect of selenium fertilization on the selenium content of soil and other environment is carefully and systematically monitored. According to the data collected to this day nothing unexpected or surprising has happened and the selenium fertilization will continue on the same way as so far.

References

1. Yläranta, T. 1985. Increasing the selenium content of cereal and grass crops in Finland. Diss. 72 p. + 125 app.

2. Anon. 1980. Selenium. Recommended dietary allowances. pp. 162-164. 9th Rev. Ed. US National Academy of Sciences. 185 p. Washington D.C.

3. Varo, P. 1988. Personal communication with P. Varo, University of Helsinki, Department of General Chemistry, SF-00710 Helsinki, Finland, January 1988.

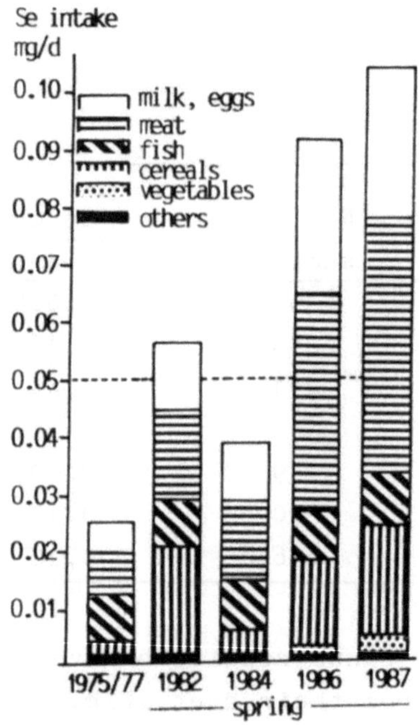

Figure 1. The daily selenium intake of the Finns, on average, since the 1970's (3).

SELENIUM TOXICOLOGY

A. PINEAU
Laboratoire de Toxicologie et d'Hygiène Industrielle
Faculté de Pharmacie, 44035 Nantes Cedex, France

The poisonous effects of selenium (Se) were recognized (7) long before its role as a nutritional trace element became apparent.

Chronic or acute intoxications can occur in different situations. Most of the compounds responsible for chronic intoxications are absorbed with food, particularly in seleniferous regions, but rarely with drinking water (7), medicines (1,6) or cosmetics (5). In the workplace, Se compounds may also lead to chronic intoxications ; however, they are, along with fatal poisoning (suicidal ingestion), the main cause of acute intoxications.

The symptomatology of Se poisoning in man shows characteristic features (6,7). Hair becomes dry and brittle, easily breaking off at scalp level. Scalp rash is associated with intolerable itching conducive to severe scratching which results in hair removal. Nails become brittle, and white spots and longitudinal streaks appear on the surface. This is followed by a break in the wall of the nail, with the broken nail dropping off in less than a month. Paronychia is common, and skin lesions occur, mainly on the limbs, with skin becoming red an swollen and then blistered and eruptive. Selenium compounds sometimes cause dermititis or allergy, with pink discoloration of the eyelids and palpebral conjunctivitis. Abnormalities of the nervous system lead to convulsions, paralysis, motor disturbances and pain in the extremities.

In diagnostic terms, the most characteristic indications of Se compound absorption by the human body are a metallic taste

Selenium in Medicine and Biology

in the mouth and a garlicky odor of the breath and sweat caused by excretion of small amounts of dimethyl selenide. This odor disappears completely within about a week after the end of exposure. It is uncertain whether there is an increase in tooth decay since results indicating a significant relationship between the Se content of body fluids and the occurence of dental caries are controversial. It would appear that further epidemiological investigations are required (4,6). Finally, chronic intoxication causes nonspecific symptoms, including digestive tract disturbances, loss of appetite, pallor, lassitude, irritability, strong headaches, giddiness and numbness. No clinical signs of hepatotoxicity are normally noted. It may be concluded that loss of hair and nails, skin lesions and garlicky breath are the main symptoms of chronic Se intoxications. To our knowledge, acute human poisoning with Se and its compounds is rather rare, as indicated by the fact that there are less than 10 publications on the subject in the literature (3). Since in most cases the compounds involved are different and the absorbed doses difficult to estimate, it is understandable that results are poorly documented and that the dose-response relationships essential to toxicology are nonexistent. The major symptomatology (3) is characterized by considerable irritation of the mucous membranes associated with nausea and more or less hemorrhagic vomiting followed by abdominal pain. Our investigations show that there is subsequent unstable blood pressure complicated by pulmonary edema due to direct lung injury or to heart failure. Intubation is required when apnea occurs. The garlicky odor is characteristic of inorganic Se ingestion. For vesicant compounds, postmortem examination shows toxic pulmonary edema, hemorrhagic and erosive mucosa with local necrosis, brain edema, diffusely swollen and dilated heart, and tubular necrosis and dilatation of the glomeruli of the kidneys. Toxicity depends largely on the kind of Se compounds involved. In toxicological terms, the Se compound itself is neither particularly dangerous nor easily absorbed by the body. However, various Se chemical compounds are strong irritants or vesicants.

and can blister and burn the skin, thus destroying cells. Hydrogen selenide induces hemolysis. Fortunately, nearly all these compounds are very unstable and may decompose to yield elemental Se or be detoxified by liver methylation. This pathway leads to production of two compounds : a volatile dimethyl selenide ($(CH_3)_2$ Se) excreted in sweat or exhaled, which is responsible for the characteristic garlicky odor ; and a hydrosoluble trimethylselenonium ($(CH_3)_3Se^+$) which is eliminated mainly by the kidneys, with small amounts excreted via feces.

Selenium has known teratogenic effects in animals (6). Observations in man remain very contradictory, although a few studies have shown a higher number of abortions and a need for controlled study of women of childbearing age exposed to Se.

For the medical scientist, it is important to be able to check the Se status of man, particularly in cases of exposure or intoxication. Blood or plasma Se can be analyzed or glutathione peroxidase activity measured. As Se is primarily excreted through the kidneys, about 30 to 70 % of the Se taken into the body leaves by urination. Analysis of total urinary Se excreted during 12 or 24 hours provides a better indication of Se content than do random samples (4). It must be emphasized that urinary Se cannot be used to monitor exposure when there is renal deficiency or exposure to irritant compounds. Analysis of sweat or feces is of little interest in view of the small amounts eliminated (5 %) and is not ordinarily performed.

It is also possible, especially in occupational medicine, to measure the amount of dimethyl selenide in breath. However, only small amounts are eliminated via the lungs (1 %). For many mineral compounds, particularly Se, head hair is not a valid biological sample since minerals are not only incorporated from blood but also from sweat, cosmetics (anti-dandruff shampoos) and the environment (water, dust, etc...). In these conditions, it is inadvisable to try to determine a correlation between Se organism impregnation and Se hair elimination (4).

The following points are recommended as a basis for diagnosis in cases of intoxication or when information about body Se intake is needed : observation of symptomatology ; interrogation of the subject about his occupational activity, place of residence, habits, drug and cosmetic use, food preferences and estimated daily Se intake from dietary sources ; body impregnation (blood Se) ; and body elimination (urinary Se).
Analytical results can then be compared with the following reference values : The current proposed adequate and safe intake of Se in food for human adults is 50 to 200 µg/day (2), although it would seem that levels of 500 to 1,000 µg/day are without danger (7). In man the oral median lethal dose is about 0,5 to 1 g of Se ($Na_2 SeO_3$, $Na_2 SeO_4$).
In drinking water, normal values are below 10 µg/l, with tolerance up to 50 µg/l.
In biological samples, normal values range from 50 to 200 µg/l in blood and are below 50 µg/l in urine. There always appear to be clear effects on nails and hair when values are above 800 µg/l in blood and 500 µg/l in urine. At blood and urine concentrations respectively of 500 and 100 µg/l, no symptoms or complaints are noted. These values are in agreement with the published data of Yang et al. (7).
No satisfactory treatment for Se toxicity is available. In chronic intoxications, the source of exposure can be eliminated ; and in acute intoxications it is practically impossible to provide more than treatment of symptoms.
In conclusion, it may be said that acute Se toxicity is infrequent and that chronic intoxication, which is caused mainly by nutritional sources, has rather marked effects. The infrequency of intoxication may be explained in part by the fact that little is known about elimination kinetics and dose-effect or dose-response relationships in man. Moreover, even though Se is toxic, it plays an important biological role as an essential trace element and also has the capacity to antagonize the poisoning effects of other minerals (arsenic, lead, cadmium, mercury) and to exert a protective action against the

adverse effects of some organic toxicants.

References

1. Anonymous. 1984. N.C.M.J. 45, 8, 525.
2. Committee on Dietary Allowances, Food and Nutrition Board. Ninth revised edition, 1980, National Academy of Sciences, Washington, D.C., p 162.
3. PENTEL, P., D. FLETCHER, J. JENTZEN. 1985. Journal Forensic Sciences. 30, 2, 556.
4. ROBBERECHT, H.J., H.A. DEELSTRA. 1984. Clin. Chim. Acta. 136, 107.
5. SHAMBERGER, R.J. 1983. In : Biochemistry of Selenium. Eds Plenum Press, New-York, p 185.
6. World Health Organization. 1987. In selenium. Environmental Health Criteria 58.
7. YANG, G., S. WANG, R. ZHOU, S. SUN. 1983. Am. J. Clin. Nutr., 37, 872.

SELENIUM TOXICITY IN INDUSTRIAL APPLICATIONS : PRESENT KNOWLEDGE

F. Baruthio

Institut National de Recherche et de Sécurité, Centre de Recherche, Service Epidémiologie, BP 27, 54501 VANDOEUVRE CEDEX (France)

Introduction

Symptoms of chronic occupational selenium poisoning were first described by Hamilton in 1917 in a copper refinery plant (1). The effects that were noticed were characterised by bronchial and pharyngeal irritations, persistant garlicky halitosis and gastro-intestinal perturbations. Selenium is used in many different industrial applications (2) but never in great quantities : it is ranked at number 60 in the world's consumption of elements (3). Elemental selenium has a relatively low level of toxicity, it is even considered to be harmless (2, 4-9). The toxicity of selenium vapour is due to the formation of selenium dioxide, which along with other compounds of selenium is dangerous for the organism.

Selenium and its compounds

Selenium exists in 3 allotropic forms (2, 5, 6, 9) : a red amorphous powder, a crystalline red monoclinic aspect, a metallic grey photo and semi conductor aspect.

Amongst the many compounds of selenium (2, 5-7) the most important are :

- selenium dioxide, SeO_2, most used in industry, formed as soon as one of the three elemental forms is heated above its melting point. On contact with water or perspiration, it is transformed into selenous acid H_2SeO_3, responsible for most of the poisonings ;
- hydrogen selenide H_2Se, a nauseous smelling gas (6, 10), is also very toxic ;
- selenium oxy-chloride, $SeOCl_2$ is an extremely vesicant yellow liquid ;
- selenium chloride $SeCl_2$ is a dark brown liquid, causing severe irritations ;

Selenium in Medicine and Biology

- selenium trioxide SeO_3 is transformed into selenic acid H_2SeO_4 by contact with water or perspiration and is very corrosive ;
- selenium hexafluoride SeF_6 is a colourless toxic gas ;
- numerous salts (selenites, selenates) and organic derivatives.

Potential risks in industrial activity

The risk of exposure to selenium and its compounds exists in numerous industrial activities by inhalation, ingestion and cutaneous contact with dust, smoke, vapours, aerosols, gas and liquids (2, 5, 7-9), sometimes containing important levels of Se O_2 and other toxic selenium compounds.

A first group of risks include extraction, treatment and refining of primary materials containing selenium (copper, lead, zinc, nickel and uranium ores, phosphates, pyrites, certain cements and slaked lime, pitch and petrol). Selenium is extracted and refined from ore concentrates and residual sludges.

The transforming industries set up a second group of risks. The electric and electronic industries are the largest consumer (melting and casting of rectifiers and photo sensitive surfaces). Glass and ceramics production (ovens, crucibles and pigments) is the second. Selenium is used as a catalyst, an additive or a component for the production of rubber, plastics, pigments, lubricators, cosmetics, animal foodstuffs, pharmacology and phytopharmacology, inks in reprography, photographic baths, etc...

Effects of exposure

2 types of effects must be distinguished, that of subacute exposure and that of chronic exposure (5-7, 9-11). The effects are shown in Table 1

No chronic or irreversible pathology was noticed (5, 9). The hepatic lesions, and renal failure found in animals submitted to subacute poisoning did not appear to affect human beings. No cancerous effects are reported.

The first symptom after exposure to selenium is a metallic taste on the tongue to which, for the most part, the subject does not pay attention. Another early sign is persistant garlicky halitosis, due to dimethylselenide, mainly eliminated by the urines (50 to 80 % of selenium inhaled) but also, in smaller proportions through perspiration or by the lungs (3 to 10 %) (6, 8). Nail bed inflammation occur rapidly. All the symptoms of chronic exposure to selenium disappear quickly and completely after seven or eight days (5-7) if the subject is removed or protected from the risk.

Table 1 : Effects of Exposure to Selenium and its Compounds

Acute exposure	Chronic exposure
Intense irritation of eyes, nose, throat, headaches, vertigo, nasal congestion, garlicky halitosis, severe dyspnea sometimes in asthma form. Pulmonary oedema. Chemical pneumonia. Severe bronchitis. Gastro-intestinal disturbances. Nail bed inflammation. Dermatitis with pruriginous papules. Sometimes allergic eruptions. Third degree burns from Se OCl_2.	Bronchial, naso-pharyngeal, skin, nail bed and eyes irritations. Gastro-intestinal disturbances. Metallic taste on the tongue, garlicky odor of the breath, sweat and urine. Increased alcaline phosphatase and transaminases. **Long term exposure** : pulmonary fibrosis, blood capillary disturbances, porphyuria. **Apparently "non specific" effects** : lassitude, irritability, asthenia, anorexia, anosmia, dizzyness. Increased risk of abortion.

Therapy

The therapeutical approach to acute or chronic selenious intoxication is mainly, if not exclusively symptomatic (6). Oxygenotherapy is necessary in the case of acute exposure. Burns are rarely severe and when well treated heal without leaving scars. The only efficient therapy known for localised effects (5, 6) on the skin, the nails and the eyes is, after a thorough washing with 10 % sodium thiosulfate solution, the application of an ointment at 10 % sodium thiosulfate. This solution reduces the irritant SeO_2 into non toxic red selenium.

Prevention

The prevention of selenious hazards necessitates a technical approach implying an atmospheric determination of selenium, to maintain the surrounding atmosphere below a threshold limit value (12) (VLE = french Exposition Limit Value or TLV = american Threshold Limit Value). At the present no threshold limit value exists for selenium and its compounds in France. Only Average Exposure Values (AEV) for hydrogen selenide and selenium hexafluoride were defined. For the other selenium compounds reference is made to the American Conference of Governmental Industrial Hygienists (ACGIH) thresholds (12). The norms of the USSR GOST 12-1-005-76 can also be used (12). These threshold values are noted in Table 2.

In France, with the exception of dyspepsia, the affections resulting from exposure to selenium and its inorganic derivatives were recognized as "occupational" diseases (9) (decree of 22 June 1984).

Table 2 : Limit Values in mg Se/m^3 Ambient Air (12)

	FRANCE	USA	USSR
Selenium + compounds		TLV 0.20	2
Selenium dioxide		TLV 0.20	0.1
Hydrogen selenide	AEV 0.08	TLV 0.20	
Selenium hexafluoride	AEV 0.20	TLV 0.20	

The necessary measures to undertake are destined to stop the irritant dusts and soluble substances from coming into contact with the skin, the eyes, the lungs and the stomach (5, 6, 9). These measures are hereby resumed : good ventilation and aspiration, individual protection equipment, no smoking, no eating and no drinking in or near the workplaces, and the use of closed process production methods, as much as possible.

The health of the workers exposed to selenium must be regularly monitored (hepatic function, hematology, respiratory function).

Special surveillance should be undertaken with subjects with the following disorders (5, 6, 9) : dyspepsia or previous peptic ulcers, cutaneous or upper respiratory tract allergies, chronic renal failure, hepatic injury, lung disease, chronic neurasthenia, anxiodepressive syndrome.

Biological monitoring of selenium in exposed subjects can be undertaken (2). An urinary threshold value (MAC = Maximum Allowable Concentration) of 0.1 mg/l of selenium has been proposed by Glover (4). If this urinary threshold value is attained, the occupational medicine should carry out all the appropriate measures to improve hygiene at the workplace.

Conclusion

Selenium is a largely used metalloid in industrial processes. Its compounds can bring about intoxications, sometimes severe. To avoid the appearance of poisoning effects, medical and biological monitoring and the use of adequate individual and collective protection are indispensable. Application of the defined norms by the industrial hygienists is recommended. With regard to the effects of long term exposure to selenium and its compounds, we noticed a lack of epidemiological studies. The relationships likely to exist between the appearance of the illnesses and the characteristics of the occupational exposure, as well as the dose-response relationship need more thorough studies.

References

1. Hamilton, A. : Industrial poisons in the United States. Macmillan Company. New-York, 1925, 303.

2. World Health Organization. Selenium. Environmental Health Criteria 58. Geneva : WHO, 1987, 306.

3. Michels, S., H.J. Einbrodt : Staub-Reinhalt Luft. 1982, 42, 147.

4. Glover, J.R. : Ann. Occup. Hyg. 1967, 10, 3.

5. Glover, J.R. : Ind. Med. 1970, 39, 50.

6. Glover, J.R., O. Levander, J. Parizek, V. Vouk. 1979. In: Handbook on the toxicology of metals (L. Friberg, G.F. Nordberg, V.B. Vouk, eds.). Elsevier North Holland Biomedical Press, p. 555.

7. Fischbein, L. 1977. In: Advances in modern toxicology - Vol. 2. Toxicology of trace elements. (R.A. Goyer and M.A. Mehlman, eds.). Halsted Press Book. p. 213.

8. Wilber, C.G. : Clin. Toxicol. 1980, 17, 171.

9. Najael : Rev. Génér. Sécurité. 1987, 63, 65.

10. Alderman, L.C., J.J. Bergin :Arch. Environ. Health. 1986, 41, 354.

11. Diskin, G.J., C.L. Tomasso, J.C. Alper, M.I. Glaser, S.E.Fliegel :Arch. Intern. Med. 1979, 139, 824.

12. Valeurs limites pour les concentrations des substances dangereuses dans l'air des lieux de travail. INRS - Cahiers de Notes Documentaires. 1987, 129, 581.

Section VI

Selenium in animals

SELENIUM DEFICIENCY IN ANIMALS.

M. Lamand

I.N.R.A. - Laboratoire des Maladies Nutritionnelles
C.R.Z.V. de Theix - 63122 Ceyrat - France.

Introduction

The toxicity of selenium has been known for a long time but its nutritional importance was demonstrated in 1957 by Schwartz and Foltz (1) : the element prevented a liver necrosis in rats. At that epoch the first studies suggested that selenium and vitamin E had very similar even identical roles. It is now known that selenium does not replace vitamin E but it is indispensable for the integrity of muscle fibres in all mammals including man (2).

The Etiology of the Myopathy in Young Ruminants

1. Selenium deficiency

A survey carried out on 150 first cut hay crops taken from permanent pasture on the principal types of soil found in France showed that the majority of these forages were deficient : 72 % had levels of less than 0.05 mg/kg DM and

Selenium in Medicine and Biology

22 % were between 0.05 and 0.1 mg/kg DM. Only 6 % were levels greater than 0.1 mg/kg (3). A subsequent survey carried out on 340 forages (4) confirmed these initial results showing that the deficiency is spread all over the country.

2. Epidemiology of myopathy

The myopathy generally appears in young animals, sometimes from the third week but most often between 1 and 2 months (5, 6). It is more uncommon in young ruminant more than one year old. The myopathy can, then, be accompanied by a myoglobinurie (7). The frequency of myopathy varies from year to year : it is most common when the milking cow has been fed hay which has grown in a very wet season (6). The intense growth of grass in humid conditions reduces the selenium content (8).

Myopathy is most common in spring, particularily in april and may. The development of deficiency as such is progressive when the selenium reserves are exhausted a release of aspartate aminotransferase and creatine kinase into plasma occurs. The myopathy is then at a pre-clinical stage which can last several weeks.

The clinical syndrom appears following a stressful situation which triggers off its full development : the trigger can be a hot thundery day, intense and unusual muscular activity or follow transport of the animal. Growth rate is also a factor. In calves suckling, a growth rate of 1400 g per day provoked an 80 % morbidity compared to 10 % when the growth rate was limited to 700 g per day. The dams of those two groups of calves had nevertheless received hay with the same selenium level (0.01 mg/kg DM).

The author has shown that in a group of farms where no selenium supplementation was practised, the milk received by the sick calves from their mothers was more unsaturated than that received by controls on the same farms (9). It is thus likely that unsaturated fatty acids favourize the development of myopathy above all when the animals are also vitamin E deficient. Moreover it is known that young grass, in spring, is rich in unsaturated fatty acids that the rumen flora cannot hydrogenate without a period of adaptation. In addition the levels in plasma of vitamin E originating from the grass rise relatively slowly in the cow thus introducing a delay in the protection of the calf (10).

Clincal Aspects of Myopathy

Symptoms

The symptoms of myopathy arise from the appearance of a lesion on several fibers or groups of muscles. Hyperthermy does not occur, except in where there is a possible microbial complication. Degeneration affects different muscle systems.

1. The locomotary muscular system. The animals cannot move about and standing is painful. They may be unable to get up suckling is then impossible. The characteristic attitude is the cud-chewing position within the back arched.
2. The respiratory muscle system. Respiration is short, rapid, and noisy ; the nasal openings dilate, the abdominal wall moves with breathing.
3. The heart. Pulsations are rapid, the two sounds may become equal with a pendular rythm. The sonority may be either metallic or buffered and distant. If the lesion is severe, sudden death may occur.

Lesions

They have no specific location (2), microscopic examination shows that the degeneration affects only a few fibers or even parts of fibers. It is rare not to find all stages of degeneration from healthy to completely broken down cells in a single zone. Degeneration progresses through two phases (11).

1. Hyaline degeneration or Zenckers'waxy degeneration. The cell loses is striation ; it may be turgid or plasmolytic. It is then rendered hypereosinophilic.
2. Inflammatory reactional phase. The degenerate cells are invaded by basophil inflammatory cells, which destroy the sparse masses of chromatin. From hyaline degeneration, the process can lead to a partial regeneration of lesioned muscular fibers. In this case newly formed nuclei invade the cell to recreate a syncytium.

The Interrelationship of Selenium and Vitamin E in Myopathy

Vitamin E deficiency can be produced in calves by feeding a skimed milk (with normal selenium levels) enriched with lard. Selenium deficiency can be produced with complete spray-dried milk originating from a selenium deficient region.

The composition of these diets is shown in table 1. The disease syndromes arising from these experimental diets were compared to calf myopathies arising spontaneously in herds

Table 1 : The Composition of Milks which Produced Experimental Deficiencies (mg/kg DM)

	Vitamin E deficiency		Selenium deficiency	
	Controls	Deficient	Controls	Deficient
Tocopherol	20	Traces	8	8
Selenium	0,1	0,1	0,25	0,02

were the forages were deficient in selenium (levels between 0.01 and 0.04 mg Se/kg DM).

Vitamin E and selenium deficiencies give rise to very similar clinical disease patterns. Nevertheless vitamin E deficiency provokes a far more acute and rapidly developing which progresses irreversibly towards death whatever the treatment. By contrast the development of the selenium deficiency syndrome is less brutal. If the muscle degeneration has not affected a vital muscle such as the heart, selenium treatment brings about a rapid recovery (about 10 days). The biochemical pathologies of vitamin E and selenium deficiencies differ (table 2) (12).

Overall lactic, pyruvic and alpha-ketoglutaric acids increase in the plasma of animals suffering from selenium deficiency whereas these acids are unaltered in vitamin E deficiency (12). Some time ago, several authors (13, 14, 15) showed that selenium was involved in the oxidation of the alpha-keto-acids of Krebs cycle. By contrast the functioning of lipoyl dehydrogenase, one of the enzymes of the pyruvate dehydrogenase complex is unaltered by selenium deficiency (12). The reasons for these apparent biochemical lesions have never been elucidated properly since. However, in the spontaneous myopathy after selenium treatment they disappear after about 10 days. The plasma selenium of vitamin E deficient calves decreases, which suggests that vitamin E has a selenium-sparing action.

Table 2 : A Comparison of the Biochemical Lesions Found in Vitamin E or Selenium Deficiency in Calf

	Spontaneous Se deficiency			Experimental Vit E deficiency			Experimental Se deficiency		
Number of animals	DEF 20		CONT 31	DEF 10		CONT 10	DEF 10		CONT 9
Selenium (µg/l)	9	*	23	12	*	20	10	*	50
Vitamin E ([1])	5.5		-	0.1	*	9.8	3.5		4.5
Pyruvic acid([1])	12.4		11.3	7.2		7.6	1.2		10.3
Lactic acid + pyruvic ([1])	373	*	207	203		163	518	*	142
α ketoglutarate[1]	6.6	*	2.6	3.2		2.7	3.4	*	1.8
Correlation α ketoglutar/Se	r = - 0.536*			r = + 0.113					

([1]) mg/l)

* $P < 0.01$ to $P < 0.05$

The Diagnosis of Myopathy

The clinical diagnosis is easy after the appearance of

locomotory, respiratory or cardiac symptoms associated with contractions. The detection of affected animals at a preclinical stage is also easy with the measurement of aspartate aminotransferase (EC 2.6.1.1) or of creatine kinase (EC 2.7.3.2.). Both these enzymes increase in the plasma at an early stage although creatine kinase goes down to normal levels in clinically sick animals and thus give a false negative result.

Overall levels greater than 100 IU/l for aspartate amino transferase or 10 IU/l for creatine kinase can be considered as indicative of a clinical or subclinical myopathy. Selenium deficiency in itself can be demonstrated by measurement of plasma levels. The deficiency level can be considered to be about 20 µg/l (16).

Erythrocyte glutathione peroxidase activity is used routinely for assess the Se status of ruminants in Danemark (17). A commercial kit is available which provides a rapid and simple method assay method (Ransel ® kit, Randox Laboratories, Northern Ireland). Such a technique will allow the comparison of result from different authors which was previously impossible with a multiplicity of technique and different methods of expression of results.

Other Conditions Linked to Selenium or Vitamin E Deficiency

While myopathy is relatively specific for young ruminant other deficiency symptoms arise in other species (2) (Table 3). Selenium deficiency reduces the fertility of ewes. According to some authors, sufficiency of Se increases the proportion of pregnant ewes and reduces the death rate of newborn lambs (18, 19). In some cases the number of lambs

Table 3 : Pathology of Selenium-Vitamin E:Deficiency States

Effects of Se-Vitamin E deficiencies	Species
Liver necrosis	Pig
Growth depression	Lamb and Calf
Exudative diathesis	Chick
Muscle dystrophy	Lamb, calf and pig
Elevated serum aspartate aminotransferase	Lamb and calf
Decreased GSH-Px	Lamb, calf and chick
Mulberry heart	Pig
Sudden death	Lamb, calf and pig
Infertility	Ewe, cow and gilt
Retained placenta	Cow
Gizzard myopathy	Chick
Pancreatic fibrosis	Chick
Peridontal disease	Sheep
Muscle calcification	Sheep and cattle.

per ewe is improved (20). By contrast, introduction of Se into the diet may have no efffect on the breeding performances of the flock (21, 22). Nevertheless, in cows, Se reduces losses of newborn calves (23).

Feeding of Se may also improve the fertility of ova destined for embryo transplantation from cows with an otherwise normal diet (24).

Some authors report that Se supplementation (25) or injections of this element (26) reduces placental retention. Other symptoms described in sheep and cattle include diarrhea (27) and poor growth of young animals

(28). Exudative diathesis of hens, characterized by muscular edema and subcutaneous hemorrhage can be prevented by feeding Se and vitamin E. The same is true for fibrosis and pancreatic atrophy.

Selenium-deficient pigs are preferentially attacked by hepatosis dietetica (28) and mulberry heart disease (29).

Selenium Levels in Animal Products of Bovine Origin

Selenium levels in milk increase with the levels of the element ingested by the cow. The response at first important becomes weaker as the dietary selenium levels exceed the usual nutritional limits.

Table 4 : Selenium Content of Milk Samples from cows receiving different levels of selenium (mg/kg DM).

Se content of diets	Number of samples	Mean	Extremes
0.01 - 0.04	35	0.058	0.027-0.1
0.2	10	0.136	0.09 -0.18

The selenium levels of meat and other abattoir material also increases with selenium intake (Table 5) (30). Muscle levels increase very moderately while the other material, liver and above all kidney, reach relatively high levels.

Table 5 : Selenium Content of Tissues from Calves Receiving Different Milks (mg/kg DM).

	Deficient calves	Normal calves	Supplemented calves
Animals number	7	9	8
Se in diet	0.03	0.09	0.4
Muscle :			
dry matter	0.25±0.10a	0.23±0.05a	0.47± 0.21b*
fresh matter	0.061±0.026a	0.055±0.010a	0.11±0.05b*
Heart :			
dry matter	0.29±0.11a	0.51±0.11b	1.13±0.20c
fresh matter	0.067±0.026a	0.111±0.014b	0.26±0.04c
Liver :			
dry matter	0.45±0.15a	0.56±0.06a	1.32±0.27b
fresh matter	0.13±0.04a	0.16±0.01b	0.40±0.09c
Kidney :			
dry matter	1.86±0.55a	3.29±0.52b	5.58±1.83c
fresh matter	0.42±0.12a	0.76±0.13b	1.24±0.36c

Means, on the same line, followed by different letters are significantly different ($P < 0.01$)

*This mean is different from the preceding ($P < 0.05$).

Conclusions

Selenium deficiency has been present, a long time in ruminants which are dependant solely upon the soil where they are raised for their forages. Animals receiving compound feedstuffs produced industrially (pigs, poultry) were supplemented when legislation permitting this was approved. Myopathy is the principal syndrome encountered, more important than reproductive problems. The systematic addition of selenium to foodstuffs has prevented the disease but at the same time the levels of selenium in animal products to which consumers are exposed has not exceeded normal limits.

Références

1. Schwarz, K., C.M. Foltz 1957. J. Am. Chem. Soc. 79, 3292-3293.

2. Gissel Nielsen, G., U.C. Gupta, M. Lamand, T. Westermarck. 1984. Adv. Agron. 37, 397-460.

3. Lamand, M. 1970. C.R. Acad. Agric. Fr. 56, 604-610.

4. Arvy, M.P. 1982. Thèse Doctorat d'Etat. Univers. Poitiers.

5. Oksanen, H.E. 1965. Acta Vet. Scand. 6, suppl. 2, 110 p.

6. Lamand, M. 1966. Ann. Nutr. Alim. 20, 13-39.

7. Allen, W.M., W.H. Parr, R. Bradley, K. Swannnack, C.R.Q. Barton, R. Tayler. 1974. Vet. Rec. 94, 373-375.

8. Gardiner, M.R. 1962. Aust. Vet. J. 38, 387.

9. Lamand, M. 1967. Ann. Zootech. 16, 5-9.

10. Mc Murray, C.H., P.K. Mc Eldowney 1977. Br. Vet. J. 133, 535-542.

11. Ferrando, R., A. Parodi. 1964. Ann. Nutr. Alim. 18, 429-460.

12. Lamand, M. 1970. C.R. Hebd. Séances Acad. Sci. 270, 417-420.

13. Bull, R.C., J.E. Oldfield. 1967. J. Nutrition 91, 237-246.

14. Connolly, J.D., K.S. Schwarz. 1963. Fed. Proc. 22, 652.

15. Connolly, J.D., K. Schwarz. 1965. Fed. Proc. 24, 623.

16. Perry, T.W., D.N. Caldwell, R.C. Peterson. 1976. J. Dairy Sci. 59, 760-769.

17. Anderson, P.H., R. Bradley, S. Berrett, D.S.P. Patterson. 1977. Br. Vet. J. 133, 160-165.

18. Godwin, K.O., R.E. Kuchel, R.A. Buckley. 1970. Aust. J. Exp. Agric. Anim. Husb. 10, 672-678.

19. Wilkins, J.F., R.J. Kilgour. 1982. Aust. J. Exp. Anim. Husb. 22, 18-23.

20. Mudd, A.J., J.L. Mackie. 1973. Vet. Rec. 93, 197-199.

21. Maxwell J.A.L. 1972. Aust. Vet. J. 48, 598-600.

22. Mitchell, D., M. Hidiroglou, K.J. Jenkins. 1975. Can. J. Anim. Sci. 55, 513-517.

23. Mace, D.T., J.A. Tucker, C.B. Bills, C.J. Ferreira. 1963. Calif. Dep. Agric. Bull. 1, 63.

24. Conrad, H.R., A.L. Moxon. 1978. Proc. Cornell Nutr. Conf. Feed Manuf., pp. 57-62.

25. Julien, W.E., H.R. Conrad, J.E. Jones, A.L. Moxon. 1977. J. Dairy Sci. 59, 1954-1959.

26. Trinder, N., R.J. Hall, C.P. Reuton. 1973. Vet. Rec. 93, 641-644.

27. Andrews, E.D., W.J. Hartley, A.B. Grant. 1968. N.Z. Vet. J. 16, 3-17.

28. Underwood, E.J. 1977. In : Trace Elements in Human and Animal Nutrition. Acad. Press New-York London.

29. Jenkins, K.J., M. Hidiroglou 1972. Can. J. Anim. Sci. 52, 591-620.

30. Lamand, M. 1972. Ann. Zootech. 21, 29-35.

WHOLE BLOOD AND PLASMA SELENIUM CONCENTRATIONS, ERYTHROCYTE AND PLASMA GLUTATHIONE PEROXIDASE ACTIVITIES IN DAMS AND LAMBS

B.A. Zachara[x], R. Zamorski[xx], K. Borowska[xx] and R. Kanarkowski[xxx]

Department of Biochemistry, Academy of Medicine[x], Department of Biochemistry, Academy of Agriculture and Technology[xx], and Medical Center of Postgraduate Education, Faculty of Pharmacy[xxx], Bydgoszcz, Poland

Introduction

Selenium /Se/ deficiency has been found in several animal species including sheep. Severe Se deficiency results in a nutritional myopathy /NM//1/. Se-responsive diseases in livestock frequently occur in countries where soils and, consequently, plants contain low levels of Se /2/. It became possible to explain the biochemical basis of the Se deficiency defects in tissues when Rotruck et al. /3/ discovered that it is a structural component of glutathione peroxidase /GSH-Px/. A linear relationship between Se intake or blood Se level and red cell GSH-Px activity has been established /4,5/. In the Bydgoszcz province as recently as 5 years ago the losses among lambs because of NM amounted to 30%. The aim of this study was to determine blood Se levels and GSH-Px activities in dams and their lambs.

Materials and Methods

Blood samples of 77 dams /Merino breed/ and their 2-4 weeks-old lambs from eight farms of the Bydgoszcz province were taken for Se and GSH-Px measurements. Whole blood and plasma Se concentration was determined by the fluorometric method of Watkinson /6/ with 2,3-diaminonaphtalene as a complexing reagent. Red blood cell and plasma GSH-Px activity was assayed using the coupled method of Paglia and Valentine /7/. A detailed desription of precision, accuracy /IAEA blood standard samples/ and statistical analysis has been published earlier /8/.

Selenium in Medicine and Biology

Results

The whole blood Se levels in dams and their lambs were 70.3±22.5 and 64.1± 19.5 ng/ml, respectively /Fig. 1/. The differences are not significant. In plasma the Se levels were 31.1±8.5 ng/ml in dams and 27.9±7.0 ng/ml in lambs /$p < 0.02$/. The data on Se concentrations in whole blood and plasma are close to the normal distribution /Fig. 2/. We found a positive significant correlation between paired samples of whole blood Se levels for dams and lambs /Fig. 3/. In plasma, however, the correlation was weaker /$r=0.26$; $p < 0.005$/. Red cell GSH-Px activity is significantly lower in lambs compared to dams /94.8±30.6 vs 110.9±35.0 U/g Hb; $p < 0.005$; Fig. 1/. The striking finding was the significantly higher plasma GSH-Px activity in lambs when compared to dams /94.2±59.0 vs 54.4±31.0 U/l; Fig. 1/. Red cell GSH-Px activity of dams correlates well with the enzyme activity of their lambs /Fig. 4/. Similar enzyme correlation has been found between plasma paired samples /$r=0.59$; $p < 0.001$/.

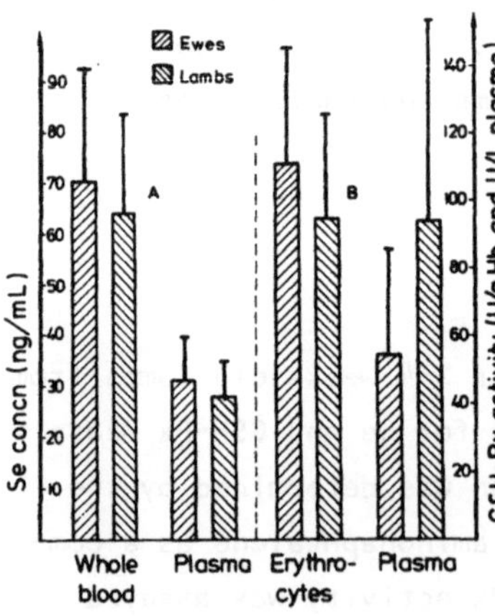

Fig. 1. Se concn in whole blood and plasma /A/ and GSH-Px activity in red cells and plasma /B/ in dams and their lambs.

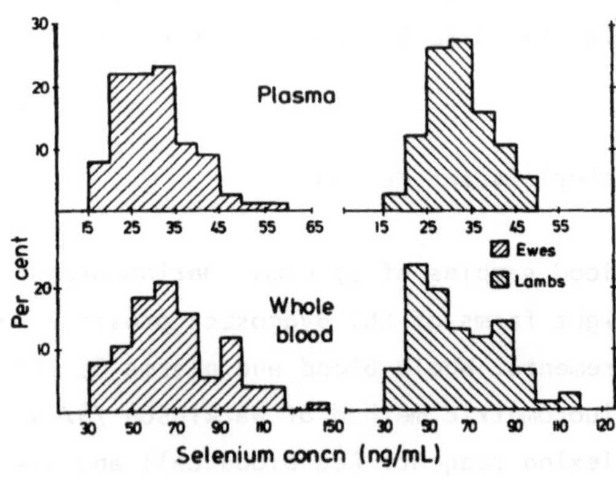

Fig. 2. Distribution of plasma and whole blood Se levels in a flock of 77 dams and lambs. Intervals of 10 ng per ml are presented.

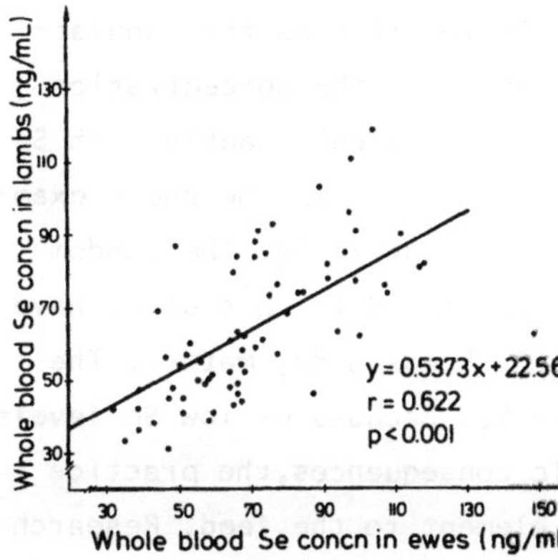

Fig. 3. Relation between paired samples of whole blood Se concn in dams and lambs

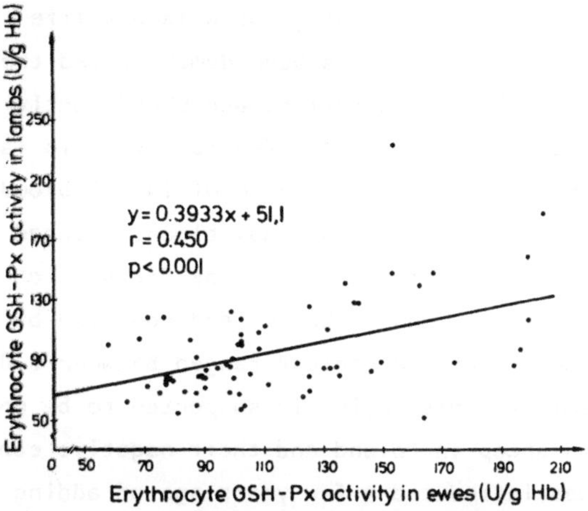

Fig. 4. Relation between paired samples of red cell GSH-Px activity in dams and lambs

Discussion

Se responsive diseases are likely to occur when the whole blood Se levels are below 75 ng/ml, and levels lower than 50 ng/ml are considered to be indicative of a deficiency state /1,9/. Our results show that about 68% whole blood samples taken both from dams and their lambs had Se concentrations lower than 75 ng/ml /Fig. 2/. In 18% of the dams and 30% of the lambs the blood samples contained less than 50 ng Se/ml. According to McDowell et al. /10/ the critical serum Se level in sheep is 30 ng/ml. As can be seen in Fig. 2, 53% of dams and 68% of the lambs covered by the present study had plasma Se levels lower than the above mentioned value. Because of low Se levels the farms in this region had a history of high mortality of lambs with classical symptoms of NM.
For most kinds of animals, including sheep, the dietary requirement of Se is said to be 0.1 ppm /1,11/. Smith and Picciano /12/ have shown that this amount is not adequate to maintain optimal maternal Se status during reproduction, especially during lactation. They proved that a dietary level of 0.2 ppm Se was necessary to maintain red cell Se level during lactation.

During the first stage of a lamb's life the only source of Se for it is maternal milk. It has been demonstrated that lactating dams show markedly lower Se concentrations and GSH-Px activities in the tissues than nonlactating controls /12/. The Se levels in lambs depend on the concentration in dams. Therefore the Se of lambs' blood correlated significantly with Se concentration in maternal blood. Reduced blood Se levels in the sheep examined by us is related to the diets which are deficient in Se. The random determinations of Se in feed consumed by dams presented in this study have shown that the content ranged between 0.03 to 0.06 mg/kg dry matter. The soil in this region is suspected to be poor in Se. Because of low Se levels in sheep in Poland and their negative economic consequences, the practice was introduced a few years ago of adding the element to the feed. Research on the effect of this dietary supplementation has only been started.

Acknowledgements

This study was supported by the Central Research Program, CPBR 10.17/IV.

References

1. The Scottish Agricultural Colleges; Scottish Agricultural Research Institutes: Trace Element Deficiency in Ruminants. 1982. Edinburgh. pp. 30-35.
2. Gissel-Nielsen, G., U.C. Gupta, M. Lamand, T. Westermarck. 1984. Adv. Agron. 37, 397-460.
3. Rotruck, J.T., A.L. Pope, H.E. Ganther, A.B. Swanson, D. Hafeman, W.G. Hoekstra. 1973. Science 179, 588-590.
4. Sankari, S., F. Atroshi. 1983. Zbl. Vet. Med. A 30, 452-458.
5. Thompson, K.G., A.J. Fraser, B.M. Harrop, J.A. Kirk. 1980. Res. Vet. Sci. 28, 321-324.
6. Watkinson, J.H. 1966. Anal. Chem. 38, 92-97.
7. Paglia, D.E., W.N. Valentine. 1967. J. Lab. Clin. Med. 70, 158-169.
8. Wąsowicz, W., B.A. Zachara. 1987. J. Clin. Chem. Clin. Biochem. 25, 409-412.
9. Andrews, E.D., W.J. Hartley, A.B. Grant. 1968. N. Zeal. Vet. J. 6, 3-17.
10. McDowell, L.R., J.H. Conrad, G.L. Ellis, J.K. Loosli. 1984. Minerals for Grazing Ruminants in Tropical Regions. University of Florida, Gainesville. pp. 1-87.
11. Ullrey, D.E. 1980. J. Anim. Sci. 51, 645-651.
12. Smith, A.M., M.F. Picciano. 1986. J. Nutr. 116, 1068-1079.

PLASMA SELENIUM CONCENTRATION AND ITS RELATIONSHIP WITH THE ACTIVITY OF GLUTATHIONE PEROXIDASE AND SOME PLASMA ENZYMES (CREATINE KINASE, LACTATE DEHYDROGENASE, ASPARTATE AMINOTRANSPHERASE) OF FIGHTING BULLS KILLED DURING A BULLFIGHT

A. Purroy, D. Revuelta
S.I.A.-I.N.I.A.(D.G.A.). Apartado 727. 50080 ZARAGOZA (Spain)

S. García-Belenguer, M. Gascón
Facultad de Veterinaria. Universidad de Zaragoza. ZARAGOZA (Spain)

J.M. Gonzalez
Hospital "Virgen de la Vega". SALAMANCA (Spain)

INTRODUCTION

Selenium (Se) concentration in animals may directly be estimated from blood and tissue analysis and indirectly by analysis of the enzyme gluthatione peroxidase (GSH-Px, E.C. 1.11.1.9) activity (1; 2; 3; 4).

Spanish bullfight supporters have been observing a worrying phenomenon in the last 20-25 years: fighting bulls have been showing a decreased aggressiveness and a lack of strength that results in numerous falls during bullfights. Those animals have been found to have very high muscular enzyme activity -Creatine Kinase (CK; E.C. 2.7.3.2), Lactate Dehydrogenase (LDH; E.C. 1.1.1.27); Aspartate Aminotranspherase (AST; E.C. 2.6.1.1) - as well as histological lesions of the cardiac and skeletal muscles (5; 6).

This work intends to determine serum Se concentration and GSH-Px activity and their correlation , as well as the relationship between these two components and different plasma enzymes (CK, LDH and AST) of bulls killed during bull fights.

MATERIALS AND METHODS

All blood samples were collected in the jugular vein of 119 4-year old bulls killed during bullfights (1987). Se concentration in serum had been found by atomic absorption obtained by hydride system (PERKIN-ELMER 1100).

Erythrocyte GSH-Px activity was determined by the method of (7) modified by (8).

Selenium in Medicine and Biology

The activities of the enzymes CK, LDH and AST were determined in an automatic analyzer (HITACHI 705).

Group A = Animals killed in spring (n = 38)
Group B = Animals killed in summer (n = 57)
Group A' = Animals of group A killed at Pamplona (n = 24)
Group B' = Animals of group B killed in Zaragoza (n = 43)
Group T = Animals of Groups A and B (n = 95)
Group S = Animals from the same herd supplemented with Se (0,25 ppm) and vitamin E (20 ppm, α-tocopherol) for at least 30 days (n = 24)

RESULTS

The mean values (± standard deviation) of plasma Se concentration, GSH-Px, CK, LDH and AST activities are shown in the following Table:

Parameter / Group	Se (µg/ml)	GSH-Px IU/gHb	CK (IU/l)	LDH (IU/l)	AST (IU/l)
A (n = 38)	0.028 (±0.013)	23.73 (±21.53)	1614 (±1362)	3296 (±876)	158 (±68)
B (n = 57)	0.053 (±0.026)	14.38 (±14.84)	2857 (±3705)	3960 (±1401)	276 (±245)
A' (n = 24)	0.031 (±0.016)	29.97 (±22.68)	1735 (±1199)	3353 (±836)	145 (±54)
B' (n = 43)	0.055 (±0.023)	13.43 (±14.72)	1959 (±1779)	3696 (±1106)	244 (±118)
T (n = 95)	0.044 (±0.025)	20.33 (±19.21)	2360 (±3046)	3694 (±1256)	229 (±203)
S (n = 24)	0.037 (±0.023)	51.28 (±31.48)	4203 (±3858)	4137 (±1479)	280 (±186)

Se concentration was higher (P<0.005) for group B than for group A, while the reverse occurred for GSH-Px activity (P<0.005). The values obtained in group T were intermediate with respect to these groups. A higher GSH-Px activity ($P<0.005$) had been found for supplemented bulls than for group T, while no increase in the Se concentration had been observed. The relationship between Se and GSH-Px had been found to be negative ($r = -0.324$, $P<0.01$) for group T, while it was positive ($r = 0.410$; $P<0.05$) for supplemented bulls.

The mean values of CK, LDH, and AST were 2360, 3694 and 229 IU/l respectively (group T). The values of CK ($P<0.1$), LDH ($P<0.05$) and AST ($P<0.005$) were higher for the animals killed in summer (B) than for those killed in spring (A). The mean values of these enzymes were higher, in absolute value (4203, 4137 and 280 IU/l), for the supplemented animals than for the others (T) although the difference was only significant for CK ($P<0.05$). These differences were all significant with respect to group A'.

The correlation coefficients between the Se concentration and the activities of CK, LDH and AST (T) were low : 0.213 ($P<0.05$); 0.221 ($P<0.05$) and 0.158 (NS) respectively. As for supplemented animals, these values were: 0.436 ($P<0.05$); 0.425 ($P<0.05$) and 0.555 ($P<0.01$). The correlation coefficients between GSH-Px activity and the activity of CK, LDH and AST, were low (NS) for group T animals: -0.137; -0.127 and -0.142, and higher and positive for supplemented animals: 0.373 (NS); 0.347 (NS) and 0.521 ($P<0.01$).

DISCUSSION

The mean Se concentration was usually lower than what is considered to be a minimum: 0.05 µg/ml (9). That may be the result of the low Se concentration in the pastures of our country (10; 11). The plasma Se concentration was higher in summer than in spring ($P<0.001$) and that agrees with the view of (12) who found that animals grazing in arid lands had higher Se concentrations than others grazing in moderate rainfall areas.

The lack of increase of plasma Se in supplemented animals does not tally with the results of other authors (4), while the increase of GSH-Px activity agrees with other results (3).

The small negative relationship between Se and GSH-Px may be due to the low levels of these components and to the fact that the animals belonged to 20 different herds which means differences in: (1) genetic origin, (2) location, (3) feeding and management, (4) season of slaughtering and (5) conditions prior to and during the bullfight.

The activities of CK, LDH and AST were much higher than those considered to be normal values (13; 14). This would confirm the existence of myodegeneration that had previously been mentioned (10; 6). Nevertheless, these activities were more important for the supplemented animals (marked lack of strength), particularly with relation to group A' (stronger animals).

Thus, our animals showed a high Se and GSH-Px deficiency paired with high values of the muscular enzymes which resulted in a lack of strength during the bullfight. GSH-Px concentration had not been a good predictor of the Se status.

Acknowledgement

This research was supported by grant no. 1239/84 (CICYT)

References

1. Anderson, P.H., S. Berrett, D.S.P. Patterson. 1978. J.Comp. Path, 88, 181.

2. Scholz R.W., L.J. Hutchinson. 1979. Am.J.Vet.Res., 40(2), 245

3. Hidiroglou, M., J. Proulx. 1985. J. Dairy Sci., 68, 57.

4. Stevens, J.B., W.G. Olson, R. Kraemer, J. Archambeau. 1985. Am.J.-Vet.Res., 46(7), 1556.

5. Purroy, A., J.M. González, S. 1985. Reprod. Nutr. Develop. 25(4A), 599.

6. Purroy, A., J.M. González, S. García-Belenguer, M. Barberán, M. Gascón. 1987. 38 Meeting European Association Animal Production. Lisbon. Sept. 28-Oct. 2.

7. Paglia, D.E., W.N. Valentine. 1967. J. Lab. Clin. Med., 70, 158.

8. Günzler, W.A., H. Kremers, L. Flohe. 1974. Z. Klin. Chem. Klin. Biochem., 10, 444.

9. Gyang, E.O., J.B. Stevens, W.G. Olson, S.D. Tsitsamis, E.A. Usenik. 1984. Am. J. Vet. Res., 45(1), 175.

10. Purroy, A., J.M. González. 1984. Bull. Acad. Vet. de France, 57, 465.

11. Montalvo, I., B. García, M. Lamand. 1982. A.Y.M.A., XXIII(21), 465.

12. Koller, L.D., P.J. South, J.H. Exon, G.A. Whitbeck. 1983. Cornell Vet., 73, 323.

13 Kaneko, J.J. 1980. In: Clinical Biochemistry of Domestic Animals. Academic Press. London. p. 831.

14. Duncan, J.R., K.W. Prasse. 1986. In: Veterinary Laboratory Medicine. Iowa State University Press. Ames, p. 285.

ALTERATIONS IN THE SEMEN QUALITY AND PLASMA ENZYMES IN BULLS. RELATION WITH SELENIUM DEFICIENCY

E.G. Capaul, A.R. Carcagno, L. Deluca

Catedra de Fisiologia Animal, Facultad de Ciencias Veterinarias, Universidad de Buenos Aires. Chorroarin 280 - 1427, Capital Federal, Argentina

Introduction

There are numerous factors for negative incidence in bulls fertility. They have been studied in detail in the Artificial Insemination Centers, and for this reason they have not been taken into account in this study. When semen abnormalities were found, a biochemical analysis was made in plasma of animals. These results are here reported as well as the effects of selenium supplementation on these indices.

Materials and Methods

- Animals : 18 bulls from an important Artificial Insemination Center.
- Feeding : Alfalfa : 4 Kgs; Plotting oat : 10 Kgs; Corn : 5 Kgs. Mineral Suppliyng (SO_4Cu : 1 Kg; SO_4Co : 200 grs; Ash bone and salt at fifty fifty till complete 100 Kgs) : ad libitum. Theoretical Se content per ration : 0.2736 ppm.
- Sampling : 2 blood and frozen semen samples at a 30 days interval, before and after selenium supplementation.
- Analysis : Plasma : Total Proteins, Albumins, Globulins, Total Bilirubins, Gamma GT; AST, CK.

  Semen : Bacteriological culture; individual motility; Vital coloration, Eosine Blue of Aniline, Urs Muller Technic; Acrosome, Watson Technic; Incubation at 37° , 1 hour and 2 hours.
- Treatment: 6.6 mg of Selenium as Sodium Selenite by injection two times a week for a total of 5 applications.

Selenium in Medicine and Biology

Results and Discussion

The results from biochemical analysis (table 1) were not relevant, except the high values for AST and CK. No hepatic alterations were observed; so, the increase of AST was infered from muscular origin, what was confirmed by the alteration in CK. After selenium treatment, the indices showed normal levels.

Table 1 : Plasma values of biochemical indices

		TBb(mg %)	γGT(mU/ml)	AST(mU/ml)	CK(mU/ml)
Before treatment	X̄	0.42 (0.32-0.54)	20 (17-24)	127 (98-180)	168 (91-262)
	SD	0.08	3	24	46
After treatment	X̄	0.39 (0.30-0.46)	18 (14-23)	45 (37-58)	20 (15-25)
	SD	0.06	2	6	3
Ref. values		0.05	25	58	20

In the semen, necrospermy and oligospermy were observed. Acrosomic studies of defrost and incubated semen samples, showed a high level of damages, with a high percentage of partial losses (table 2). After treatment samples normalization is observed.

Table 2 : Values in semen

		MOTILITY (IMP) (% Live X vigour)			VITAL COLORATION (% Live)			ACROSOMES (% Intact)		
		Inmed	1st h	2nd h	Inmed	1st h	2nd h	Inmed	1st h	2nd h
Before treatment	X̄	161	60	42	43	23	6	44	22	8
	SD	14	12	9	5	4	3	8	3	2
After treatment	X̄	305	253	155	66	41	27	86	79	69
	SD	21	28	36	5	5	4	3	5	4
Ref. values		250	200	150	65	45	35	65	55	45

Bull seminal plasma contains appreciable amounts of the Selenoenzyme Glutathione Peroxidase (GSH-Px). In case of selenium deficiency, spermatozoa die without defense against peroxides. Protective effect of selenium consists in the peroxide destruction before they attack the lipids of the spermatic membrane (1). In the tail of sperm, there is a selenopeptide. In Se deficiency, a fracture of the tail near of the middle part has been reported, altering motility (2, 3). This alteration cannot be prevented by vitamin E, but by Se (4).

Conclusions

The simultaneous occurence of established parameters for Se deficiency in plasma and in semen suggests a deficiency state in the animals in spite of the lack of specific analysis. The diagnostic is further demonstrated by the normalization of altered parameters after treatment.

Acknowledgement

This work was supported by "Programa 1987-1988 : Investigacion y Desarrollo". Secretaria de Ciencia y Tecnica U.B.A. (UBACYT) Subsidio VE-028.

References

1. Jones, R., T. Mann. 1977. J. Reprod. Fert. 50, 261.
2. Pallini, U., F. Bacci. 1979. J. Submicro. Cytol. 2, 165.
3. Pond, F., M.J. Tripp., S.H. Wu, P.D. Whanger, J.E. Olfred. 1980. In : 9° Congreso Internacional de Reproduccion Animal e Inseminacion Artificial. Madrid. Vol. V p. 157.
4. Segerson, E., B. Johnson. 1981. J. Anim. Sc. 51, 2, 395.
5. Smith, D.G. et coll. 1979. Biol. Reprod. 20, 377.
6. Watson, P. 1975. Vet. Record 97 (1) 12.

SELENIUM IN MARINE WADERS

A.A.Goede
Delft University of Technology, Interfaculty Reactor Institute
Mekelweg 15, 2629 JB Delft, the Netherlands

Introduction

In marine waders from a large estuarine area, the Dutch Wadden Sea, selenium organ concentrations up to 110 mg/kg dry weight have been observed (1). Concentrations at this level have been reported to be highly toxic to wildfowl reproduction (2). In marine mammals such high concentrations have been described in association with high mercury levels (3). The significance or possible threat of selenium to the marine birds is currently investigated.

Material and Methods

Breeding populations of Oystercatcher, Haematopus ostralegus, have been sampled on four locations: along the Dutch Wadden Sea coast (marine, polluted), on Dutch grasslands (freshwater, unpolluted) and on a Norwegian island (marine unpolluted and freshwater unpolluted). Blood and egg samples have been collected and trace element concentrations were determined by neutron activation analysis (4).

Results and Discussion

Selenium levels in the red blood cells (RBC) appear to be very

Selenium in Medicine and Biology

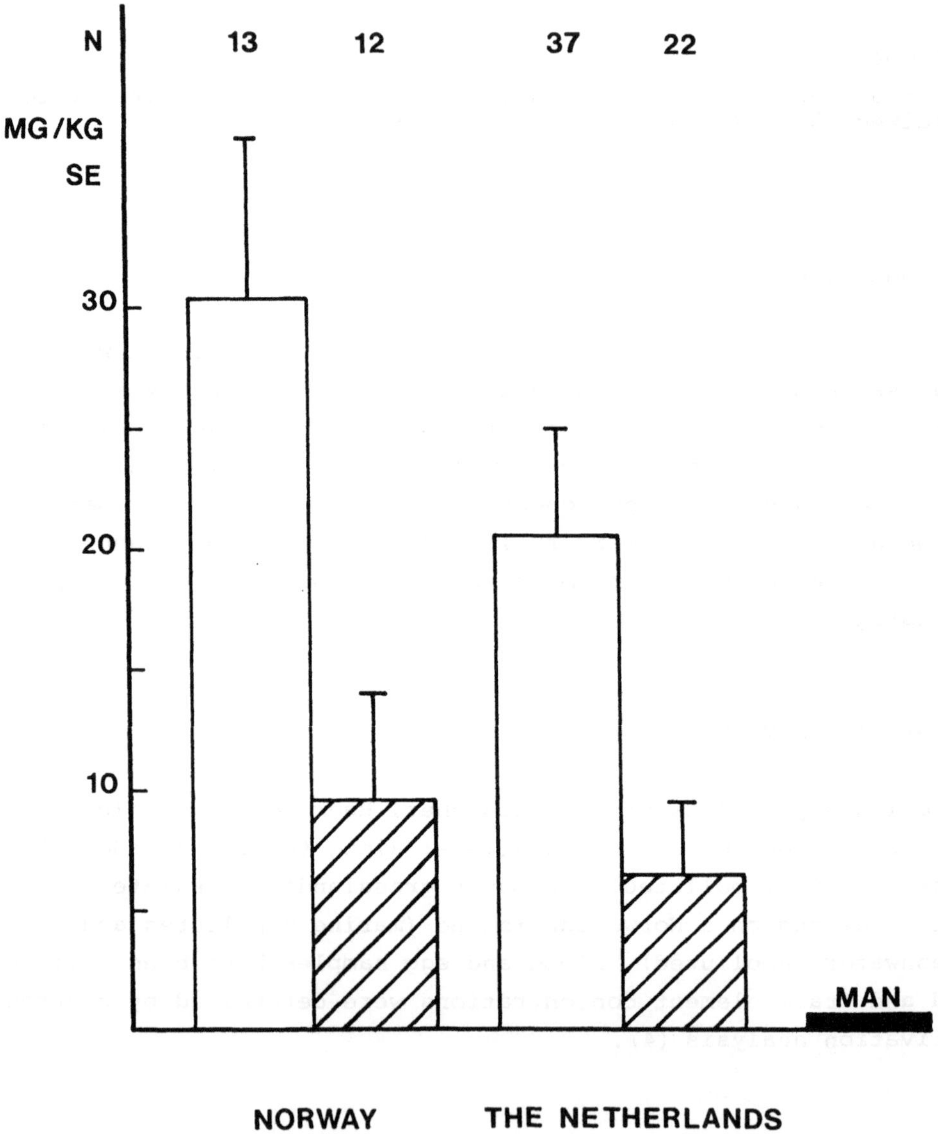

Figure 1.
Selenium Concentrations (Mean ± S.D. dry weight) in Red Blood Cells of Marine (▭) and Freshwater (▨) Feeding Oystercatchers. N = sample size.

high in the marine feeding birds. Though levels in freshwater feeding birds are much lower, they are still high compared to levels in man (Figure 1). Concentrations in Figure 1 are given in mg/kg dry weight. When concentrations are converted to µg/g Hb, differences in selenium levels become visible between Norwegian and Dutch birds : resp. 39.1 and 19.5 (marine samples), 12.4 and 5.7 (freshw.samples) µg Se/g Hb. The differences are attributed to differences in dry matter: resp. 1.3 and 0.9 g/g Hb in Norwegian and Dutch birds. The haemoglobin level in whole blood is similar in all samples and similar to man: circa 150 g Hb/l.

Despite these highly different RBC selenium levels in marine and freshwater feeding birds, egg selenium concentrations do not differ (Table I).

Table I
Egg Selenium Concentrations (Mean ± S.D in mg/kg dry weight) in Coastal and Inland Breeding Populations of the Oystercatcher in the Netherlands

location	yolk	white	n
marine, polluted	1.8 ± 0.4	3.3 ± 0.7	30
freshw., unpolluted	1.9 ± 0.3	2.9 ± 0.6	18

In California, where selenium intoxication in wildfowl (waders included) was reported (2), selenium levels up to 110 mg/kg were measured.

Selenium is able to prevent toxic effects of high mercury, cadmium or arsenic intake and gets accumulated with these elements. In the Oystercatcher no relation is found between selenium and mercury, cadmium or arsenic concentrations in either RBC or egg tissues. Furthermore, there is no difference in RBC selenium concentration between the birds from the polluted and unpolluted marine environment.

Conclusions

The high selenium concentrations in waders is a marine characteristic (confirmed by (unpubl.) kidney data). In contrast to California wildfowl, the high selenium levels are not toxic to the reproduction, nor can they be attributed to a detoxifying function in relation with mercury, cadmium or arsenic exposure. As these possibilities have been excluded, the high selenium concentrations in marine waders have no precedent. Further research will be focussed on the possible essential role of the element and the chemical speciation.

References

1. Goede, A.A. 1985. Environ. Pollut. 37, 287.
2. Ohlendorf, H.M., D.J. Hoffman, M.K. Saiki, T.W. Aldrich. 1986. Sc. Total Environ. 52, 49.
3. Koeman, J.H., W.S.M. van de Ven, J.J.M. de Goeij, P.S. Tjioe, J.L. van Haaften. 1975. Sc. Total Environ. 3, 279.
4. de Bruin, M., P.J.M. Korthoven, P. Bode. 1982. J.Rad.anal. Chem. 70, 497.

EFFECT OF SELENIUM ON THE METABOLISM OF A NOVEL METHANOGEN

Shanthi Krishnan, N. Vasanthy, K. Lalitha
Department of Chemistry, Indian Institute of Technology, Madras 600 036, India

Introduction

Methanogens of the archaebacterial kingdom are phylogenetically distinct group among the procaryotic organisms. Intensive studies on methanogens in recent years have revealed unique and novel biological and biochemical features (1,2). Addition of selenium stimulates the growth of Methanococcus vannielii (3) and methanosarcina barkeri (4). The occurence of selenium in the seleno enzyme, formate dehydrogenase, in the chemical form of selenocysteine residues has been reported in Methanococcus vannielli. The metabolic significance of selenium is thus related to its role as an integral part of many enzyme of bacterial origin and the mammalian glutathione peroxidase (5). In an attempt to understand the importance of selenium so far reported in methanogenic species and its role in anoxic environment, the effect of selenium on the growth and metabolism of a novel methanol utilising methanogen is described.

Results

The methanogen isolated from the anaerobic Leucaena leucocephala degrading medium was maintained on a basal medium that contained the following in grams/litre : K_2HPO_4 (1.4) KH_2PO_4 (0.75), $MgCl_2$ (0.2) and NH_4Cl (0.9) and methanol (5.0 ml) devoid of any added mineral nutrient. Cells were gram positive short rods with an average size of 0.57 µm in diameter and 1.7 µm length (scanning electron micrographs available from the author). Growth of the cells corresponded with the measured fluorescence intensity of the methanogonic cofactor F_{420}.

Selenium in Medicine and Biology

Cells grew optimally at a methanol concentration of 162 mM at 37° C and at pH 7.4 yielding 14 mmoles methane in 13 days. Appreciable growth was also noted with 2 propanol, limited with acetate while no growth was observed with formate. The fluorescence intensity measurements of F_{420} showed that addition of selenium (Se) as sodium selenite (0.1 µM) increased growth by 22 % and nickel (Ni) as nickel chloride (5µM) by 25%, while addition of both Ni and Se at the same concentration to the culture medium enhanced the growth by 50 % when compared to control.

Aspartate aminotransferase (AAT), Alanine aminotransferase (ALAT) and ATPase were monitored in both soluble and membrane fraction of the sonicated cell extract. Specific activity of AAT was 60 % higher in the sonicated supernatant while that of ALAT was higher by about 8 % when compared to their respective activities measured in membrane fractions left after sonication. Specific activity of ATPase was 2 fold high in sonicated cell extract (Table 1).

Table 1 : Aspartate aminotransferase, Alanine aminotransferase and ATPase activities in the methanol utilising methanogen

	Soluble fraction		Membrane fraction	
Parameters	µ/mg protein	µ/g cells	µ/mg protein	µ/g cells
Aspartate aminotrans-ferase	0.75	2.08	0.47	0.6514
Alanine aminotrans-ferase	0.91	2.42	0.84	5.45
ATPase	3.84	10.24	1.77	11.40

Cell soluble and cell membrane fraction were obtained by sonication.

AAT activity : µmoles pyruvate formed/10 min at 37° C;
ALAT activity : µmoles pyruvate formed/10 min at 37° C;
ATPase activity : µmoles Pi liberated/10 min at 37° C.

Addition of Se (0.1 µM) resulted in significant changes in the electrophoretic protein profiles of the soluble and membrane fractions and were different from those induced by Ni (5.0 µM). Of the membrane bound proteins, about 90 % were solublised by Triton treatment whereas in the presence of Se only 37 % could be released. The detection of one anionic and one cationic isoenzyme of AAT in this isolate is of significance. Soluble glutathione peroxidase (GSHPx) activity of this isolate was enhanced by 45 % in the presence of 0.1 µM selenite.

Discussion

The organism isolated appears to be novel and hitherto unreported. The detection of isoenzymes of AAT and GSHPx activity is significant in consideration of specific metabolic roles recognised for these enzymes in higher organisms. The isoenzymes of AAT are widely distributed in all living systems and have metabolic and regulatory significance. Anionic soluble and cationic mitochondrial isoenzymes of AAT are well characterised in many animal species (6). The detection of isoenzymes of AAT in the methanogenic isolate emphazises the importance and probable evolution at a very early stage in life processes. The detection of GSHPx activity in this species reported for the first time requires further inquiry. The presence of ATPase in methanogens have also been reported by other workers (7,8). These preliminary investigations indicate the presence of the enzyme in both soluble and membrane fractions. Further studies are being carried out as to the extent of solublisation by different methods, the actual levels of the enzymes and more detailed investigations on the level of selenium required and the effect thereof.

References

1. Balch, W.E., G.E. Fox, L.J. Magrum, C.R. Woese, R.S. Wolfe. 1979. Microbiol. Rev. 43, 260-296.

2. Daniels, L., R. Sparling, G.D. Sprott. 1984. Biochim. Biophys.Acta 768, 113-163.

3. Jones, J.B., T.C. Stadtman. 1977. J. Bacteriol. 130, 1404-1406.

4. Scherer, P., H. Sahm. 1981. Eur. J. Appl. Microtriol. Biotechnol. 12, 28-35.

5. Stadtman, T.C. 1980. Annu. Rev. Biochem. 49, 93-110.

6. A.E. Braunstein. 1973. In : The Enzymes, vol. 9 (P.D. Boyer, ed.) A.P. Newyork, pp. 393-440.

7. Roth, R., R. Duft, A. Binder, R. Bachofen. 1986. System, Appl. Microbiol. 7, 346-348.

8. Inatomi, K.I. 1986. J. Bacteriol. 167, 837-841.

THE ADDITION OF SELENIUM TO FEEDSTUFFS AND THE LEGAL RESTRICTIONS RELATED TO THIS PRACTICE

M. Lamand
I.N.R.A. - Laboratoire des Maladies Nutritionnelles
C.R.Z.V. de Theix - 63122 Ceyrat (France)

The toxicity of selenium is well known and it is highly desirable that levels of the element in animal products should not be abnormal, thus in most countries, restrictive legislation limits its usage in animal foods.

Since selenium accumulates in the liver and kidney of animals, a cumulative toxicity can also develop in the consumers of this material from the carcases. Moreover the Nelson group (1) produced hepatic tumours in rats given selenium. However these results have remained unconfirmed and selenium has been since, shown to be anticarcinogenic(2).

In France the risk of toxicity is low, there is no area where either soil or plants are rich in selenium. The water from the spa at La Roche Posay contains a lot of selenium (32 to 48 µg per litre) (3) and some kaolin quarries in this area produce forage plants which are usually high in selenium, but this is exceptional (Table 1) (4).

The governments of the various countries generally adopt a very prudent position with respect to legislation controlling the usage of selenium as a dietary additive to prevent deficiency.

Some therapeutic forms have been given market authorization. These are injectable solutions or slow release tablets

Selenium in Medicine and Biology

based on selenium and iron oxide. It is certain that the administration of these forms could lead to overdosage. To avoid this danger the employment of medicament containing selenium must be under veterinary supervision.

Table 1 : The Selenium Levels of Some Plants Growing on an Old Kaolin Clay Quarry (mg/kg dry matter) (seed stage - Renoir - La Roche Posay).

	Number of samples	Level	Range
Lotus corniculatus L.	8	10,6 ± 8,2	0,9 - 19,7
Trifolium subterraneum L.	8	9,4 ± 6,3	3,5 - 21,2
Trifolium repens L.	6	3,3 ± 1,8	0,4 - 5,7
Festuca ovina L.	7	0,6 ± 0,1	0,4 - 0,7
Taraxacum dens leonis Desf	2	2,8	1,1 - 4,6
Medicago sativa L.	1	11,9	

The supplementation of foods with selenium is safe provided the quantities added are supervised and follow the following criteria :

- the addition of selenium should be in response to a deficiency in the forage.
- the levels in the prepared mineral mixes should be tightly supervised.
- the mixes should be homogeneous.
- the supplementation should put neither animal nor carcase consumer at risk.

widespread and of economic significance particularly with respect to myopathy of young ruminants, French legislation allows the systematic addition of 0.1 mg Se/kg DM to all compound feeds but does not authorise levels in excess of 0.5 mg Se/kg DM (5). The legislation specifies that only compound feeds can be enriched and the sale of mineral mixes containing high levels of selenium to the general public is forbidden. Obviously the animals appetite limits compound feed ingestion whereas they could easily take in excessive amounts presented in other forms.

The Economic Community has adopted similar measures (6) and authorises a concentration of no more than 0.5 mg Se/kg of the foodstuff. By contrast the Food and Drug Administration is less restrictive since this body has authorized the sale of mineral mixes rich in selenium intended for addition to (and dilution by) the compound foodstuff. Selenite or selenate can be used and the recommendations are the following :

1. Poultry : in complete feed of growing chicken up to 16 weeks of age at a level not to exceed 0.1 ppm ; in complete feed of turkeys at a level not to exceed 0.2 ppm.
2. Swine : in complete feed, the Se supplementation for pigs from weaning to 50 lb (22.7 kg) b.w. is 0.3 ppm. At 22.7 kg b.w., Se supplementation should revert back to 0.1 ppm Se.
3. Sheep : in complete feed, at a level not to exceed 0.1 ppm ; in a feed supplement for limited feeding at a level not to exceed an intake of 0.23 mg/head/day.
4. Beef cattle : in complete feed at a level not to exceed 0.1 ppm ; in a feed supplement for limited feeding at a level not to exceed an intake of 1 mg/head/day, up to 20 ppm in a salt-mineral mixture for free choice feeding at a rate not to exceed an intake of 1 mg/head/day.

5. Dairy cattle : in completed feed at a level not to exceed 0.1 ppm (7).

Animals kept intensively receive foodstuffs from the compounders whose levels of Se have been corrected. The risk of meat dangerously high in selenium is low. The frequent consumption of other material such as liver or kidney, could considerably increase the selenium intake but the danger remains slight provided the animal foodstuffs remain below 0.5 mg Se/kg. Some countries such as New Zealand (8) or Finland (Kumpulainen, personal communication) have authorized the addition of selenium to phosphate fertilisers. The progressive accumulation of selenium in the soil is a possibility and constant surveillance of animal products will be necessary.

References

1. Nelson A.A., O.G. Fitzhugh, H.O.Calvery 1943. Cancer Res. 3, 230-236.
2. Scott, M.L. 1973. J. Nutr. 103, 803-810.
3. Lepeintre, C. 1964. Thèse Doctorat 3e cycle Univers. Paris.
4. Arvy, M.P. 1982. Thèse Doctorat d'état Univers. Poitiers.
5. Journal Officiel de la République Française 1973, 105, 13338-13345.
6. Journal Officiel des Communautés Européennes 1970. L 270, 14 XII 1970 p. 1, artic. 4.
7. Ullrey, D.E. 1980. J. Anim. Sci. 51, 645-650.
8. Gissel-Nielsen, G., U.C. Gupta, M. Lamand, T. Westermarch 1984. Adv. Agron. 37, 397-460.

TOXICOLOGY OF SELENIUM IN VETERINARY MEDICINE

G. Keck

Laboratoire de Toxicologie, Ecole Nationale Vétérinaire de Lyon, B.P. 31, 69752 CHARBONNIERES CEDEX

Deficiencies in selenium have been identified for many years in domestic animals and are corrected either in food, either by parenteral administration. However, selenium is a fairly toxic oligo-element for domestic animals, on acute or chronic mode. These aspects will be presented in the light of the fieldcases treated by the Veterinary Antipoison Centre at the Veterinary School of Lyon.

1) Acute toxicity

On a general basis, the lethal oral dose of selenium is very low, in the range of 1-5 mg/kg.
Selenites are more toxic than selenates, selenides and elementary sélenium. Following intra-muscular injection, the DL 50 of sodium selenite in sheep is 0,5 mg/kg (Blodgett and Bevill, 1987). The susceptibility depends on the status in selenium : deficient animals seem much more susceptible than those with normal levels. Stress (by contention) or cobalt defiency could also favorize the toxicity.

-Several acute accidents following parenteral injection of selenium formulations, usually in association with vitamine E have been reported at our Centre or in the literature.

Therapeutic doses are in the range of 0,03 mg/kg. However, over-concentrations can be easily observed in commercial preparations : the decantatation of the aqueous phase can lead to concentrations 50 times higher than nominal concentrations (Krieger et al., 1986). In Australia, oral administration of selenite has also raised number of acute accidents (Cooper, 1987).

Selenium in Medicine and Biology

-The symptoms of acute intoxication are intense and appear 2-3 hours after administration : anorexia, prostration, progressively severe dyspnea with cyanosis and nasal discharge.

-The lesions affect mainly the cardio-respiratory tract : pulmonary oedema, hydrothorax, degeneration of the myocardium. The heart seems to be the target-organ of selenium toxicity and shows the highest levels of selenium with the liver (Blogdett and Bevill, 1977).

-The local tolerance of I.M administration of selenium preparations is not good (muscular necrosis or abcess).

2) Sub-acute/chronic toxicity

The recomended and toxic levels for domestic animals are presented in Table I (Buck et al., 1986).

These levels show large variations following the chemical form, the diet (contents in proteins and sulfates), age and individual susceptibility.

Species	Recomended levels	Non-toxic levels	Toxic levels
Cattle	0.10	2.0	8.0
Sheep	0.10	?	10.0
Swine	0.10	2.5	7.0
Poultry	0.15	5.0	15.0

Table 1 - Recomended and toxic levels of selenium in food (as ppm of ration)

-Modalities of sub-acute/chronic intoxications

Sub-acute or chronic intoxications can be due either to plants either to excessive supplementation in food.

-Intoxications by seleneferous plants

are known in some regions of USA (Dakota, Kansas) and in Australia ; they seem very rare in Europe, especially in France.

Clinical manifestations have lead to a sub-division into 2 syndroms :

- "blind-staggears" syndrom, due to the ingestion of typically seleniferous plants (Astragalus ...), with anorexia, colics, ataxia loss of view, dyspnea
- "alcali-disease" syndrom, more chronic, due to plants containing lower levels : loss of hair, deformation of the hoofs are associated to progressive emaciation.

The lesions, in both syndroms, are : hepatic cirrhosis, chronic nephritis, myocardial degeneration.

-Sub-acute or chronic intoxications

due to excessive supplementation of selenium in food are not unfrequent. Very often, the errors of incorporation are in a range of 10 to 100 times the normal level. The excess can also be due to a poor cleaning of an apparatus after the fabrication of selenium-rich condiment.

The clinical manifestations associate generally : anorexia loss of hair, dyspnea, hepatic and renal degeneration, cardiac degeneration.

3) Analytical diagnosis

The levels of selenium found in tissues after acute or chronic intoxication are given in Table II.

Acute intoxication	liver-kidney	10-25 ppm
Chronic intoxication	liver-kidney	1-5 ppm
	hoof	8-20 ppm

Table II - Levels of selenium in tissues after acute or chronic intoxications.

4) Embryo-toxicity

Selenium crosses easily the placenta and can perturbate the foetal development.

Deformation of the legs, chetivity of new-borns and sterilities have been described in seleniferous regions of the USA (Buck et al.).

Saws receiving 1 ppm of selenium in their food gave birth to piglets of low weight and deformed hoofs.

5) Mechanisms of toxicity

The toxicity of selenium seems to be linked with the inhibition of respiratory enzymes, especially deshydrogenases, maybe by interference with sulfur (Cooper, 1987).

6) Treatment of intoxications

As for most of intoxications by cumulative elements, the treatment is very difficult and deceiving.

Some dispositions coulb may be reduce reduce the toxic impact :

- a high level of protein in the ration, especially casein with SO_4^{--}
- supplementation with arsenic (Bartik and Pisak, 1981)

Conclusion

The susceptibility of domestic animals to acute or chronic toxicity of selenium in very high, as is their susceptibility to deficiencies in selenium.

Many factors of variations play a major role in practical conditions, which make necessary exchanges between toxicologists and biochemists, but also between medicals and veterinarians.

LIST OF PARTICIPANTS

Mr AIME Christian
Chemin de la tuille
64250 CAMBO FRANCE

Mme AKESSON Anita
University hospital Dept. of Rheumatology
S-22185 LUND SWEDEN

Mr AKESSON Bjorn
University Hospital
Dept. Clin. Chem.
S-22185 LUND SWEDEN

Mme ALCARAZ Arlette
C.H.R.U.G.
Laboratoire Biochimie C
BP 217 X 38043 GRENOBLE CEDEX FRANCE

Mr ALLAIN Pierre
C.H.U.
Laboratoire Pharmacologie
rue Larrey 49040 ANGERS FRANCE

Dr ALNOR K.C.
Kirkevejen 20
Teestrup
4690 HASLEV DENMARK

Mme ANDERSEN Helve Raun
Klingstrupvanget 103 1tv
5230 ODENSE DENMARK

Mme ANDRE Dominique
C.H.R.U.G.
Laboratoire Biochimie C
BP 217 X
38043 GRENOBLE CEDEX FRANCE

Dr ARCAN J. Claude
Laboratoire MILLOT SOLAC
16, avenue Georges V
75008 PARIS FRANCE

Mle ARNAUD Josiane
C.H.U.R.G. Laboratoire Biochimie C
BP 217 X
38043 GRENOBLE CEDEX FRANCE

Mr AUGER Michel
1150, bd St Joseph Est
HIJIL5 MONTREAL CANADA

Dr BARDET J.Pierre
Laboratoire LA ROCHE POSAY
29, avenue des Fontaines
86270 LA ROCHE POSAY FRANCE

Mr BAREAU Xavier
6, rue Francis Poulenc
19100 BRIVE FRANCE

Mr BARRILLON Dominique
Hôpital PASTEUR F2 - BP 69
06002 NICE CEDEX FRANCE

Mr BARUTHIO François
INRS
avenue de Bourgogne BP 27
54501 VANDOEUVRE LES NANCY FRANCE

Dr BECQUE Claudine
14, rue de l'Argenterie
66000 PERPIGNAN FRANCE

Pr BEHNE Dietrich
HAHN MEITER
Institut Postfach 39 01 28
Glieniker Strasse 100 D
100 BERLIN 39 FRG

Mme BELIN Nicole
C.H.R.U.G. Laboratoire Biochimie C
BP 217 X
38043 GRENOBLE CEDEX FRANCE

Mr BERTHELOT Alain
Faculté de Médecine Pharmacie
Laboratoire de Physiologie Pharmacie
25030 BESANCON FRANCE

Dr BEYEN Katharina
Med. Institut Fur Umwelthygiene an der Universitat Dusseldorf Postfach 5634
4000 DUSSELDORF 1 FRG

Mr BIENVENU Pierre
CRSSA
Division de Radiobiochimie
rue du Lieutenant Raoul BATANY
92140 CLAMART FRANCE

Mme BIRON Martine
Laboratoire AGUETTANT
1, avenue J. Carteret
69007 LYON FRANCE

Dr BOGDEN John D.
Uni. of Medicine and Dentistry of N.J.
Dept. of Preventive Medicine and Comm. Health
NEWARK NEW JERSEY 07103 2757
U.S.A

Mr BORTOLI Doti
Industrie Ricerche Chimiche IRCA - ED Albano SPA
87, via Tonale
24061 ALBANO S. ALESSANDRO BG
ITALY

Mr BOUGLE Dominique
2, rue Robert Tournière
14000 CAEN FRANCE

Dr BOURDON Raymond
25, rue Croulebarbe
75013 PARIS FRANCE

Mme BUNEAUX Françoise
19, avenue Rabelais
92160 ANTONY FRANCE

Mr BUREAU François
C.H.R.U. Laboratoire Biochimie A
avenue de la Cote de Nacre
14040 CAEN CEDEX FRANCE

Mr BURNEL Daniel
Faculté de Médecine A et B
Service de Chimie BP 184
54505 VANDOEUVRE LES NANCY
FRANCE

Dr CALESTREME Ph
C.H.U.
Laboratoire
47023 AGEN FRANCE

Mme CARGAGNO Araceli R.
Laguna 853
1714 ITUZAINGO (Buenos Aires)
ARGENTINA

Mr CARRIERE
JOHNSON MATTHEY S.A.
13, rue de la Perdrix
BP 50240 - ZI Paris Nord II
95956 ROISSY CDG Cédex FRANCE

Mle CARTIER-MILLON Françoise
C.H.R.U.G.
Laboratoire Biochimie C
BP 217 X 38043 GRENOBLE CEDEX
FRANCE

Dr CERVILLA Rodriguez Jose
Hosp. Gen. Gallicia Dpto. Pediatria
Apartado 149 SANTIAGO
COMPOSTELA SPAIN

Mr CHAPPUIS Philippe
Hôpital LARIBOISIERE
Laboratoire Central de Biochimie
2, rue Ambroise Paré 75475 PARIS
FRANCE

Mme CHARLOT Colette
Faculté de Pharmacie
Laboratoire de Chimie Ana. et Toxicol.
11, avenue Charles Flahaut
34060 MONTPELLIER FRANCE

Mr CHARRE Olivier
BIOTROL
FRANCE

Mle CHARTON Emmanuelle
SANOFI Recherche
195, route d'Espagne
31036 TOULOUSE FRANCE

Mr CHAUMELLE Pierre
21, rue du Repos
75020 PARIS FRANCE

Mr CHAUSSE
Laboratoire d'Analyses Médicales
40, avenue Gambetta 12700
CAPDENAC FRANCE

Dr CHRISTIAENS Emile
Institut Medico-Chirurgical
Chaussée de Saint Amand 80
7500 TOURNAI BELGIUM

Mr CIAPELLANO Salvatore
Facoltà Agraria
via Celoria n°2
20133 MILANO ITALY

Pr CLAUSEN Jorgen
Roskilde Universitetscenter Marbjergvej
35 Postbox 260
DK-4000 ROSKILDE DENMARK

Mr COUDERC Rémi
112, avenue Maréchal Joffre
92340 BOURG LA REINE FRANCE

Mme DARDELET Brigitte
C.H.R.U.G.
Laboratoire Biochimie C
BP 217 X
38043 GRENOBLE CEDEX FRANCE

Mme DAUBIGNARD Corinne
Sté DIG DIAGNOSTIC et Services
124, rue des Rosiers
93400 ST OUEN FRANCE

Mr DAVIES Stephen
The Stone House
9, Wey Mouth St.
WIN 3FF LONDON UNITED KINGDOM

Mr DE LA TULLAYE Jean
SANOFI Recherches
37, avenue Pierre 1er de Serbie
75008 PARIS FRANCE

Mr DEBERTOLIS
Laboratoires LABCATAL
7, rue Roger Salengro
92541 MONTROUGE CEDEX FRANCE

Dr DECKERS Yves
Laboratoire UNDA
118, avenue Jules Bordet
1140 BRUXELLES BELGIUM

Mme DELANNOY
CAHIERS DE NUTRITION ET DE
DIETETIQUE SEPAIC
42, rue du Louvre
BP 551
75027 PARIS CEDEX 01 FRANCE

Mme DELREZ Evelyne
39, avenue des Fontaines
86270 LA ROCHE POSAY FRANCE

Mme DEMICHEL Patricia
Groupement Régional d'Etudes et de
Recherches en Hydrologie Thermale
3, place de la Victoire
33076 BORDEAUX CEDEX FRANCE

Pr DERBIN C.
Université Paris Nord
Laboratoire de Neurocytologie
74, rue Marcel Cachin
93012 BOBIGNY CEDEX FRANCE

Mme DESRAME Madeleine
69, rue d'Ezy
27540 IVRY LA BATAILLE FRANCE

Mr DESSALLES Pierre Henri
C.H.U.
Services des Urgences
24019 PERIGUEUX CEDEX FRANCE

Dr DRATWA Max
Hôpital Brugman
4, place Vangehuchten
1020 BRUXELLES BELGIUM

Mme DUBOIS Françoise
C.H.R. de Nancy Brabois
Laboratoire de Biochimie Service du
Prof PAYSANT
54500 VANDOEUVRE FRANCE

Mme DUCOS-FONFREDE Simone
Faculté des Sciences Pharmaceutiques
Chimie Bioinorganique
2 bis, bd Tonnellé
37042 TOURS CEDEX FRANCE

Mle DUCROS Véronique
C.H.R.U.G.
Laboratoire Biochimie C
BP 217 X
38043 GRENOBLE CEDEX FRANCE

Mle DUMOLARD Laurence
Laboratoire HERBAXT
5, bis rue Chappon BP 145
77335 MEAUX CEDEX FRANCE

Dr DYARD François
Laboratoire MILLOT SOLAC
16, avenue Georges V
75008 PARIS FRANCE

Mr EPINAT Rémy
2, place Jules Rabourdin
42300 VILLEREST FRANCE

Mr ESCHNAUER Heinz
Gelnhauser Str.
6463 FREIGERICHT FRG.

Mr FAIVRE B.
PERKIN ELMER
1, rue Francklin
BP 67
78390 BOIS D'ARCY FRANCE

Dr FAJGELJ Ales
J. STEFAN Institut
Jamova 39
61000 LJUBLJANA YOUGOSLAVIA

Mr FAUGERON
Société PHIRA
15, rue de Champagne
77400 BUSSY ST MARTIN FRANCE

Mr FAURE Henri
C.H.R.U.G
Laboratoire Biochimie C
BP 217 X
38043 GRENOBLE CEDEX FRANCE

Mr FAVIER Alain
C.H.R.U.G.
Laboratoire Biochimie C
BP 217 X
38043 GRENOBLE CEDEX FRANCE

Mme FONTAINE Jeanine
Institut de Pharmacie
Laboratoire Pharmacologie CP 20517
1050 BRUXELLES BELGIUM

Mr FORGEOT Marcel
IRB
29, rue de Noisy
78870 BAILLY FRANCE

Mr FRAGA J.M.
Fac. Medicina Apartado
149 SANTIAGO COMPOSTELA SPAIN

Mme FUSELIER
Laboratoire AGUETTANT
1, avenue J. Carteret
69007 LYON FRANCE

Mme GARCY Christiane
15, rue D'Orléans
31000 TOURS FRANCE

Mr GARIN
VG INSTRUMENTS Lyon
9, rue St Eusèbe
69003 LYON FRANCE

Mr GARRET Christian
SPIRAL Recherche et Développement
3, rue des Mardors Couternon
21560 ARC SUR TILLE FRANCE

Mme GERBER Ghislaine
Laboratoire HOLISTICA
56, Zone Industrielle
13790 ROUSSET FRANCE

Dr GERBER Mariette
Centre Paul Lamarque Cliniques Saint Eloi
34094 MONTPELLIER CEDEX FRANCE

Mr GIROUD
Société JOBIN YVON
16-18, rue du Canal
91163 LONGJUMEAU CEDEX FRANCE

Mme GNAT Danièle
Hôpital Saint Pierre
Départ. Chimie Médicale
322, rue Haute
1000 BRUXELLES BELGIUM

Dr GOEDE A.A.
Interuniversitair Reactor Institut
Mekelweg, 15
2629 JB DELFT HOLLAND

Mr GRONBAEK Henning
Institut for Neurobiology Arhus Universitet
8200 ARHUS N - DENMARK

Dr GROTH Ulrich
Aukammallee 33
D-6200 WIESBADEN FRG

Mle GUIGUES Martine
Laboratoire des GRANIONS
14, avenue des Frères Crovetto
98000 MONACO

Mr GUILLARD Olivier
Hôpital J. BERNARD
Laboratoire Central de Biochimie Cité
Hosp. de la Milétrie - BP 577
86021 POITIERS CEDEX FRANCE

Dr GUTTERIDGE John M.C.
National Institut for Biological Stand.
Holly Hill Hampstead
NW3 6RB LONDON ENGLAND

Dr HACHACHE
C.H.R.U.G.
AGDUC - BP 217 X
38043 GRENOBLE CEDEX FRANCE

Dr HAWKER Felicity
Royal Prince Alfred Hospital
Intensive Care
2050 SYDNEY NSW AUSTRALIA

Mme HINCAL Filiz
MEDHEALTH TRADING LTD
8, avenue de Chantelle
1206 GENEVE SWITZERLAND

Mme HOUOT Odile
Centre de Médecine Préventive
2, avenue J. Parisot
54500 VANDOEUVRE FRANCE

Mme HUTIN Marie-France
2, rue des Glacis
54000 NANCY FRANCE

Mr HUTTUNEN Jussi
NAT. PUBLIC HEALTH Institut
Mannerheimintie 166
SFOO280 HELSINKI FINLAND

Mr INFANTE Francisco
Hôpital Cantonal Division de
Gastroentérologie et Nutrition
1211 GENEVE 4 - SWITZERLAND

Mme JAUDON Marie-Chantal
26, rue Dagorno
75012 PARIS FRANCE

Dr KACIMI G.
9, Rampe Louni Arezki
B.E.O. ALGER ALGERIA

Dr KASTENMAYER Peter
FINNIGAN MAT GmbH
Barkhausenstr. 2
D2800 BREMEN FRG

Mr KECK Gérard
Ecole Nat. Vétérinaire de Lyon
69370 CHARBONNIERES FRANCE

Mme KIA Dominique
C.H.R.U.G. Laboratoire Biochimie C
BP 217 X
38043 GRENOBLE CEDEX FRANCE

Mr KIEFFER Félix
C/O WANDER A.G.
Monbijoustrasse 115 Postfach 2747
3001 BERN SWITZERLAND

Mr KLINKHAMER Adrianus Cornelis
Burg Jacoblaan 8
1401 BM BUSSUM
THE NETHERLANDS

Mr KLUG Zvergen
Kvhredder 55-61
2000 2-HAMBOURG 65 FRG

Mr KOUKAY Nizam
C.H.R.U.G.
Laboratoire Biochimie C
BP 217 X 38043 GRENOBLE CEDEX
FRANCE

Mr KUUSNIEMI Ossi
Huhtamaki Oy Novamed PO Box 415
20101 TURKU FINLAND

Mr LADROITTE Philippe
Diététique et Santé
BP 106 31250 REVEL FRANCE

Mr LAFOND Jean-Luc
C.H.R.U.G.
Laboratoire Biochimie C
BP 217 X 38043 GRENOBLE CEDEX
FRANCE

Mr LAGARDE Claude
Laboratoire d'Analyses Médicales
40, avenue Gambetta
12700 CAPDENAC FRANCE

Dr LAMAND Michel
INRA, CRZV de Theix
Labo. Des Maladies Nutritionnelles
63110 BEAUMONT FRANCE

Mme LASSEN Karin
Laessoegade 122
5230 ODENSE M. DENMARK

Mr LEROUX
VG Instruments
3,rue du Maréchal de Lattre de Tassigny
78150 LE CHESNAY FRANCE

Mr LEWIS Norman
CELL LIFE INTERN. INC.
9373 Activity Road Ste 1
CA 92126 SAN DIEGO U.S.A.

Dr LHOTE Bruno
7, rue Custine
75018 PARIS FRANCE

Dr LOMBECK I.
Medizinische Einrichtungen der Universt. Dusseldorf Kinderklinik und Poliklinik Moorenstrasse
5 D-4000 DUSSELDORF 1 FRG

Mr MAHFOUZ Ryad
Laboratoires MILLOT SOLLAC
16, avenue Georges V
75008 PARIS FRANCE

Mr MAIER Eddie
EHICAS Laboratoire de Chimie Minérale
1, rue Blaise Pascal
67008 STRASBOURG FRANCE

Mr MALVY Denis
C.H.R. BRETONNEAU
Service de Néphrologie
37044 TOURS CEDEX FRANCE

Mr MAN Fernando
Hôpital ST LAZARD Inserm U 290
107, rue du Faubourg St Denis
75010 PARIS FRANCE

Mme MARCHAND Martine
Hôpital BRETONNEAU
2, rue Carpeaux
75018 PARIS FRANCE

Mr MASSOL Michel
Université Paul SABATIER
Laboratoire de Chimie Bioinorganique
38, rue des 36 Ponts
31400 TOULOUSE FRANCE

Dr MENZER Bruno
RP SCHERER SA
67930 BEINHEIM FRANCE

Mr METZGER Jean Marc
25, quai Mullenheim
67000 STRASBOURG FRANCE

Mr MISSIAEN Irwin
REVOGAN Moerstraat 58
B-9810 GENT-DRONGEN BELGIUM

Mr MOESGAARD Sven
Jellingvej 9
7100 VEJLE DENMARK

Pr MOLLE Léopold
Université Libre de Bruxelles
Campus Plaine 205-1
B-1050 BRUXELLES BELGIUM

Mr MORAND
Société PHILIPS Division Sciences Industrie
105, rue de Paris
93002 BOBIGNY CEDEX FRANCE

Mr MOURGEON Gérard
Société GIF DIAGNOSTIC ET SERVICES
124, rue des Rosiers
93400 SAINT OUEN FRANCE

Mle MOYNIER Isabelle
Faculté de Pharmacie
avenue G. Flahault
34060 MONTPELLIER FRANCE

Mr NAHMANI
Laboratoire RICHELET
6, rue de la Paix
75002 PARIS FRANCE

Mr NALBONE Gilles
INSERM U 130
10, avenue Viton
13009 MARSEILLE FRANCE

Mr NARBONNE Jean-François
Université Bordeaux I
avenue des Facultés
33405 TALENCE FRANCE

Dr NAVEZ Bernard
313, Les Marlières
4155 VILLERS LE TEMP BELGIUM

Mr NEVE Jean
Université Libre de Bruxelles
Laboratoire de Toxicologie
Campus Plaine 205-1
B 1050 BRUXELLES BELGIUM

Mr N'GUEYEN
L'OREAL
188, rue P. Hochard Chevilly Larue
94150 RUNGIS FRANCE

Mr NICOLS
Laboratoires LABCATAL
7, rue Roger Salengro
92541 MONTROUGE CEDEX FRANCE

Dr NIELSEN GISSEL Gunnar
Riso Nat. Laboratory
PO Box 49
DK 4000 ROSKILDE DENMARK

Mr NIELSEN GISSEL Mogens
University Aarhus Zoological Lab.
DK 8000 AARHUS C. DENMARK

Dr OSTER Oskar
Universtatsklinik Mainz-Langenbeckstr
1 Abteilung fur Klinische Chemie und
Laboratorium Medizin
D 6500 MAINZ FRG

Mr PAUQUAY François
La Mouderie
4561 DALHEM BELGIUM

Mr PELEN
Laboratoires LABCATAL
7, rue Roger Salengro
92541 MONTROUGE CEDEX FRANCE

Dr PENGLOAN
C.H.U. BRETONNEAU
Service de Néphrologie
37044 TOURS CEDEX FRANCE

Mme PERETZ Anne
Université Libre de Bruxelles
Hôpital ST PIERRE
rue Haute, 322
1000 BRUXELLES BELGIUM

Mr PERONNET Alain
Laboratoire HERBAXT
5, bis rue Chappon
BP 145
77335 MEAUX CEDEX FRANCE

Mr PINEAU Alain
Faculté des Sciences Pharmaceutiques
Centre de Dosage des Elements Traces
1, rue G. Weil - BP 1024
44035 NANTES CEDEX FRANCE

Mme POINSOT Simone
RHONE ALPES CONGRES
28, rue du Lieut. Colonel Girard
69007 LYON FRANCE

Mr POTIE Jean-Claude
Laboratoire d'Analyses Médicales
46, cours Vallier
38160 SAINT MARCELLIN FRANCE

Mr PRADIER François
l'OREAL
A 341
41, rue Martre
92117 CLICHY CEDEX FRANCE

Mr PURROY Antonio
Sia Inia (DGA) Appartado 727
50080 SARAGOSA SPAIN

Dr REVEILLARD Pierre
Clinique LAFARGUE
rue Gentil Ader
64100 BAYONNE FRANCE

Mr REY Jean-Marc
MULTIPHARMA
67, bd Vital Bouhot
92200 NEUILLY/SEINE FRANCE

Mr RICHARD Alain
Laboratoires LA ROCHE POSAY
86270 ROCHE POSAY FRANCE

Mme RICHARD Marie-Jeanne
C.H.R.U.G.
Laboratoire Biochimie C
BP 217 X
38043 GRENOBLE CEDEX FRANCE

Mr ROCIPON Jacques
L.A.C.B.
39, avenue du Granier
38240 MEYLAN FRANCE

Mr RODUIT Claude
Laboratoire MGD
18, bd des Philosophes
1205 GENEVE SWITZERLAND

Mr ROGUET Roland
l'OREAL
1, avenue E. Schueller
93600 AULNAY SOUS BOIS FRANCE

Mr ROQUES Christian
Société LE SAFFRE
137, rue Gabriel Péri
BP 202
59703 MARC EN BAROEUIL FRANCE

Mme RUAL Monique
C.H.R.U.G.
Laboratoire Biochimie C
BP 217 X
38043 GRENOBLE CEDEX FRANCE

Dr RUDOLPH Nathan
State University of New York Health
Science Center at Brooklyn
450, Clarkson avenue Box 49
NY11203 BROOKLYN USA

Mr RUFFIEUX Daniel
C.H.R.U.G.
Laboratoire Biochimie C
BP 217 X
38043 GRENOBLE CEDEX FRANCE

Dr SAINT GEORGES Martine
48, bd Richard Lenoir
75011 PARIS FRANCE

Mme SAUNIER Brigitte
PF Industrie
46 bis, rue Albert 1er
81106 CASTRES FRANCE

Mme SAVOURE
UFR Médicale Villejean
35043 RENNES CEDEX FRANCE

Mr SCHRAUZER Gerhard
University of California
Dept of Chemistry
92093 CA SAN DIEGO U.S.A.

Mr SEALES P.
WASSEN INTERNATIONAL LTD
14, The Mole Business Park
Leatherheat
KT22 7BA SURRY ENGLAND

Mr SHEEHAN Timothy
Regional Toxicology Laboratory
Dudley Road Hospital
B18 7QH BIRMINGHAM UNITED
KINGDOM

Mme SHEVLANE Nicola
Orchard Road
5G 85HE ROYSTON. HERTS UNITED
KINGDOM

Mr SHOONBROOD Jacques
8, rue St Jean
6200 GOSSELIES BELGIUM

Mme SIMONOFF Monique
CNRS
Laboratoire de Chimie Nucléaire UA 451
33170 BORDEAUX-GRADIGNAN
FRANCE

Mr SOTTO
Société PHILIPS Division Sciences
Industrie
105, rue de Paris
93002 BOBIGNY CEDEX FRANCE

Mr SOUDANT Etienne
l'OREAL
188, rue Paul Hochart BP 553
Chevilly Larue
94152 RUNGIS CEDEX FRANCE

Dr STIEFEL Thomas
GN PHARM
Nobelstrasse 15
D-7000 STUTTGART 80 FRG

Mme THAUVIN Estelle
C.H.R.U.G. Laboratoire Biochimie C
BP 217 X 38043
GRENOBLE CEDEX FRANCE

Dr THERON Patrice
Hôpital BRETONNEAU
Laboratoire Biochimie
2, rue Carpeau
75018 PARIS FRANCE

Mr THOMASSEN Y.
Institute of Occupational Health Gydas
Vei 8
PO BOX 8149 DEP
OSLO 1 NORWAY

Dr THUNUS L
Institut de Pharmacie
3, rue Fusch
4000 LIEGE BELGIUM

Mme TIBERGHIEN Marie-Pascale
RESEARCH LABORATORY TALKIN
CA8ILE BRAMPTON
UNITED KINGDOM

Mr TOLONEN Matti
Oppibranten 10B4
SF 02210 ESBO FINLAND

Mr TRUONG HO MINH
LABM RISSE 23, rue D'Auberton
75005 PARIS FRANCE

Dr VALLON Daniel
Les Villas du Golf
Chemin de la Valmasque
06600 ANTIBES FRANCE

Mme VAN BEERS Danièle
Hôpital BRUGMANN
Sérologie Virologie
1020 BRUXELLES BELGIUM

Mme VAN ERUM Suzy
Hôpital ST PIERRE
Départ. de Chimie Médicale
rue Haute 322
1000 BRUXELLES BELGIUM

Mme VAN HALEWYCK Karin
8, Rykswegl
2670 PUURS BELGIUM

Mr VANDEWIELLE Jean Marc
315, rue Verte
59170 CROIX FRANCE

Mr VANSLEMBROUCK André
REVOGAN
Moerstraat 58
B-9810 GENT-DRONGEN BELGIUM

Dr VIEGA G. CRESPO Anna Maria
Fac. Ciencias dept Zoologia
R. ESC. PO1
1200 LISBOA PORTUGAL

Mle VISO Anne Catherine
INSERM U 303
BP 3
O6230 VILLEFRANCHE SUR MER
FRANCE

Mr WALLIN Eli
Jellingvej 9
7100 VEJLE DENMARK

Mr WEIGERT Peter
94 AM Schlachtensee
D-1000 BERLIN (W) 38 FRG

Mme WILKE Bérénice Cunha
2, allée de Meppel
54500 VANDOEUVRE LES NANCY
FRANCE

Mr WOLFFRAM Siegfried
Institut Für Veterinar Physiologie
University of Zurich Winterhurerstrasse
260 8057 ZURICH SWITZERLAND

Mr YLARANTA Toivo
Agricultural Research Centre
31600 JOKIOINEN FINLAND

Mr ZACHARA Broniseaw
A. Academy of Medicine Depart. of Biochemistry
85092 BYDGOSZ KARKOWICZA 24
POLAND

Mme ZAGO Susanna
Monte Cervino 5
20047 BRUGNERIO ITALY

Mr ZAWISLAK Richard
Faculté de Médecine
Labo. de l'Institut de Chimie Organique
11, rue Human
67085 STRASBOURG CEDEX FRANCE

Dr ZIMMERLI D.
Bundesamt fur Gesundheitswesen
Bollwerk 27
Postfach 2644
3001 BERN SWITZERLAND

Mr ZUMBIHL
Laboratoire HERBAXT
5 bis, rue Chappon BP 145
77335 MEAUX CEDEX FRANCE

AUTHORS INDEX

SUBJECT INDEX

Bei Fragen zur Produktsicherheit wenden Sie sich bitte an:
If you have any questions regarding product safety,
please contact:

Walter de Gruyter GmbH
Genthiner Straße 13
10785 Berlin
productsafety@degruyterbrill.com